Student Interactive Workbook

Biology
The Unity and Diversity of Life

THIRTEENTH EDITION

Cecie Star

Ralph Taggart

Christine Evers

Lisa Starr

Prepared by

Kathy Hecht
Nassau Community College

Tara-Jo Homlberg
Nowthwestern Connecticut Community College

BROOKS/COLE
CENGAGE Learning

Australia • Brazil • Japan • Korea • Mexico • Singapore • Spain • United Kingdom • United States

ISBN-13: 978-1-111-98700-8
ISBN-10: 1-111-98700-9

Brooks/Cole
20 Davis Drive
Belmont, CA 94002-3098
USA

Cengage Learning is a leading provider of customized learning solutions with office locations around the globe, including Singapore, the United Kingdom, Australia, Mexico, Brazil, and Japan. Locate your local office at: **www.cengage.com/global**

Cengage Learning products are represented in Canada by Nelson Education, Ltd.

To learn more about Brooks/Cole, visit **www.cengage.com/brookscole**

Purchase any of our products at your local college store or at our preferred online store **www.cengagebrain.com**

Printed in the United States of America
2 3 4 5 6 7 15 14

CONTENTS

PREFACE

Tell me and I will forget, show me and I might remember, involve me and I will understand.
—Chinese Proverb

The proverb outlines three levels of learning in increasing order of effectiveness. The writer of the proverb understood that humans learn most efficiently when they involve themselves in the material to be learned. This student workbook is like a tutor; when properly used it increases the efficiency of your study periods. The interactive exercises actively involve you in the most important terms and central ideas of your text. Specific tasks ask you to recall key concepts and test your understanding of the facts. As you complete the exercises, you will discover items that you need to reexamine or clarify. Your performance on these tasks provides an estimate of your next test score based on specific material. Most importantly though, this biology student workbook, together with the text, helps you make informed decisions about matters that affect your own well-being and that of your environment. In the years to come, human survival on planet Earth will demand administrative and managerial decisions based on an informed background in biology.

HOW TO USE THIS STUDENT WORKBOOK

Following this preface, you will find an outline that will show you how the student workbook is organized and will help you use it efficiently. Each chapter in the student workbook begins with a title, chapter introduction, and list of focal points. The Interactive Exercises follow, wherein each chapter is divided into sections of one or more of the main (1st level) headings labeled 1.1, 1.2, and so on. *For easy reference to an answer or definition, each question and term in this unique student workbook is accompanied by the* appropriate text page number(s). The Interactive Exercises begin with a list of Selected Words (other than boldfaced terms) chosen by the authors as those most likely to enhance understanding of the material in the chapter. This is followed by a list of Boldfaced, Page-Referenced Terms that appear in the text. These terms are essential to understanding each student workbook section of a particular chapter. Space is provided by each term for you to formulate a definition in your own words. Next is the Interactive Exercises which is a series of exercises designed to enhance your understanding of the main topics in the chapter. These exercises include questions types like matching, fill-in-the-blank, true/false, and labeling.

A Self-Test immediately follows the Interactive Exercises. This test is composed primarily of multiple-choice questions. Any wrong answers in the Self-Test indicate portions of the text you need to reexamine. A series of Chapter Objectives / Review Questions follow each Self-Test. These serve as a quick review of the tasks that you should be able to accomplish if you have understood the assigned reading in the text. The final part of each chapter is Integrating and Applying Key Concepts. It invites you to try your hand at applying major concepts to situations in which there is not necessarily a single correct answer. Therefore, answers are not provided in the answer section at the end of the student workbook. Your text generally will provide enough clues to get you started on an answer, but this part is intended to stimulate your thinking and provoke group discussions.

A person's mind, once stretched by a new idea, can never return to its original dimension.
—Oliver Wendell Holmes

STRUCTURE OF THIS STUDENT WORKBOOK

The following outline shows how each chapter in this student workbook is organized.

Chapter Number ─────────▶

Chapter Title ─────────▶

32

ANIMAL TISSUES AND ORGAN SYSTEMS

Introduction ─────────▶

This includes a brief overview to the subject matter of the chapter.

Chapter 32 starts with a description of how cells are organized and connected to form organs, organ systems, and multicellular animals. The four types of animal tissues are described: epithelial, connective, muscle, and nerve. Vertebrate organ systems are introduced along with a more detailed description of how tissues are integrated into a specific organ system—the integumentary system.

Focal Points ─────────▶

- Figure 32.2 [p.540] illustrates how cells are attached to each other to form coherent tissues.
- Figure 32.4 [p.541] shows various types of epithelium.
- Figure 32.5 and 32.6 [pp.542-543] illustrate connective tissues.
- Figure 32.8 [p.544] has images of muscle tissues.
- Figure 32.9 [p.545] shows a typical neuron.
- Figure 32.11 and 32.12 [pp.546-547] describe anatomical terms and outline the major human organ systems.
- Figure 32.13 [p.568] diagrams human skin structure.

Interactive Exercises ─────────▶

The Interactive Exercises are divided into numbered sections by titles of main headings and page references. Each section begins with a list of author-selected words that appear in the chapter. This is followed by a list of important boldfaced, page-referenced terms from each section of the chapter. Each section ends with interactive exercises that vary in type and require constant interaction with the important chapter information

Self-Test ─────────▶

This is a set of questions designed to provide a quick assessment of how well you understand the information from the chapter.

Chapter Objectives / Review Questions ─────────▶

This section provides a list of concepts that you need to master before proceeding to the next chapter. Page numbers from the text are provided should you be unable to answer the questions.

Integrating and Applying Key Concepts ─────────▶

These represent "big-picture" or "real-life" applications of the concepts presented in the chapter.

Answers to Interactive Exercises and Self-Test ─────────▶

Answers for all the Interactive Exercises and the Self-Test can be found at the end of the student workbook. The answers are arranged by chapter number and main heading.

CREDITS AND ACKNOWLEDGMENTS

This page constitutes an extension of the copyright page. We have made every effort to tract the ownership of all copyrighted material and to secure permission from copyright holders. In the event of any question arising as to the use of any material, we will be pleased to make the necessary corrections in future printings. Thanks are due to the following authors, publishers, and agents for permission to use the material indicated.

Chapter 2
p. 16 (17): Lisa Starr with PDV ID:IBNA; H.R. Drew, R>M. Wing, T. Takano, C. Broka, S. Tanaka, K. Itakura, R.E. Dickerson; Structure of a B-DNA Dodecamer. Conformation and Dynamics, PNAS.
p. 17 (2): Lisa Starr with PDB ID:IBNA; H.R. Drew, R.M. Wing, T. Takano, C., Broka, S. Tanaka, K. Itakura, R.E> Dickerson; Structure of B-DNA Dodecamer. Conformation and Dynamics, PNAS.

Chapter 3
p. 32 (1): PDB files fro Klotho Biochemical Compounds Declarative Database.

Chapter 4
p. 42 (19-40): Lisa Starr

Chapter 22
p. 219 (1-6): D.J. Patterson / Seaphot Limited: Plant Earth Pictures.

Chapter 23
p. 233 (1-7): Lisa Starr
p. 226 (photo): A.&E. Boomford, Ardea, London; Lee Casebere

Chapter 24
p. 248 (28): Carolina Biological Supply Company

Chapter 28
p. 292 (1-7): Carolina Biological Supply Company
p. 295 (photo, right, top): C.E. Jeffree, et al., *Planta*, 172(1):20-37, 1987. Reprinted by permission of C.E. Jéffree and Springer-Verlag.
p. 295 (photo, right, bottom): Jeremy Burgess / SPL / Photo Researchers, Inc.
p. 296 (8-10): Chuck Brown

Chapter 29
p. 305 (13-15): Micrograph Chuck Brown

Chapter 30
p. 312 (1-4): © Gary Head
p. 317: Patricia Schulz

Chapter 36
p. 393: Yokochi and J. Rohen, Photographic Anatomy of the Human Body, 2nd Ed., Igaku-Shoin, Ltd., 1979

Chapter 47
p. 521: D. Robert Franz; art After Paul Hertz.

1

INVITATION TO BIOLOGY

INTRODUCTION

Chapter 1 introduces you to the study of biology as well as presenting terminology that help scientists interpret patterns in nature. This chapter discusses the levels of organization in nature and current classification systems (taxonomy). It also introduces ideas that help explain the diversity of organisms today. Finally, it introduces the concepts of scientific theory and critical thinking, emphasizing the steps of the scientific method.

STUDY STRATEGIES

- Read all of Chapter 1 with the goal of familiarizing yourself with the boldface terms.
- Explain the levels of organization of nature.
- Compare and contrast the movement of energy and nutrients through ecosystems.
- Describe how organisms are grouped based on traits.
- Identify the processes by which life becomes diverse.
- Explain and be able to recognize the steps of the scientific method and how scientific information is analyzed and interpreted.

FOCAL POINTS

- Nature is comprised of many levels of organization, beginning at the level of atoms and extending to the level of the biosphere [pp.4-5, Figure 1.2].
- Energy flows through a system while nutrients cycle through a system [p.6, Figure 1.3]
- Organisms grow and reproduce, passing along inherited traits encoded in their DNA [p.7]
- Organisms can be grouped into kingdoms based on primary characteristics [pp.8-9, Figure 1.5]
- Taxonomy classifies organisms based on their shared traits [p.10, Figure 1.6]
- Scientific investigation involves asking specific questions in nature, creating a hypothesis about a single variable, making observations, analyzing data, and drawing conclusions which can then become part of a theory [pp.12-18, Table 1.1]

INTERACTIVE EXERCISES

1.1. THE SECRET LIFE OF EARTH [p. 3]

Boldfaced Terms

biology _____

_____

Dichotomous Choice [p.3]

Choose the correct one of the two possible answers given between parentheses in each statement.

1. _____ (Biology / Taxonomy) is the scientific study of life.

2. _____ Approximately (2000 / 20) species become extinct every minute in the world's rainforests.

3. _____ Humans (are / are not) intimately connected with the world around us.

1.2. LIFE IS MORE THAN THE SUM OF ITS PARTS [pp.4-5]

Boldfaced Terms

atoms _____

_____

molecules _____

_____

cell _____

_____

organism _____

_____

tissue _____

_____

organ _____

_____

organ system _____

_____

population _____

_____

community _____

_____

ecosystem _____

_____

biosphere _____

_____

Matching [pp.4-5]

Match each of the following levels of biological organization with its correct definition.

1. _____ molecule
2. _____ cell
3. _____ community
4. _____ ecosystem
5. _____ organ system
6. _____ organ
7. _____ population
8. _____ biosphere
9. _____ tissue
10. _____ atom
11. _____ multicelled organism

A. Interaction of two or more tissues to perform a common task

B. All the regions of Earth that hold organisms

C. The smallest unit of life that can survive and reproduce on its own

D. Interaction of organs physically or chemically to perform a common task

E. Two or more atoms bonded together

F. All populations of all species occupying a given area

G. The fundamental unit of matter

H. The interaction of a community and the physical environment

I. Group of individuals of the same species occupying a specific area

J. Organized arrays of cells that interact for a specific task

K. Individual composed of different types of cells

Sequence [pp.4-5]

Arrange the following levels of organization in nature in the correct hierarchical order. Write the letter of the **least inclusive** level next to 12. Write the letter of the **most inclusive** level next to 22.

12. _____
13. _____
14. _____
15. _____
16. _____
17. _____
18. _____
19. _____
20. _____
21. _____
22. _____

A. community
B. tissue
C. cell
D. organ
E. organ system
F. atom
G. biosphere
H. molecule
I. population
J. multicelled organism
K. ecosystem

Fill-in-the-blank [pp.4-5]

Complex properties emerge from smaller parts combining together. The new characteristics are referred to as (23)_____. Organization of life begins at the level of the (24)_____ which then join together into (25)_____. A (26)_____ is the smallest unit of life. One or more of these may work together to organize an (27)_____. Within multicellular organisms, the (28)_____ are made up of specialized cells that have an organized pattern. (29)_____ are made up of various tissues working together and work together to form (30)_____.

Labeling [pp.4-5]

Label the following pictures with the correct level of organization.

A. Atom
B. Molecule
C. Cell
D. Tissue
E. Organ
F. Organ system

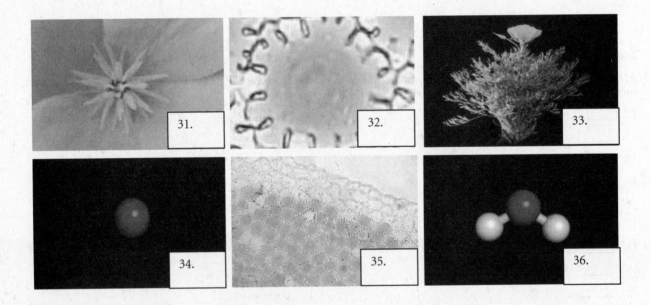

1.3. HOW LIVING THINGS ARE ALIKE [pp.6-7]

Boldfaced Terms

energy _____

_____

nutrient _____

producers _____

_____

photosynthesis _____

_____

consumers _____

_____

homeostasis _____

_____

DNA _____

_____

development _____

_____

growth _____

_____

reproduction _____

_____

inheritance _____

_____

Fill-in-the-Blanks [p.6]

Energy refers to the capacity to do (1) _____. Organisms spend a great deal of time acquiring energy and (2) _____, which are essential atoms or molecules that an organism cannot make for itself. If an organism is able to acquire its energy and raw materials from the environment, and subsequently, make their own food, they are called (3) _____. Plants are capable of making their own food through the process of (4) _____, in which they obtain energy from sunlight and convert it to a usable form. On the other hand, (5) _____ cannot make their own food and must acquire it by eating plants and other organisms.

Matching [p.7]

6. _____ homeostasis   8. _____ development
7. _____ reproduction   9. _____ inheritance

10. _____ growth
11. _____ DNA

A. The actual transmission of traits (through DNA) from parents to offspring

B. The molecule within an organism that guides metabolic activity

C. Maintenance of a constant internal environment

D. The orderly transformation of the first cell of a new individual into an adult

E. The increase in cell number, size, and volume

F. The mechanism by which parents transmit DNA to offspring

1.4. HOW LIVING THINGS DIFFER [pp.8-9]

Boldfaced Terms

biodiversity _____

_____

nucleus _____

_____

bacteria _____

_____

archaea _____

_____

eukaryotes _____

_____

protists _____

_____

fungi _____

_____

plants _____

_____

animals _____

_____

Choice [pp.8-9]

For each description in questions 1-7, choose the correct domains.

     a. Bacteria   b. Archaea   c. Eukarya

1. _____ May be located in extreme environments like hydrothermal vents or frozen desert rocks

2. _____ Includes plants, animals, and fungi

3. _____ Includes all multicelled organisms

4. _____ Single-celled, lack nucleus; most closely related to more derived multicelled organisms

5. _____ Includes protists

6. _____ The most numerous organisms on Earth

7. _____ Contains a nucleus in each cell

Matching [pp.8-9]

8. _____ bacteria

9. _____ archaea

10. _____ protist

11. _____ fungi

12. _____ plants

13. _____ animals

A. multicellular photosynthesizers

B. unicellular organisms that tend to live in extreme environments

C. active multicellular consumers

D. the most numerous organisms on Earth

E. extremely diverse eukaryotes ranging from multicellular to unicellular organisms

F. eukaryotic decomposers, parasites, or pathogens

1.5. ORGANIZING INFORMATION ABOUT SPECIES [pp.10-11]

Boldfaced Terms

species _____

_____

taxonomy _____

_____

genus _____

_____

specific epithet _____

_____

taxon _____

_____

traits _____

_____

Dichotomous Choice [p.10-11]

Choose the correct one of the two possible answers given between parentheses in each statement.

1. _____ The current system of naming organisms was developed by (Linnaeus / Darwin).

2. _____ Organisms within the same species share a unique set of characteristics known as (chemistry / traits).

Critical Thinking [p.10-11]

3. Why do most scientists consider the "biological species concept" useful but artificial?_____

_____

_____

1.6. THE SCIENCE OF NATURE [pp.12-13]

1.7. EXAMPLES OF EXPERIMENTS IN BIOLOGY [pp.14-15]

Boldfaced Terms

critical thinking _____

_____

science _____

_____

hypothesis _____

_____

inductive reasoning _____

_____

prediction _____

_____

deductive reasoning _____

_____

model _____

_____

experiments _____

_____

variables _____

independent variable _____

dependent variable _____

experimental group _____

control group _____

data _____

scientific method _____

Sequence [pp.12-13]

Arrange the following steps of the scientific method in correct chronological order. Write the letter of the first step next to 1, the letter of the second step next to 2, and so on.

1. _____
2. _____
3. _____
4. _____
5. _____
6. _____
7. _____

A. Develop a hypothesis
B. Repeat the tests or devise new ones
C. Devise ways to test the accuracy of predictions drawn from the hypothesis (use of observations, models, and experiments)
D. Make a prediction, using the hypothesis as a guide; the "if-then" process
E. If tests do not provide expected results, check to see what might have gone wrong
F. Objectively analyze and report the results from tests and the conclusions drawn
G. Observe some aspect of nature and research what others have found out about it

Matching [pp.12-15]

Using Table 1.3 from your text as a guide, match each of the following aspects of a sunflower experiment with its correct description.

8. _____ hypothesis
9. _____ control group
10. _____ report
11. _____ question
12. _____ observation
13. _____ experimental group

14. _____ assessment
15. _____ prediction

A. Fertilizer may help my sunflowers grow taller

B. Does fertilizer help my sunflowers grow taller

C. Ten plants are not treated with fertilizer

D. Submit the results and conclusions of my sunflower fertilizer experiments to the scientific community for review and publication

E. If fertilizer helps my sunflowers grow, then plants treated with fertilizer will grow taller than plants not treated with fertilizer

F. Compile results from sunflower fertilizer experiments and draw conclusions fro them

G. Ten plants are treated with fertilizer

H. I noticed that many people buy fertilizer for their sunflowers

Fill-in-the-Blank [pp.12-13]

(16) _____ is a way of evaluating information before accepting it based on quality. This process is one of the foundations of (17) _____ which is a systematic method of studying the observable world. Scientists begin the process of study by making observations, asking questions, and then forming a (18) _____which is a testable explanation for a natural phenomenon. Scientists then make a (19) _____, usually in the form of an "if- then" statement. An (20) _____ is a specialized test used to examine a cause-and-effect relationship but is not always used if a model is more suitable. Scientists control their experiment by using (21) _____ and (22) _____ variables. By manipulating these two types of variables, scientists set up two types of groups: an (23) _____ group and a (24) _____ group. The results, or (25) _____, are then analyzed and discussed with other scientists to come to a conclusion.

True/False [pp. 12-13]

If the statement is true, write a "T" in the blank. If the statement is false, make it correct by changing the underlined word and writing the correct word in the blank.

26. _____ The process of forming a hypothesis, testing and evaluating the hypothesis, and developing a conclusion is referred to as <u>an experiment</u>.

1.8. ANALYZING EXPERIMENTAL RESULTS [pp.16-17]

Boldfaced Terms

sampling error _____

_____

probability _____

_____

statistically significant _____

_____

Short Answer [pp.16-17]

1. Why do researchers use large sample sizes and repeat their experiments?_____
   _____

   _____

2. Why do researchers try to design single-variable experiments, rather than examining multiple variables all at once? _____

   _____

   _____

3. What does "statistically significant" mean for scientists? _____

   _____

   _____

True/False [pp. 16-17]

If the statement is true, write a "T" in the blank. If the statement is false, make it correct by changing the underlined word and writing the correct word in the blank.

4. _____ The difference between results obtained from testing a subset of a group and results from test a whole group is termed <u>sampling error</u>.

5. _____ <u>Statistically significant</u> refers to the measure of the chance that a particular outcome will occur.

Dichotomous Choice [pp. 16-17]

Choose the correct one of the two possible answers given between parentheses in each statement.

6. _____ A (small / large) subset of a larger group is preferred to test via experimentation or survey via observation.

7. _____ Statistical significance (does / does not) refer to the result's importance.

8. _____ Humans, and scientists, are by nature (subjective / objective) but the process of the scientific method serves to minimize potential bias.

1.9. THE NATURE OF SCIENCE [pp.18-19]

Boldfaced Terms

scientific theory _____

_____

law of nature _____

_____

Short Answer [pp.18-19]

1. How does a "scientific theory" differ from the everyday use of the word "theory"? _____

    _____

    _____

2. How would you define a "law of nature"? _____

    _____

    _____

3. What are some of the limits to science? _____

    _____

    _____

    _____

SELF-TEST

____ 1. Two or more atoms bonded together are
         called a _____.
         a. molecule
         b. organism
         c. population
         d. tissue

____ 2. All populations of all species occupying a
         given area are called a _____.
         a. biosphere
         b. organ system
         c. community
         d. multicelled organism

____ 3. A(n) _____ is a characteristic of a system
         that do not appear in any of the system's
         individual parts.
         a. atoms
         b. complex property
         c. emergent property
         d. quality of life

____ 4. The most comprehensive of all levels of
         organization of life is a(n) _____
         a. biosphere
         b. organ system
         c. community
         d. multicelled organism

____ 5. An example of an organism that uses
         photosynthesis to obtain energy from
         sunlight is a _____.
         a. mouse
         b. bird
         c. tree
         d. mushroom

____ 6. Two of the essentials components for all
         forms of life are:
         a. energy and nutrients
         b. sunlight and carbon dioxide
         c. energy and molybdenum
         d. gamma rays and nutrients.

____ 7. Organisms _____ changes in the
         environment and _____.
         a. sense; grow
         b. sense; respond
         c. taste; grow
         d. think about; respond

____ 8. The only domain that includes
         multicellular organisms with a nucleus is
         _____.
         a. Archaea
         b. Eubacteria
         c. Bacteria
         d. Eukarya

____ 9. Which of the following is not a eukaryote?
         a. bacteria
         b. plant
         c. fungi
         d. animal

____ 10. The orderly transformation of the first cell
          of a new individual into an adult is _____.
          a. development
          b. reproduction
          c. inheritance
          d. evolution

11. The process of naming and classifying organisms is:
   a. ecology
   b. scientific method
   c. taxonomy
   d. botany

12. The first step in the scientific method is:
   a. observe
   b. ask a question
   c. make a hypothesis
   d. analyze data

13. An example of a control group would be:
   a. A group that is composed only of males
   b. A group that does not receive the treatment of the experimental variable
   c. A group that starts the experiment later than other groups
   d. A group that receives two treatment variables instead of three

14. If an experiment is not appropriate or feasible based on the question being asked, a _____ may be substituted.
   a. schematic
   b. equation
   c. drawing
   d. model

15. Energy may be defined as _____.
   a. maintenance of a constant internal environment
   b. the ability to withstand photosynthesis
   c. the capacity to do work
   d. the ability to pass nutrients from parents to offspring

16. Different species of Galapagos Island finches have different beak types to obtain different kinds of food. One species removes tree bark with sharp beak to forage for insect larvae and pupae, whereas another species has a large, powerful beak capable of crushing and eating large, heavy, coated seeds. _____ selection is responsible for the evolution of beaks in these finches.
   a. Adaptive
   b. Natural
   c. Artificial
   d. Discriminate

17. Which of these statements is true?
   a. Nutrients cycle within a system, while energy flows through a system.
   b. Both nutrients and energy flow through a system.
   c. Both nutrients and energy cycle within a system.
   d. Nutrients flow through a system, while energy cycles within a system.

18. Which of these must scientists take care to reduce or eliminate in their research?
   a. bias
   b. probability
   c. improbable hypotheses
   d. critical thinking

19. If a field of science has tested a particular hypothesis over many years or decades, it is supported by vast amounts of data, and other phenomena have been explained by the hypothesis, the hypothesis may then be referred to as a _____.
   a. result
   b. conclusion
   c. theory
   d. supported guess

CHAPTER OBJECTIVES/REVIEW QUESTIONS

This section lists general and detailed chapter objectives that can be used as review questions. You can make maximum use of these items by writing answers on a separate sheet of paper. Fill in answers where blanks are provided. To check for accuracy, compare your answers with information given in the chapter or glossary.

1. Diagram the levels of organization in nature, beginning with the *atom* and ending with the *biosphere*. [pp.4-5]
2. Distinguish between the terms *energy* and *nutrient*. [p.6]
3. Describe how producers are different from consumers [p.6]
4. Explain why homeostasis is important for living organisms. [p.7]
5. Describe the differences among inheritance, reproduction, and development. [p.7]
6. Distinguish among the three domains: Eukarya, Bacteria, and Archaea. [p.8]
7. Explain how organisms are organized and named. [p.10]
8. Define *scientific theory* and give an example of one. [p.12]

10. List and explain the general steps used in the scientific method. [p.12]
11. Distinguish between a control group and experimental group in a scientific test. [p.13]
12. Define *sampling error* and give an example of one. [p.16]

INTEGRATING AND APPLYING KEY CONCEPTS

1. Humans have the ability to maintain body temperature very close to 37°C.
   a. What conditions would tend to make body temperature drop?
   b. What measures do you think your body takes to raise body temperature when it drops?
   c. What conditions would cause body temperature to rise?
   d. What measures do you think your body takes to lower body temperature when it rises?
2. If a theory has not been proven false in 100 years of testing, can we conclude that it is a scientific truth? Why or why not?
3. What would happen if scientists failed to use control groups in all of their experiments? How might this change our interpretation of their results?
4. Can we use the scientific method to answer the following question: "Is it wrong for me to be angry when squirrels take birdseed out of my birdfeeder?" Justify your answer.
5. Re-read the potato chip experiment described on page 14 of your text. Name two ways that you might improve the quality of the experiment if you were to complete it again.

2
LIFE'S CHEMICAL BASIS

INTRODUCTION

This chapter looks at the basic chemistry needed to understand biology. Topics covered include atoms and their structure, chemical bonding, properties of water, and acids, bases, and buffers. Many topics discussed here will reappear throughout the book, so it is important that you understand these concepts before moving on.

STUDY STRATEGIES

- Read all of chapter 2 with the goal of familiarizing yourself with the boldface terms
- Describe the components and composition of an atom, the charges of subatomic particles, and terminology of elements and atoms.
- Identify the applications of radioisotopes in medicine and research.
- Explain how the number of electrons changes the energy of the atom and why they are the basis of chemical behavior.
- Describe the various types of chemical bonds and how they are formed.
- Identify the various types of representations of chemical bonds.
- Explain the importance of hydrogen bonds and the various properties of water that make it important for life.
- Distinguish between acids and bases and recognize the importance of buffers to living systems.

FOCAL POINTS

- Figure 2.2 [p.24] identifies the subatomic particles and their relative positions within an atom.
- Figure 2.3 [p.24] illustrates the periodic table of the elements.
- Figure 2.5 [p.26] demonstrates how to form shell models which reveal electron vacancies in atoms.
- Figure 2.6 [p.27] demonstrates how ions are formed.
- Figure 2.8 [p.29] indicates the various kinds of covalent bonds found in biological molecules.
- Figure 2.11 [p.31] shows how hydrogen bonds are formed and how, while weak individually, collectively they can stabilize the structure of large molecules.
- Section 2.6 [pp.32-33] and Figure 2.12 [p.32] describe acids, bases, and buffers.
- Table 2.2 [p.34] summarizes the most important concepts of the chapter.

INTERACTIVE EXERCISES

2.1. MERCURY RISING [p.23]

2.2. START WITH ATOMS [pp.24-25]

Boldfaced, Page-Referenced Terms

atoms _Fundamental building block of all matter._

protons _Positively charged subatomic particle that occurs in the nucleus of all atoms._

neutrons _Uncharged subatomic particle in the atomic nucleus._

nucleus _of an atom; core area occupied by protons and neutrons._

electrons _Negatively charged subatomic particle._

charge _Electrical property._

atomic number _Number of protons in the atomic nucleus._

elements _A pure substance that consists only of atoms with the same number of protons._

periodic table _Tabular arrangement of all known elements by their atomic number._

isotope _Form of an element that differs in the number of neutrons its atom carries._

mass number _of an isotope, this total number of protons + neutrons in the atomic nucleus._

radioisotope _Isotope with an unstable nucleus._

radioactive decay _Process by which atoms of a radioisotope emit energy and/or subatomic particles when their nucleus spontaneously breaks up._

tracer _A substance that can be traced via its detectable component._

Matching [pp.24-25]

Choose the most appropriate answer for each term.

1. __K__ atoms
2. __C__ protons
3. __O__ nucleus
4. __D__ PET
5. __L__ neutrons
6. __B__ electrons
7. __F__ atomic number
8. __I__ mass number
9. __J__ elements
10. __E__ isotopes
11. __G__ radioisotopes
12. __A__ tracers
13. __M__ radioactive decay
14. __H__ Dmetry Mendeleev
15. __N__ Melvin Calvin

A. Compounds that have a radioisotope attached to allow determination of the pathway or destination of a substance

B. Subatomic particles with a negative charge

C. Positively charged subatomic particles within the nucleus

D. Positron-emission tomography; obtains images of particular body tissues

E. Atoms of a given element that differ in the number of neutrons

F. The number of protons in an atom

G. A form of isotope that contains an unstable nucleus that emits energy and particles in an attempt to stabilize its structure

H. Creator of the Periodic Table of the Elements

I. The number of protons and neutrons in the nucleus of one atom

J. Fundamental forms of matter that occupy space, have mass, and cannot be broken down into something else

K. Smallest units that retain the properties of a given element

L. Subatomic particles within the nucleus carrying no charge

M. the process of nuclear breakdown of radioisotopes

N. Used radioisotopes to identify specific reaction steps in photosynthesis

O. The atoms central core

True/False [pp.24-25]

If the statement is true, write a T in the blank. If the statement is false, rewrite it to make it correct.

16. __T__ Ions are formed when an atom gains or loses neutrons

17. __T__ The mass number of each element refers to the total number of protons and neutrons in its nucleus

18. __T__ Atoms of an element can differ structurally from one another.

19. __T__ An atom of lead has 82 protons

20. __T__ Radioactive decay is dependent on external forces such as temperature and pressure

21. __T__ Electrons carry a negative charge.

22. __F__ Molecules of life carry a high proportion of nickel atoms.

2.3. WHY ELECTRONS MATTER [pp.26-27]

Boldfaced, Page-Referenced Terms

shell models <u>helps us visualize how electrons populate atoms.</u>

free radicals <u>Atoms that have unpaired electrons.</u>

Matching [pp.26-27]

Choose the most appropriate answer for each term.

1. __B__ free radicals
2. __D__ shell model
3. __E__ lowest energy level
4. __A__ orbitals
5. __F__ higher energy levels
6. __C__ ion

A. Regions of space around an atom's nucleus where electrons are likely to be at any one instant

B. Unstable atoms that have a strong tendency to interact with other atoms

C. An atom with a different number of electrons and protons

D. A graphic representation of the distribution of electrons in their energy levels

E. Energy of electrons in the orbital closest to the nucleus

F. Energy of electrons farther from the nucleus than the first orbital

Complete the Table [pp.26-27]

7. Referring to the Periodic Table in the Appendix, enter the information for each of the following elements: *atomic number, atomic mass, number of protons, number of neutrons, and number of electrons*

a. sodium (Na) <u>atomic # = 11   atomic mass = 23</u>
   <u>P = 11   E = 11   N = 12</u>

b. bromine (B) <u>atomic # = 5   atomic mass = 10</u>
   <u>P = 5   E = 5   N = 5</u>

c. carbon (C) <u>atomic # = 6   atomic mass = 12</u>
   <u>P = 6   E = 6   N = 6</u>

d. hydrogen (H) <u>atomic # = 1   atomic mass = 1</u>
   <u>P = 1   E = 1   N =</u>

e. oxygen (O) _Atomic # = 8   atomic mass = 16_
_P = 8   E = 8   N = 8_

f. neon (Ne) _Atomic # = 10   atomic mass = 20_
_P = 10   E = 10   N = 10_

g. chlorine (Cl) _Atomic # = 17   atomic mass = 35_
_P = 17   E = 17   N = 18_

Fill-in-the-Blanks [pp.26-27]

An (8) _____ is uncharged only when it has as many electrons as protons. An (9) with different

numbers of electrons and protons is called an (10) _____. An (10) is formed when an atom gains or loses (11)

_____. Some atoms are chemically (12) _____ because their outermost shell is filled. Other atoms have

unpaired electrons, are not very stable, and are known as (13) _____.

Identification [pp.26-27]

14. The following example demonstrates a shell model for the element helium. In Figures a–f, the number of
protons and the number of neutrons are shown within the nucleus. For each energy level, indicate the number
of electrons present, using the form 2e, 3e, and so on. In the space under each diagram, indicate the name of the
element.

He

a. _H_   b. _C_   c. _N_

d. _O_   e. _P_   f. _S_

2.4. CHEMICAL BONDS: FROM ATOMS TO MOLECULES [pp.28-29]

Boldfaced, Page-Referenced Terms

chemical bond _An attractive force that arises between two atoms when their electrons interact._

molecule _results when two or more atoms bond together_

compound _Molecule that has atoms of more than one element_

ionic bond _Type of chemical bond in which a strong mutual attraction links ions of opposite charge_

polarity _separation of charge into positive and negative regions._

electronegativity _Measure of the ability of an atom to pull electrons away from other atoms._

covalent bond _Two atoms share a pair of electrons, so that each atom's vacancy becomes partially filled._

Matching [pp.28-29]

Choose the most appropriate answer for each term.

1. _G_ molecule
2. _D_ compound
3. _E_ chemical bond
4. _A_ ionic bond
5. _C_ covalent bond
6. _B_ electronegativity
7. _F_ polarity

A. The mutual attraction of two atoms due to opposite charges

B. Measure of an atom's ability to pull electrons from other atoms

C. A bond involving the sharing of electrons

D. Types of molecules composed of two or more different elements in proportions that never vary

E. An attractive force that arises between two atoms when their electrons interact

F. A separation in charge due to ionization

G. Results when two or more atoms bond together

Short Answer [pp.28-29]

8. Referring to the following example of hydrogen gas, complete the two diagrams below it by placing electrons (as dots) in the outer shells to identify the covalent bonding that forms oxygen gas and water molecules.

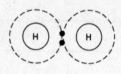

H₂

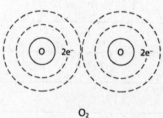

O₂

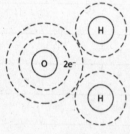

H₂O

9. Distinguish between a nonpolar covalent bond and a polar covalent bond. Cite an example of each.

_____

_____

_____

10. Explain the importance of hydrogen bonds in establishing the structure of DNA.

_____

_____

_____

11. Distinguish between a "molecule" and a "compound". Cite an example of each.

_____

_____

_____

2.5. HYDROGEN BONDS AND WATER [pp.30-31]

Boldfaced, Page-Referenced Terms

hydrogen bond _____

_____

cohesion _____

_____

evaporation _____

_____

temperature _____

_____

solvent _____

_____

hydrophillic _____

_____

salt _____

_____

solute _____

_____

solution _____

_____

mixture _____

_____

hydrophobic _____

_____

Fill-in-the-Blanks [pp.30-31]

The (1) _____ of water molecules allows them to hydrogen-bond with each other. Water molecules hydrogen-bond with polar molecules, which are (2) _____ (water-loving) substances. Polarity causes water to repel oil and other nonpolar substances, which are (3) _____ (water-dreading). (4) _____ is a measure of the molecular motion of a given substance. Liquid water changes its temperature more slowly than air because of the great amount of heat required to break the high number of (5) _____ bonds between water molecules; this property helps (6) _____ temperature in aquatic habitats and cells. (7) _____ occurs when large energy inputs increase molecular motion to the point where hydrogen bonds stay broken, releasing individual molecules from the water surface. Below 0°C, water molecules become locked in the lattice-like bonding pattern of (8) _____, which is less dense than water. Water is an excellent (9) _____, in which ions and polar molecules readily dissolve. Substances dissolved in water are known as (10) _____. Collective hydrogen bonding creates a high tension on surface water molecules, resulting in (11) _____, the property of water that explains how long, narrow water columns rise to the tops of tall trees.

Dichotomous Choice [pp.30-31]

Choose the correct one of the two possible answers given between parentheses in each statement.

12. _____ Hydrogen bonds are the (strongest / weakest) chemical bonds.

13. _____ Temperature is a measure of the (heat / motion) of an atom.

14. _____ Ice can act as an (insulator / air conditioner) for aquatic environments in the winter.

15. _____ Mixtures are an (intermingling / chemical combination) of substances.

16. _____ Solutions are (heterogeneous / homogeneous) mixtures.

2.6. ACIDS AND BASES [pp.32-33]

Boldfaced, Page-Referenced Terms

concentration _____

_____

pH_____

_____

acids _____

_____

bases _____

_____

buffer _____

_____

Matching [pp.32-33]

Choose the most appropriate answer for each term.

1. _____ acids
2. _____ pH scale
3. _____ alkaline
4. _____ H$^+$
5. _____ bases
6. _____ OH
7. _____ buffer system
8. _____ salt

A. Hydroxide ion

B. Substances that accept H$^+$ when dissolved in water

C. Used to represent H$^+$ concentration in fluids

D. Partnership between a weak acid and the base that forms when it dissolves in water; counteracts slight pH shifts

E. Hydrogen ion or proton

F. Substances that donate H$^+$ when dissolved in water

G. A term used to describe a basic solution

H. A compound that dissolves easily in water and releases ions other than H$^+$ and OH$^-$

Ordering [pp.32-33]

9. Refer to Fig 2.12 and order the following materials from most acidic to least acidic. Include the pH of each (round the pH to the nearest whole number):

ammonia, beer, black coffee, blood, hair remover, lemon juice, pure water, seawater, acid rain, drain cleaner, milk,

_____

_____

_____

_____

_____

_____

_____

Short Answer [pp. 32-33]

10. If pH changes from 1 to 3, what happens to the hydrogen ion concentration? _____

_____

_____

11. What is occurring when you take an antacid after a meal of tomato sauce-covered pasta and soda? _____

_____

_____

12. Explain how buffers maintain homeostasis. _____

_____

_____

_____

SELF-TEST

___ 1. Each element has a unique _____ which refers to the number of protons present in its atoms.
   a. isotope
   b. mass number
   c. atomic number
   d. radioisotope

___ 2. An ionic bond is one in which _____.
   a. electrons are shared equally
   b. electrically neutral atoms have a mutual attraction
   c. two charged atoms have a mutual attraction due to a transfer of electrons
   d. electrons are shared unequally

___ 3. A covalent bond is one in which _____.
   a. electrons are shared
   b. electrically neutral atoms have a mutual attraction
   c. two charged atoms have a mutual attraction due to a transfer of electrons
   d. electrons are lost

___ 4. Isotopes are atoms that differ by the number of
   a. neutrons
   b. protons
   c. electrons
   d. atoms

___ 5. If neon has an atomic number of 10 and an atomic mass of 20, it has _____ neutron(s) in its nucleus.
   a. one
   b. two
   c. five
   d. ten

___ 6. Radioactivity releases
   a. vibration
   b. odors
   c. radiation
   d. sounds

___ 7. Substances that are nonpolar and repelled by water are _____.
   a. hydrolyzed
   b. nonpolar
   c. hydrophilic
   d. hydrophobic

___ 8. A nonpolar covalent bond implies that
_____.
  a. one negative atom bonds with a
     hydrogen atom
  b. it is a double bond
  c. there is no difference in charge at
     the ends (the two poles) of the bond
  d. atoms of different elements do not
     exert the same pull on shared
     electrons

___ 9. The periodic table was developed by
  a. Melvin Calvin
  b. Andrew Benson
  c. Dmitry Mendeleyev

___ 10. This type of bond contributes to the 3-D
shape of large molecules.
  a. hydrogen
  b. ionic
  c. covalent
  d. inert
  e. single

___ 11. In the shell model of the atom, the _____
level contains the most energy.
  a. innermost
  b. outermost
  c. middle

___ 12. Ions are atoms that differ by the number of
  a. neutrons
  b. protons
  c. electrons
  d. atoms

___ 13. A solution with a pH of 2 is_____
times more acidic than one with a pH of 5.
  a. 3
  b. 10
  c. 30
  d. 100
  e. 1,000

___ 14. The universal solvent is
  a. salt
  b. water
  c. blood
  d. seawater

___ 15. A control that minimizes unsuitable pH
shifts is a(n) _____.
  a. hydrophilic compound
  b. salt
  c. base
  d. acid
  e. buffer

___ 16. Which of the following properties of water
protects aquatic organisms during a long,
cold winter?
  a. Cohesion
  b. Solvent properties
  c. Temperature-stabilizing
  d. None of the above

CHAPTER OBJECTIVES/REVIEW QUESTIONS

1. Define *element*. [p.24]
2. List and describe the three types of subatomic particles. [p.24]
3. Distinguish between the mass number and atomic number of an element and tell what subatomic particles contribute to the value of each. [pp.24-25]
5. Explain why helium, neon, and argon chemically stable. [p.27]
6. What information may be obtained about an atom from a shell model? [pp.26-27]
7. Explain the difference between molecules and compounds. [p.28]
8. How does a mixture differ from a compound? [pp.28, 31]
9. Explain why H_2 is an example of a nonpolar covalent bond and why H_2O has two polar covalent bonds. [pp.28-29]
10. Explain the biological importance of hydrogen bonds. [p.30]
11. What is meant by the temperature of an object at the molecular level? [p.30]
12. What occurs at the molecular level during the process of evaporation? [p.30]
13. Describe what happens when a substance is dissolved in water. [p.31]
14. Explain the difference between acids and bases in terms of their relationship to hydrogen ions (H^+). [pp.32-33]
15. Define *buffer system*; cite an example and describe how buffers operate. [pp.32-33]

INTEGRATING AND APPLYING KEY CONCEPTS

1. Explain what would happen if water were a nonpolar molecule instead of a polar molecule. Would water be a good solvent for the same kinds of substances? Would the nonpolar molecule's heat regulation likely be higher or lower than that of water? How about the cohesive nature? Would the ability to form hydrogen bonds change? Is it likely that the nonpolar molecules could form unbroken columns of liquid? What implications would that hold for trees?

2. What would be the implications for life on Earth if water did not have temperature-stabilizing effects? What would be different within individual organisms? What would change in ecosystems such as ponds and rivers? What would change on a global (biosphere) level? Why then do scientists search for evidence of water on other planets?

3. Interrelate the following terms in a concept map: atom, atomic number, electron, element, mass number, negative charge, neutron, nucleus, isotope, ion, positive charge, and proton.

4. Why are buffers important to biological molecules? Do all buffers maintain the same pH? Why is this important?

3

MOLECULES OF LIFE

INTRODUCTION

Often students question why a course in the biological sciences needs to start with a discussion of chemistry. Yet, chemistry is the foundation of life. Your body is powered by chemical reactions that convert food into energy and raw materials for the building of cells and tissues. In this chapter you will examine the basic principles of organic chemistry, which focuses on the element carbon. You will then examine the biologically important organic compounds, namely the carbohydrates, fats, proteins, and nucleic acids.

STUDY STRATEGIES

- Read all of Chapter 3 with the goal of familiarizing yourself with the boldface terms.
- Describe the basic structure of organic molecules and the importance of carbon to life.
- Explain the role of functional groups for organic molecules.
- Define condensation and hydrolysis reactions.
- Describe simple, short-chain, and complex carbohydrates and explain the various roles of carbohydrates in living organisms.
- Compare and contrast saturated and unsaturated fatty acids and explain the various roles of lipids in living organisms.
- Describe the structure of a protein, identify the various levels of folding of a protein, and explain the various roles of proteins in living organisms.
- Describe the structure of nucleic acids and explain the various roles of nucleic acids in living organisms.

FOCAL POINTS

- Figure 3.5 [p.40] illustrates condensation and hydrolytic reactions, which are the cornerstone of biochemistry. The principles are the same for all of the biologically important molecules.
- Table 3.1 [p.41] identifies the common functional groups found in biological molecules.
- Figure 3.8 [p.43] shows the difference in structure between starch, cellulose, and glycogen, all polymers of glucose. The different structures are due to different monomer bonding patterns.
- Figure 3.11 [p.44] shows the formation of a triglyceride and the structure of a phospholipid.
- Figure 3.14 [p.46] shows the generalized structure of an amino acid.
- Figure 3.16 [p.47] shows the different levels of protein structure and how they contribute to a three-dimensional molecule. Proteins are the working molecules of the cell. Their function is based on their shape.
- Figure 3.18 [p.49] shows the structure of nucleic acids.

INTERACTIVE EXERCISES

3.1. FEAR OF FRYING [p.37]

3.2. ORGANIC MOLECULES [pp.38-39]

3.3. MOLECULES OF LIFE - FROM STRUCTURE TO FUNCTION [pp.40-41]

Boldfaced, Page-Referenced Terms

organic _____

_____

condensation_____

_____

enzyme _____

_____

functional group _____

_____

hydrocarbons _____

_____

hydrolysis _____

_____

metabolism _____

_____

monomers_____

_____

polymers _____

_____

reaction_____

_____

Fill-in-the-Blanks [pp.38-39]

The molecules of life are (1) _____ compounds, which are defined as containing the element (2)

_____ and at least one (3) _____ atom. The term is a holdover from a time when chemists thought

"organic" substances were the ones made naturally in living (4) _____ only, as opposed to (5) _____

substances that formed abiotically. The term persists, although scientists now synthesize organic compounds in (6)

_____ and have reason to believe that organic compounds were present on Earth before organisms were.

Carbon's importance to life starts with its versatile (7) _____ behavior. Each carbon atom can covalently bond with as many as (8) _____ other atoms. The (9) _____ consist only of hydrogen atoms covalently bonded to (10) _____. Examples are (11) _____ and other fossil fuels. Each organic compound has one or more (12) _____ groups, which are particular atoms or clusters of atoms covalently bonded to (13) _____. The hydroxyl functional group has a (14) _____ characteristic, which allows it to form (15) _____ bonds and dissolve quickly.

True/False [pp.37-39]

If the statement is true, write a T in the blank. If the statement is false, rewrite it to make it correct.

16. _____ Enzymes are a class of proteins that make reactions proceed faster than they would otherwise.

17. _____ All functional groups are polar.

18. _____ Hydrocarbons are the simplest organic molecule.

19. _____ Structural details give clues to how molecules function.

20. _____ Carbon only makes up a small percentage of living organisms.

Dichotomous Choice [pp.37-39]

Choose the correct one of the two possible answers given between parentheses in each statement.

21. _____ Carbon atoms typically form (ionic / covalent) bonds.

22. _____ A (structural formula / space-filling model) indicates the overall shape of a molecule.

23. _____ Organic molecules are made up primarily of (hydrogen / oxygen) and carbon.

Matching [pp.40-41]

Match each of the following functional groups with its correct description.

24. _____ sulfhydryl group

25. _____ hydroxyl group

26. _____ carbonyl group

27. _____ phosphate group

28. _____ carboxyl group

29. _____ amine group

A. Stabilizes the structure of proteins

B. Found within the structure of ATP and DNA

C. The functional group of the alcohols

D. Used in the building of fats and carbohydrates

E. A key component of the amino acids and fatty acids

F. Contains nitrogen and is important in some nucleotide base

Labeling [p.41]

For each of the following molecules, shade in all of the atoms associated with the indicated functional group. Use Table 3.1 as a reference.

30. methyl

31. carbonyl

32. carboxyl

33. carbonyl

34. hydroxyl

35. amino

36. phosphate

Matching [p.40-41]

Choose the most appropriate answer for each term.

37. _____ metabolism

38. _____ condensation reaction

39. _____ monomers

40. _____ hydrolysis

41. _____ polymers

42. _____ functional-group transfer

43. _____ cleavage

44. _____ rearrangement

45. _____ electron transfer

A. Activity by which cells acquire and use energy for cellular process

B. A type of reaction that splits molecules using water

C. The individual subunits of organic molecules

D. Any reaction that splits a molecule into two smaller molecules

E. The type of chemical reaction that moves electrons between molecules

F. The movement of functional groups between molecules

G. The formation of a covalent bond by the removal of –OH and H functional groups, forming water

H. Long chains of subunits, sometimes consisting of millions of individual subunits

I. A change in the internal bond structure of a molecule

Short Answer [pp.37-41]

46. Why is carbon such an important element to life?

_____

_____

_____

3.4. CARBOHYDRATES [pp.42-43]

Boldfaced, Page-Referenced Terms

carbohydrate_____

_____

cellulose _____

_____

disaccharide _____

_____

glycogen _____

_____

monosaccharide _____

_____

polysaccharide _____

_____

starch _____

_____

Identification [p.42-43]

1. In the diagram, identify condensation reaction sites between the two glucose molecules by circling the components of the water molecule whose removal allows a covalent bond to form between the glucose molecules (Figure 3.7, p.42). Note that the reverse reaction is hydrolysis and that both condensation and hydrolysis reactions require enzymes in order to proceed efficiently.

Choice [pp.42-43]

Choose the class of carbohydrates (a–c) associated with the terms in items 2–9.

    a. oligosaccharides   b. polysaccharides c. monosaccharides

2. _____ "complex" carbohydrates

3. _____ chitin

4. _____ disaccharides

5. _____ ribose and deoxyribose

6. _____ lactose, sucrose, and maltose

7. _____ glucose and fructose

8. _____ starch and glycogen

9. _____ cellulose

Fill-in-the-Blanks [p.42-43]

(10) _____ is a structural carbohydrate associated with plant cell walls. The chains of (11)

_____ used to form this molecule are stabilized by (12) _____ bonds. (13) _____ is also a polymer

of glucose and is the primary energy source for plants. (14) _____ is a third class of glucose polymer which

provides back up energy stores for animals.

Matching [pp.42-43]

Match each of the following carbohydrates with its correct function.

15. _____ sucrose

16. _____ chitin

17. _____ glucose

18. _____ ribose

19. _____ cellulose

20. _____ amylose

21. _____ glycogen

A. Instant energy source for most organisms; precursor of many organic molecules; serves as building block for larger carbohydrates

B. A form of starch produced by plants to store excess sugars

C. Most plentiful sugar in nature, formed from glucose and fructose

D. Animal starch that is stored in liver and muscle tissue of mammals

E. Main structural material in some external skeletons and other hard body parts of some animals and fungi

F. Structural material of plant cell walls

G. Five-carbon sugar occurring in DNA and RNA

3.5. GREASY, OILY—MUST BE LIPIDS [pp.44-45]

Boldfaced, Page-Referenced Terms

fat _____

_____

fatty acid _____

_____

lipid_____

_____

lipid bilayer _____

_____

phospholipid_____

_____

saturated fatty acid_____

_____

steroid_____

_____

unsaturated fatty acid_____

_____

wax_____

_____

Labeling [pp.44-45]

1. In the appropriate blanks, label the molecules shown as *saturated* or *unsaturated*. For the unsaturated molecules, circle the regions that make them unsaturated.

a. _____

b. _____

c. _____

a.   oleic acid

b.   stearic acid

c.   linolenic acid

Choice [pp.44-45]

For questions 2-17, choose from the answers below. Some answers may be used more than once.

a. triglycerides   b. phospholipids   c. waxes   d. steroids

2. _____ The sex hormones are formed from this class
3. _____ Richest source of body energy
4. _____ The lipid found in honeycombs
5. _____ Cholesterol belongs to this class
6. _____ This class may have either saturated or unsaturated structure
7. _____ This molecule has a phosphate functional group in place of a fatty acid chain
8. _____ This is the only class of lipids that lack fatty acid tails
9. _____ The primary component of cell membranes
10. _____ All possess a rigid backbone of four fused carbon rings
11. _____ Found in the cuticles of plants
12. _____ Precursors of vitamin D and bile salts
13. _____ This class is made from three fatty acids combined with a unit of glycerol
14. _____ Vegetable oil belongs to this class
15. _____ Used by vertebrates for insulation
16. _____ Furnishes protection and lubrication for hair, skin, and feathers
17. _____ The neutral fats belong to this class

Dichotomous Choice [pp.44-45]

Choose the correct one of the two possible answers given between parentheses in each statement.

18. _____ Most vegetable oils are (unsaturated / saturated).
19. _____ Unsaturated fatty acids have one or more (single / double) bonds that make them less flexible.
20. _____ Triglycerides are (hydrophobic / hydrophilic).

Short Answer [pp.44-45]

21. Why are saturated fatty acids considered to be much less healthy than unsaturated fatty acids.

_____

_____

_____

22. What characteristics of phospholipids make them excellent for building cell membranes?

_____

_____

_____

3.6. PROTEINS—DIVERSITY IN STRUCTURE AND FUNCTION [pp.46-47]

3.7. WHY IS PROTEIN STRUCTURE SO IMPORTANT? [p.48]

Boldfaced, Page-Referenced Terms

amino acid _____

_____

peptide bond_____

_____

polypeptide _____

_____

protein _____

_____

denature _____

_____

prion_____

_____

Labeling [p.46]

1. In the following model of an amino acid, label the R group, amino group, and carboxyl group.

a. _____

b. _____

c. _____

Matching [pp.46-48]

Choose the most appropriate statement for each term.

2. _____ amino acid

3. _____ peptide bond

4. _____ polypeptide chain

5. _____ primary structure

6. _____ secondary structure

7. _____ tertiary structure

8. _____ domain

9. _____ quaternary structure

10. _____ mutation

11. _____ amino acid substitution

12. _____ denaturation

A. Coils or twists in amino acids caused by hydrogen bonds

B. Three or more amino acids joined in a linear chain

C. An event that can alter protein structure enough to block or enhance its function

D. The type of covalent bond linking one amino acid to another

E. Globular proteins and hemoglobin are examples of this level of protein structure

F. The unwinding of protein structure causing a change in shape

G. The lowest level of protein structure consisting of a linear, unique sequence of amino acids

H. A small organic compound having an amino group, an acid group, a hydrogen atom, and an R group

I. The level of organization determined by interacting domain

J. A mutation that results in replacement of the correct amino acid with an incorrect amino acid

K. A structurally stable unit of a polypeptide chain

Fill-in-the-Blanks [pp.46-48]

What is the take-home lesson? A protein's (13) _____ dictates its function. Hemoglobin, hormones, (14) and _____ transporters—such proteins help us survive. Twists and folds in their (15) _____ chains form anchors, or (16) _____-spanning barrels, or jaws that grip enemy agents in the body. (17) _____ can alter the chains enough to block or enhance an anchoring, transport, or defensive function. Sometimes the consequences are awful. Yet changes in sequences and functional (18) _____ also give rise to variation in (19) _____—the raw material for (20) _____.

Choice [p.45]

For questions 21-27, choose from the answers below. Answers may be used more than once. Refer to Figure 3.17.

a. primary protein structure

b. secondary protein structure

c. tertiary protein structure

d. quarternary protein structure

21. _____ Each protein's unique sequence of amino acids

22. _____ A structurally stable unit known as a "domain" comprises this level of protein structure

23. _____ Involves polypeptide chains forming sheets and/or helices

24. _____ Held in place by hydrogen bonds

25. _____ Two or more polypeptide chains associated as one molecule

26. _____ Results in a functional protein

27. _____ Leads to the formation of fibrous proteins that are involved in structure and organization of cells and tissues

Identification [pp.41, 46]

28. Study the structural formulas of the two adjacent amino acids. Identify the enzyme action causing formation of a covalent bond and a water molecule (through a condensation reaction) by circling an H atom from one amino acid and an –OH group from the other amino acid. Also circle the covalent bond that formed the dipeptide.

amino acid   amino acid   dipeptide

Short Answer [pp.46-48]

29. Why are prions such a concern and how does protein folding play a role? Name two prion diseases.

_____

_____

_____

3.8. NUCLEIC ACIDS [p.49]

Boldfaced, Page-Referenced Terms

ATP _____

_____

DNA _____

_____

nucleic acid _____

_____

nucleotide _____

_____

RNA _____

_____

Labeling [p.49]

1. In the following diagram of a nucleotide, label the phosphate groups, nitrogenous base, and five-carbon sugar subunits.

a. _____

b. _____

c. _____

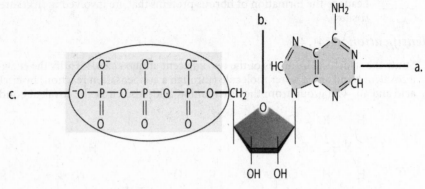

Matching [p.49]

Choose the most appropriate answer for each term.

2. _____ adenosine triphosphate

3. _____ nucleic acids

4. _____ double helix

5. _____ RNAs

6. _____ DNA

A. A molecule that can transfer phosphate groups, making molecules reactive

B. Single-stranded molecules made of ribonucleotides

C. Chains of nucleotides in which the sugar of one is bonded at the phosphate group of the other

D. The signature molecule of life, made from deoxyribonucleotides

E. Two nucleotide chains twisted together to form a DNA molecule

True/False [p.49]

If the statement is true, write a T in the blank. If the statement is false, rewrite it to make it correct.

7. _____ Nucleotides are small, inorganic molecules.

8. _____ Nucleotides can function as energy carriers, information carriers, and enzyme helpers.

9. _____ Nucleic acids are composed of a sugar, phosphate group and nucleotide base.

10. _____ All species have the same pattern of DNA.

11. _____ ATP acts as an information molecule.

12. How does DNA guide the development of an organism?

_____

_____

_____

_____

SELF-TEST

Choice

For questions 1-10, choose the class of organic molecule to which the item belongs. Some answers may be used more than once.

a. lipids   b. nucleic acids   c. proteins   d. carbohydrates

1. _____ glycoproteins

2. _____ phospholipids

3. _____ glycogen

4. _____ adenosine triphosphate

5. _____ sucrose and maltose

6. _____ triglycerides

7. _____ DNA and RNA

8. _____ cholesterol

9. _____ glycogen and starch

10. _____ waxes

Multiple Choice

___ 11. Amino acids are linked by _____ bonds to form the primary structure of a protein.
     a. disulfide
     b. hydrogen
     c. ionic
     d. peptide

___ 12. Carbohydrates are made up of.
     a. carbon, hydrogen, and oxygen
     b. carbon, hydrogen, and nitrogen
     c. hydrogen, oxygen, and sulfur
     d. carbon and hydrogen

___ 13. Proteins _____.
     a. are weapons against disease-causing bacteria and other invaders
     b. are composed of amino acid subunits
     c. may act as hormones
     d. may function as enzymes
     e. all of the above

___ 14. A long chain of carbohydrates is called a
     a. polypeptide
     b. lipid
     c. protein
     d. polysaccharide

___ 15. The primary storage form of energy in plants is
     a. glycogen
     b. starch
     c. cellulose
     d. protein

___ 16. Which of the following does not belong to the lipid class of organic molecules?
     a. sterols
     b. waxes
     c. phospholipids
     d. glycoproteins
     e. triglycerides

17. DNA _____.
   a. is one of the adenosine phosphates
   b. is one of the nucleotide coenzymes
   c. is double-stranded
   d. is composed of monosaccharides

18. Denaturation is a change in _____ shape.
   a. lipid
   b. carbohydrate
   c. protein
   d. nucleic acid

19. Of what is a triglyceride composed
   a. 3 fatty acids
   b. one glycerol with 3 fatty acids
   c. 3 glycerol with one fatty acid
   d. 3 glycerols

20. Carbon is a part of so many different substances because _____.
   a. carbon generally forms two covalent bonds with a variety of other atoms
   b. a carbon atom generally forms four covalent bonds with a variety of atoms
   c. carbon ionizes easily
   d. carbon is a polar compound

21. Which of the following levels of protein structure is not correctly linked to its description?
   a. Primary—the linear sequence of amino acids
   b. Secondary—coiling of a polypeptide due to the action of hydrogen bonds
   c. Tertiary—interactions between the domains of a protein
   d. Quaternary—chemical interactions between multiple polypeptide chains
   e. All of the above are correct

22. _____ are molecules used by cells as structural materials, as energy transport molecules, or as storage forms of energy.
   a. lipids
   b. nucleic acids
   c. carbohydrates
   d. proteins

23. Hydrolysis could be correctly described as the _____.
   a. heating of a compound in order to drive off its excess water and concentrate its volume
   b. breaking of a long-chain compound into its subunits by adding water molecules to its structure between the subunits
   c. linking of two or more molecules by the removal of one or more water molecules
   d. constant removal of hydrogen atoms from the surface of a carbohydrate
   e. prime example of a condensation class of reactions

24. The functional group(s) associated with amino acids is/are _____.
   a. hydroxyl
   b. carbonyl
   c. amide/amine
   d. carboxyl
   e. both amide/amine and carboxyl

25. Which is not a function of carbohydrates in general?
   a. structural support
   b. energy storage
   c. enzymatic catalysis

CHAPTER OBJECTIVES/REVIEW QUESTIONS

1. Distinguish between organic and inorganic compounds. [p.38]
2. Explain why carbon is important in the study of biological chemistry. [p.38]
3. Explain the role of a functional group. [pp.40-41]
4. Be able to identify select functional groups. [pp.40-41]
5. Describe the importance of enzymes to chemical reactions. [p.40]
6. Define and distinguish between *condensation reactions* and *hydrolysis*; cite a general example. [p.40]
7. Give the general characteristics of a carbohydrate. [p.42]
8. Name and generally define the three classes of carbohydrates. Give an example of each. [pp.42-43]
9. For the complex carbohydrates, indicate which is associated with animals, which with plants, and which with fungi. [pp.42-43]
10. Give the general characteristics of lipids and list their general functions. [pp.44-45]
11. Distinguish a saturated fatty acid from an unsaturated fatty acid. [p.44]
12. Describe the structure of triglycerides. [p.44]

13. Describe the structure of phospholipids and give their biological importance. [p.45]
14. Describe how waxes are used by organisms. [p.45]
15. Give the general characteristics of steroids and describe how their structure differs from the other classes of lipids. [p.45]
16. List the functions of steroids. [p.45]
17. Describe the structure of an amino acid and how amino acids form peptide bonds. [p.46]
18. Describe how the primary, secondary, tertiary, and quaternary structure of proteins results in complex three-dimensional structures. [pp.46-47]
19. Define *denaturation*. [p.48]
20. List the three parts of every nucleotide. [p.49]
21. Distinguish between DNA and RNA. [p.49]

INTEGRATING AND APPLYING KEY CONCEPTS

1. Humans can obtain energy from many different food sources. Do you think this ability is an advantage or a disadvantage in terms of long-term survival? Why?
2. Modern diets often focus on proteins as a primary energy source. What are the preferred energy molecules of the cell? What is the overall role of proteins? Do you think that the metabolic pathways have evolved for long-term protein use for energy?
3. Condensation reactions are sometimes called dehydration synthesis reactions. Explain why this term is also appropriate.
4. In the early 20th century, scientists thought that proteins might be the hereditary material. What characteristics of proteins would have supported this idea? Why wouldn't carbohydrates or lipids be good candidates for hereditary material?
5. A common statement in nutrition is "you are what you eat." Explain why this statement is both true and false.

4

CELL STRUCTURE AND FUNCTION

INTRODUCTION

For biologists, the cell is where it all begins—it is considered to be the fundamental unit of life. In this chapter you will begin to explore some of the basic principles of cell structure. As you will see, cells are intricate, efficient, biological machines. Even the bacteria, which some call "simple cells," are complex in their interactions with the environment. Your primary goal for this chapter is to understand the terminology used to describe cellular structures and their functions. Future chapters of the textbook will explore in greater detail how cells operate in their environment. Understanding the interrelated structure and function of cellular organelles will provide you with the foundation needed later on in the textbook.

STUDY STRATEGIES

- Read all of Chapter 4 with the goal of familiarizing yourself with the boldface terms.
- Describe the basic structure of all living cells.
- Explain why cells need to be small.
- Identify the four basic tenets of the cell theory.
- Explain the basic structure of a compound light microscope.
- Discuss the difference between a scanning electron microscope and a transmission electron microscope.
- Identify the basic structure and function of prokaryotic cells.
- Discuss the importance of biofilms to ecosystems.
- Identify the basic structure and function of eukaryotic cells.
- Discuss the structure and function of the following organelles: nucleus, endomembrane system (ER, Golgi apparatus, peroxisome, and vacuole), lysosome, mitochondrion, plastids, chloroplasts, chromoplasts, amyloplasts, and cytoskeleton (microtubules, microfilaments, intermediate filaments, motor proteins, and centrioles).
- Explain the importance of the following cell surface specializations: extracellular matrix, primary and secondary cell walls, lignin, cuticle, and cell junctions (plasmodesmata, tight junctions, adhering junctions, and gap junctions).
- Discuss the characteristics of life.

FOCAL POINTS

- Figure 4.2 animated [p.54] shows the difference between prokaryotic and eukaryotic cells.
- Figure 4.3 animated [p.54] illustrates the basic structure of the cell membrane.
- Figure 4.4 animated [p.55] demonstrates the relationship between cell surface area and cell volume.
- Figure 4.5 animated [p.56] compares the structure of light and electron microscopes.
- Figure 4.6 [pp.56-57] depicts relative sizes of objects and the visual tools necessary to see them.
- Figure 4.8 animated [p.58] shows the general body plan of a prokaryote.
- Table 4.2 [p.60] summarizes eukaryotic organelles and their functions.
- Figure 4.12 animated [p.61] is an excellent visual tool for studying plant and animal cellular structure and function.
- Figure 4.14 animated [p.63] shows how the structure of the nuclear envelope is related to its function.
- Figure 4.15 animated [p.64] illustrates the interrelationships between components of the endomembrane system.
- Figure 4.17 [p.67] details the structure of a mitochondrion.

- Figure 4.18 animated [p.67] reveals the internal structure of the chloroplast.
- Figure 4.19 animated [p.68] details the various cytoskeletal elements.
- Figure 4.21 animated [p.69] demonstrates the mechanism behind movement of cilia and flagella.
- Figure 4.22 animated [p.70] demonstrates the development of plant cell wall structure.
- Figure 4.24 animated [p.71] illustrates the different types of cellular junctions.
- Table 4.4 [p.73] is an excellent comparison of the cell components of prokaryotic and eukaryotic cells and their functions in those cells.

INTERACTIVE EXERCISES

4.1. FOOD FOR THOUGHT [p.53]

4.2. CELL STRUCTURE [pp.54-55]

4.3. HOW DO WE SEE CELLS? [pp.56-57]

Boldfaced, Page-Referenced Terms

plasma membrane_____

_____

cytoplasm _____

_____

organelles _____

_____

nucleus _____

_____

surface-to-volume ratio _____

_____

cell theory _____

_____

Matching [pp.54-55]

Choose the most appropriate answer for each term.

1. _____ prokaryotic cells
2. _____ plasma membrane
3. _____ cytoplasm
4. _____ cell theory
5. _____ nucleus
6. _____ eukaryotic cells
7. _____ surface-to-volume ratio
8. _____ lipid bilayer
9. _____ organelles
10. _____ cell

A. Structures that carry out specialized functions in a cell

B. The structural foundation of all cell membranes

C. The type of cell that lacks a nucleus

D. A physical relationship that constrains increases in cell size and shape

E. The smallest unit of life that retains all the properties of life

F. Theory composed of 4 generalization that explains observations of cells

G. The thin outermost membrane of cells that separates metabolic activities from events outside the cell

H. In eukaryotic cells, the membranous sac that contains the DNA

I. The semifluid material between the plasma membrane and the region of DNA

J. A type of cell possessing internal membranes that divide the cytoplasm into compartments

Short Answer [p.55]

11. List the four basic principles of the cell theory.

_____

_____

_____

_____

_____

_____

Matching [pp.56-57]

Choose the most appropriate statement for each term relating to microscopes.

12. _____ micrograph

13. _____ transmission electron microscope

14. _____ wavelength

15. _____ compound light microscope

16. _____ scanning electron microscope

17. _____ fluorescence microscope

18. _____ contrast

A. Glass lenses bend incoming light rays to form an enlarged image of a cell or another specimen

B. The distance from one wave's peak to the peak of the wave behind it

C. A narrow beam of electrons moves back and forth across the surface of a specimen coated with a thin metal layer

D. A photograph of an image formed with a microscope

E. Difference between light and dark sections of stained specimens; allows greater detail to be seen

F. Electrons pass through a thin section of cells to form an image of its internal details

G. A cell or molecule is the source of emitted light when a laser is focused on the specimen

Matching [p.55]

Match the scientist(s) with the statement that best describes his contribution to science.

19. _____ Antoni van Leeuwenhoek

20. _____ Robert Hook

21. _____ Robert Brown

22. _____ Matthias Schleiden and Theodor Schwann

23. _____ Rudolf Virchow

A. Discovered that cells divide into descendant cells

B. The first to identify a plant cell

C. First articulated the cell theory

D. First coined the term "cellulae" (cell)

E. The first to observe living microscopic "animalcules" and "beasties"

Completion [p.56]

Fill in the blanks to complete the following table.

24. a. 1 _____ = 1/100 Meter; b. 1 Meter = _____ cm

25. a. 1 _____ = 1/1000 Meter b. 1 Meter = _____ mm

26 a. 1 _____ = 1/1,000,000 Meter b. 1 Meter = _____ μm

27. a. 1 _____ = 1/1,000,000,000 Meter b. 1 Meter = _____ nm

4.4. INTRODUCING "PROKARYOTES" [pp.58-59]

4.5. INTRODUCING EUKARYOTIC CELLS [pp.60-61]

Boldfaced, Page-Referenced Terms

ribosomes _____

_____

plasmids _____

_____

nucleoid _____

_____

cell wall _____

_____

flagella _____

_____

pili_____

_____

biofilm _____

_____

Matching [pp.58-59]

Match each of the following prokaryotic structures with the correct description.

1. _____ cell walls
2. _____ capsule
3. _____ plasma membrane
4. _____ flagella
5. _____ ribosomes
6. _____ pili
7. _____ nucleoid
8. _____ plasmid
9. _____ biofilm

A. Sites where polypeptide chains are built

B. Permeable to dissolved substances; composed of protein or peptidoglycan

C. Small circles of DNA containing a few genes

D. Selectively controls the movement of materials in and out of the cytoplasm

E. Slender cellular structures involved in mobility

F. Made of polysaccharides; allows bacteria to attach to surfaces

G. Community of prokaryotes living in a shared, sticky mass of polysaccharides and glycoproteins

H. Protein filaments that project from surface of bacterial cells; used for attachment

I. Region of bacterial cell where a single, circular DNA molecule is located

Matching [pp.60-61]

Select the function that most closely fits the eukaryotic cellular structures listed.

10. _____ lysosome
11. _____ Golgi body
12. _____ nucleus
13. _____ chloroplast
14. _____ endoplasmic reticulum (ER)
15. _____ cell wall
16. _____ mitochondrion
17. _____ cytoskeleton
18. _____ centriole

A. Protects DNA from damaging reactions in cytoplasm

B. Modifies new polypeptide chains; synthesizes lipids

C. Protects and structurally supports plant cells

D. Intracellular digestion; recycles materials

E. Produces ATP by aerobic respiration

F. Produces and organizes the microtubules

G. Finishes, sorts, and ships proteins, lipids, and enzymes

H. Photosynthetic organelle

I. Internal structural support for the cell

Labeling [p.61]

Identify each indicated part of the accompanying illustrations.

19. _____

20. _____

21. _____

22. _____

23. _____

24. _____

25. _____

26. _____

27. _____

28. _____

29. _____

30. _____

31 _____

32. _____

33. _____

34. _____

35. _____

36. _____

37. _____

38. _____

39. _____

40. _____

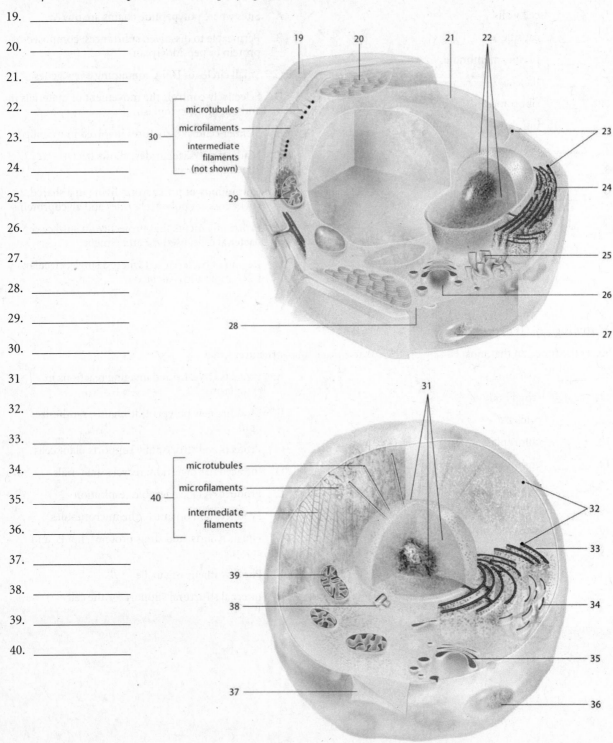

Labeling [p.58]

Identify each indicated part of the accompanying illustrations.

41. _____

42. _____

43. _____

44. _____

45. _____

46. _____

47. _____

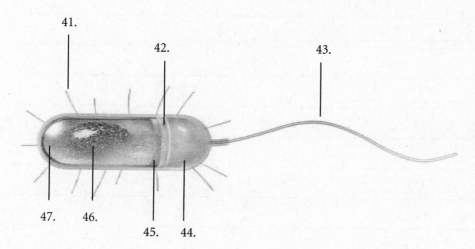

Short Answer [p.58-61]

48. Why do scientists compare a cell to a factory?

_____

_____

_____

4.6. THE NUCLEUS [pp.62-63]

4.7. THE ENDOMEMBRANE SYSTEM [pp 64-65]

4.8. LYSOSOME MALFUNCTION [p.66]

Boldfaced, Page-Referenced Terms

nuclear envelope _____

_____

nucleolus _____

_____

nucleoplasm _____

_____

chromosome _____

_____

chromatin _____

_____

endomembrane system _____

_____

endoplasmic reticulum _____

_____

ER _____

_____

vesicles _____

_____

peroxisomes _____

_____

lysosomes _____

_____

vacuoles _____

_____

central vacuoles _____

_____

Golgi body _____

_____

Short Answer [p.66]

1. Explain how a genetic mutation may cause an error in cell metabolism resulting in Tay-Sachs disease.

_____

_____

_____

Matching [pp.62-66]

Match each of the following terms with the corresponding description.

2. _____ chromosome
3. _____ nucleolus
4. _____ smooth ER
5. _____ nuclear envelope
6. _____ rough ER
7. _____ chromatin
8. _____ peroxisome
9. _____ lysosome
10. _____ Golgi body
11. _____ ribosome
12. _____ sarcoplasmic reticulum
13. _____ vesicles
14. _____ vacuoles
15. _____ nucleoplasm

A. Site of ribosome assembly
B. Finishing, sorting, and packaging of proteins and lipids
C. Viscous fluid within the nucleus
D. Contains enzymes for intracellular digestion
E. Folds polypeptides into tertiary-structured proteins
F. A type of smooth ER found in muscle cells
G. A double membrane that encloses the DNA
H. Makes lipids for the cell membrane
I. Membrane-bound intracellular trash cans
J. Synthesizes polypeptide chains
K. Small, membrane-enclosed saclike organelles
L. One DNA molecule and associated proteins
M. All the DNA and associated proteins in the nucleus
N. Contains enzymes to digest fatty acids, amino acids, alcohol, and hydrogen peroxide

Sequence [pp.64-65]

Place the numbers 1-7 in the spaces before each statement to put into proper sequence these events in the organelles of the endomembrane system.

16. _____ Rough ER folds a polypeptide chain into a protein with tertiary structure

17. _____ RNA moves through nuclear pores into the cytoplasm

18. _____ Golgi body adds the finishing touches to a protein and packages it for transport

19. _____ Using instructions from RNA, ribosomes synthesize polypeptide chains from amino acids

20. _____ DNA transcribes protein-making instructions into RNA

21. _____ A transport vesicle leaves the Golgi and merges with the plasma membrane

22. _____ A transport vesicle leaves the ER and travels to the Golgi body

4.9. OTHER ORGANELLES [pp.66-67]

Boldfaced, Page-Referenced Terms

mitochondrion _____

_____

plastids _____

_____

chloroplasts _____

_____

Choice [pp.66-67]

For questions 1-15, choose from the following:

    a. mitochondria   b. chloroplasts   c. amyloplasts   d. central vacuole   e. chromoplasts

1. _____ Organelle specialized for photosynthesis in eukaryotic cells
2. _____ Organelle that makes ATP during aerobic respiration
3. _____ Organelle that lacks pigments and stores starch grains
4. _____ More plentiful in cells that have big demands for energy
5. _____ Organelle with an abundance of carotenoids but no chlorophylls
6. _____ Fluid pressure in this organelle keeps plant cells firm
7. _____ Organelles that convert sunlight energy into chemical energy
8. _____ Abundant in starch-storing cells of stems, potato tubers, and seeds
9. _____ Stores amino acids, sugars, ions, and toxic wastes in its fluid-filled interior
10. _____ Found in nearly all eukaryotic cells
11. _____ Function as gravity-sensing organelles
12. _____ Uses sunlight to synthesize ATP
13. _____ Builds carbohydrates from carbon dioxide and water
14. _____ Resemble bacteria in size, form, and biochemistry
15. _____ Stores pigments that give red or orange color to fruits, flowers and roots

Short Answer [p.66]

16. Define endosymbiosis. _____

_____

_____

17. Why are mitochondrial disorders so serious? _____

_____

_____

4.10. THE DYNAMIC CYTOSKELETON [pp.68-69]

4.11. CELL SURFACE SPECIALIZATIONS [pp.70-71]

Boldfaced, Page-Referenced Terms

cytoskeleton _____

_____

microtubules_____

_____

microfilaments _____

_____

cell cortex _____

_____

intermediate filaments _____

_____

motor proteins _____

_____

cilia _____

_____

centriole _____

_____

basal body _____

_____

pseudopods _____

_____

extracellular matrix (ECM) _____

_____

primary wall _____

_____

secondary wall _____

_____

lignin _____

_____

cuticle _____

_____

cell junctions _____

_____

plasmodesmata _____

_____

tight junctions_____

_____

adhering junctions _____

_____

gap junctions_____

_____

Matching [pp.68-69]

Select the statement that best fits each component of the cytoskeleton.

1. _____ microtubules
2. _____ microfilaments
3. _____ cell cortex
4. _____ intermediate filaments
5. _____ motor protein
6. _____ kinesin
7. _____ flagella
8. _____ cilia
9. _____ centriole
10. _____ 9 + 2 array
11. _____ pseudopods
12. _____ dynein

A. Arrangement of microtubules lengthwise through a flagellum or cilium

B. A reinforcing mesh of microfilaments under the plasma membrane

C. Long hollow cylinders made of subunits of tubulin

D. A motor protein used to bend a flagellum

E. The "false feet" eukaryotic cells use in locomotion and obtaining food

F. Long slender structure used for motion; cells have one or a few of them

G. Fibers composed mainly of subunits of actin

H. A barrel-shaped organelle responsible for the growth of structures made of microtubules

I. The most stable part of the cytoskeleton; strengthen and maintain tissue structures

J. Short, slender structures used for coordinated motion; cells have many of them

K. A type of accessory protein that uses ATP to move cell parts along microtubule and microfilament tracks

L. A motor protein used to move chloroplasts toward or away from a light source

Matching [pp.70-71]

Choose the most appropriate statement for each term.

13. _____ primary wall
14. _____ secondary wall
15. _____ tight junctions

16. _____ adhering junctions
17. _____ gap junctions
18. _____ plasmodesmata

19. _____ middle lamella

20. _____ pectin

21. _____ cellulose

22. _____ lignin

23. _____ cuticle

24.. _____ extracellular matrix (ECM)

A. Seal abutting cells that line the outer surfaces and internal cavities of animals so no fluids can pass between them

B. The main constituent of a plant's primary cell wall

C. A strong, waterproof component of the secondary cell walls of older stems and roots

D. Layers of firm material secreted onto the primary cell wall's inner surface that strengthen the wall and maintain its shape

E. Thin, pliable structure made of polysaccharides that allows a plant cell to grow and enlarge

F. Nonliving mixture of fibrous proteins and polysaccharides surrounding the cells that secrete it

G. A sticky layer in between the primary walls of adjoining cells

H. Substance a young plant cell first secretes onto the outer surface of its plasma membrane

I. Cytoplasmic channels that cross the cell walls and join neighboring plant cells

J. A protective body covering made of cell secretions

K. Anchor cells to each other and to the extracellular matrix

L. Open channels that connect the cytoplasm of neighboring animal cells

Labeling [p.70-71]

Identify each indicated cell junction from the accompanying illustration.

25. _____

26. _____

27. _____

28. _____

28.

25.

26.

27.

SELF-TEST

___ 1. Which of the following statements most correctly describes the relationship between cell surface area and cell volume?
  a. As a cell expands in volume, its surface area increases at a rate faster than its diameter does
  b. Volume increases with the square of the diameter, but surface area increases only with the cube
  c. If a cell were to grow four times in diameter, its volume of cytoplasm increases 16 times and its surface area increases 64 times
  d. Volume increases with the cube of the diameter, but surface area increases only with the square

___ 2. The cellular structure in animals that is involved in the process of intracellular digestion is the _____ .
  a. lysosome
  b. rough endoplasmic reticulum
  c. microtubules
  d. mitochondria

___ 3. Which of the following is not found in a prokaryotic cell?
  a. DNA
  b. plasma membrane
  c. cytoplasm
  d. nucleus

___ 4. The nucleolus is the site where _____.
  a. the protein and RNA subunits of ribosomes are assembled
  b. the chromatin is formed
  c. chromosomes are bound to the inside of the nuclear envelope
  d. chromosomes duplicate themselves prior to cell division

___ 5. The _____ is free of ribosomes and curves through the cytoplasm like connecting pipe. It is the site of lipid synthesis and some detoxification.
  a. lysosome
  b. Golgi body
  c. smooth ER
  d. rough ER

___ 6. Which of the following is not present in all cells?
  a. cell wall
  b. plasma membrane
  c. ribosomes
  d. DNA molecules

___ 7. A part of the endomembrane system the _____ finishes, packages, and sorts lipids and proteins.
  a. endoplasmic reticulum
  b. Golgi bodies
  c. peroxisomes
  d. lysosomes

___ 8. Chloroplasts _____
  a. are specialists in oxygen-requiring reactions
  b. function as part of the cytoskeleton
  c. trap sunlight energy and produce organic compounds
  d. assist in carrying out cell membrane functions

___ 9. Aerobic respiration occurs in the _____.
  a. chloroplast
  b. mitochondria
  c. peroxisome
  d. cytoplasm

___ 10. The component of the cell that is involved with cellular movement is the _____.
  a. nucleolus
  b. cell wall
  c. chromatin
  d. cytoskeleton

___ 11. The _____ controls construction of parts of the cytoskeleton in animals.
  a. mitochondrion
  b. endoplasmic reticulum
  c. centriole
  d. ribosome

___ 12. _____ allow communication between the cytoplasm of adjacent cells.
  a. tight junctions
  b. adhering junctions
  c. gap junctions
  d. all of the above

Cell Structure and Function   **59**

13. The _____ creates an image by passing a beam of electrons back and forth across the surface of a specimen coated with a thin layer of metal.
   a. compound light microscope
   b. fluorescence microscope
   c. transmission electron microscope
   d. scanning electron microscope

14. _____ is the sticky substance that holds young plant cells together.
   a. pectin
   b. cellulose
   c. lignin
   d. chitin

15. Which of the following is the energy supplier for the cell?
   a. lysosome
   b. mitochondria
   c. Golgi apparatus
   d. nucleus

16. The modern Cell Theory states that all cells come from
   a. non-living material
   b. nothingness
   c. other existing cells

17. The organelle responsible for storing water in the plant cell is the
   a. Golgi body
   b. endoplasmic reticulum
   c. vacuole
   d. ribosome

18. What is one major feature that plant cells have the animal cells do not?
   a. lysosome
   b. cell wall
   c. plasma membrane

19. Anton von Leeuwenhoek _____.
   a. looked at cork under the microscope
   b. stated that all animals are made of cells
   c. all plants are made of cells
   d. used a simple light microscope and saw microorganisms

20. Which of the following uses a laser to look for reflecting light off of a specimen?
   a. compound light microscope
   b. scanning electron microscope
   c. transmission electron microscope
   d. florescence microscope

21. Which of the following correctly orders the measurement from smallest to largest units?
   a. meter, nanometer, centimeter, millimeter
   b. nanometer, millimeter, centimeter, meter
   c. millimeter, meter, centimeter, nanometer
   d. centimeter, nanometer, millimeter, meter

22. Which of the following is NOT a eukaryote?
   a. bacteria
   b. fungus
   c. plant
   d. animal

CHAPTER OBJECTIVES/REVIEW QUESTIONS

1. Describe the very basic components of a cell. [p.54]
2. Explain how the surface-to-volume ratio of a cell determines its size. [p.56-57]
3. Distinguish among the different forms of microscopes including their uses and limitations. [pp.56-57]
4. Describe the basic structure of a prokaryotic cell. [pp.58-59]
5. Describe the differences between prokaryotic and eukaryotic cells. [pp.58-60, 73]
6. Explain the advantage of complex compartmentalization (organelles) in eukaryotic cells. [p.60]
7. Briefly describe the cellular location and function of the organelles typical of most eukaryotic cells: nucleus, ribosomes, endoplasmic reticulum, Golgi body, various vesicles, mitochondria, and the cytoskeleton. [pp.60, 73]
8. List the two main functions of a nucleus. [p.62]
9. Describe the structure of a eukaryotic nucleus. [pp.62-63]
10. Distinguish between a chromosome and chromatin. [p.63]
11. List the organelles that are part of the endomembrane system and their functions. [pp.64-65]
12. Explain the variety of tasks that lysosomes and peroxisomes perform for a eukaryotic cell. [p.66]
13. Understand the role of the mitochondria. [p.66]
14. Give the general function of the following plant organelles: chloroplasts, chromoplasts, amyloplasts, and the central vacuole. [p.66-67]

15. Distinguish between microtubules, microfilaments, and intermediate filaments with regard to structure and function. [p.68-69]
16. Explain the three basic mechanisms of eukaryotic cell movement. [p.69]
17. Understand the purpose of a cell wall and the types of eukaryotic cells that utilize a cell wall. [p.70-71]
18. Explain the difference in function between gap junctions, adhering junctions, and tight junctions. [p.71]
19. Identify the primary properties of living organisms. [p.72]

INTEGRATING AND APPLYING KEY CONCEPTS

1. Which parts of a cell constitute the minimum for keeping the simplest of living cells alive?
2. How did the existence of a nucleus, compartments, and extensive internal membranes confer selective advantages on cells that developed these features?
3. Cells are frequently described as biological factories. Compare a typical eukaryotic cell to a modern automobile factory. Describe the function of the nucleus, endoplasmic reticulum, mitochondria, Golgi body, ribosomes, vesicles, lysosomes, and peroxisomes in relation to a factory.
4. Dental plaque is a biofilm that supports the growth of many types of bacteria including Streptococcus mutans, a major factor in the formation of cavities. Consider the nature of a biofilm and explain the benefits of frequent brushing and flossing of teeth.
5. Suppose that you wanted to develop a drug that inhibited the ability of a eukaryotic cell to export proteins. Explain what possible organelles you would target and the basic mechanism by which your drug would work.

5

GROUND RULES OF METABOLISM

INTRODUCTION

This chapter introduces energy and describes some of its properties. It also describes enzyme structure and function and the relationship between enzymes and energy. The chapter then discusses how enzyme reactions are put together in an organism's metabolism.

STUDY STRATEGIES

- Read all of Chapter 5 with the goal of familiarizing yourself with the boldface terms.
- Identify the Laws of Thermodynamics and explain their importance to biological systems.
- Define *energy* and identify the various forms.
- Compare and contrast endergonic and exergonic reactions and define reactants and products.
- Discuss metabolism and feedback inhibitions importance for living organisms.
- Summarize redox reactions.
- Relate cofactors to metabolism and explain the ATP-ADP cycle.
- Identify the main components of cell membranes and their functions.
- Explain the processes of diffusion and osmosis.
- Discuss the aspect of semipermeability in membranes and why that characteristic is so important.
- Compare and contrast passive transport, active transport, endocytosis, and exocytosis.

FOCAL POINTS

- The introductory paragraphs and final pages [pp.77, 96-97] look at some of the roles of alcohol dehydrogenase.
- Figure 5.2 [p.78] illustrates the concept of entropy.
- Section 5.2 [pp.78-79] describes energy and its characteristics.
- Figure 5.4 [p.79] demonstrates the one-way flow of energy into and out of living organisms.
- Figure 5.8 [p.81] demonstrates the role of activation energy in a chemical reaction.
- Sections 5.4-5.6 [pp.82-87] look in detail at enzyme structure and function.
- Figure 5.18 [p.87] illustrates the structure of ATP and the ADP/ATP cycle.
- Figure 5.20 [pp.88-89] show the structure of the fluid mosaic model of biological membranes
- Table 5.2 [p.88] provides a detailed explanation of the various functions of membrane proteins.
- Figure 5.22 [p.90] illustrates the concept of selectively permeability.
- Figure 5.23 [p.91] looks at the importance of osmosis in biological systems.
- Figures 5.24-5.26 [pp.92-93] provides illustrations of passive, active, and cotransport.
- Figure 5.27 [p.94] provides a detailed overview of endocytosis and exocytosis.

INTERACTIVE EXERCISES

5.1. A TOAST TO ALCOHOL DEHYROGENASE [p.77]

5.2. ENERGY IN THE WORLD OF LIFE [pp.78-79]

Boldfaced, Page-Referenced Terms

energy _____

_____

kinetic energy _____

_____

first law of thermodynamics _____

_____

entropy _____

_____

second law of thermodynamics _____

_____

potential energy _____

_____

Fill-in-the-Blanks [p.77]

(1) _____ molecules move quickly from the stomach to the bloodstream. The enzyme (2) _____ is found in the liver and helps rid the body of alcohol. One possible outcome of liver cell death is (3) _____ which is characterized by inflammation and destruction of the liver tissue. The liver has many important functions; it (4) _____, (5) _____, and (6) _____. (7) _____ is defined as consuming five alcoholic drinks in a two hour period. More than 1400 students die as a result of (7) _____ every year.

Choice [pp.78-79]

In the blank preceding each item, indicate if the first law of thermodynamics (I) or the second law of thermodynamics (II) is best described.

8. _____ Apple trees absorbing energy from the sun and storing the energy in the chemical bonds of starch and sugar

9. _____ A hydroelectric plant at a waterfall, producing electricity

10. _____ A cup of hot coffee cooling over time

11. _____ The glow of an incandescent bulb following the flow of electrons through a wire

12. _____ Earth's sun continuously losing energy to its surroundings

13. _____ The movement of a gasoline-powered automobile

14. _____ Humans running the 100-meter dash following usual food intake

15. _____ The death and decay of an organism

Short Answer [pp.78-79]

16. Why is it said that the total amount of energy available for doing work in the universe is always decreasing?

   _____

   _____

17. Define *entropy*._____

   _____

   _____

18. Describe the First and Second Laws of Thermodynamics._____

   _____

   _____

19. What is the difference between potential and kinetic energy? _____

   _____

   _____

True/False [pp.78-79]

If the statement is true, write a T in the blank. If the statement is false, make it correct by changing the underlined word(s) and writing the correct word(s) in the answer blank.

20. _____ The first law of thermodynamics states that entropy is constantly increasing in the universe.

21. _____ A molecule of glucose has less entropy than the individual carbon, hydrogen, and oxygen atoms that go into making it.

22. _____ The amount of low-quality energy in the universe is decreasing.

23. _____ No energy conversion can ever be 100 percent efficient.

24. _____ In our world, energy can flow in two directions: from the sun, to producers, to consumers, and back again.

5.3. THE ENERGY IN THE MOLECULES OF LIFE [pp.80-81]

5.4. HOW ENZYMES MAKE SUBSTANCES REACT [pp.82-83]

Boldfaced, Page-Referenced Terms

reactants _____

   _____

products _____

_____

endergonic _____

_____

exergonic_____

_____

activation energy_____

_____

catalysis_____

_____

substrates _____

_____

active site_____

_____

transition site _____

_____

induced fit model _____

_____

Fill-in-the-Blanks [pp.80-81]

In most reactions, (1) _____ of the reactants differs from (1) of the products. Reactions in which the reactants have less energy are (2) _____. The opposite is true in (3) _____ reactions. Regardless of the type of reaction, (4) _____ is needed to start the reaction. All reactions have (4) but different reactions require different amounts. Cells (5) _____ free energy by building organic compounds. Cells (6) _____ free energy by breaking bonds of organic compounds.

Choice [pp.80-81]

Classify each of the following reactions as *endergonic* or *exergonic*.

7. _____ Burning wood at a campfire

8. _____ The products of a chemical reaction have more energy than the reactants

9. _____ Glucose + oxygen = carbon dioxide + water plus energy

10. _____ The reactants of a chemical reaction have more energy than the product

11. _____ The reaction releases energy

Dichotomous Choice [pp.80-81]

Choose the correct one of the two possible answers given between parentheses in each statement.

12. _____ Every chemical bond holds (energy/light).

13. _____ (Activation/potential) energy is the minimum amount needed to start chemical reactions.

14. _____ (Photosynthesis/respiration) is an example of an endergonic reaction.

15. _____ Molecules entering into a reaction are (reactants/products).

Matching [pp.80-83]

Match the most appropriate letter to its number.

16. _____ exergonic reaction   A. Substances able to enter into a reaction

17. _____ endergonic reaction   B. Reactions that release energy

18. _____ reactants (substrates)   C. Proteins (usually) that catalyze reactions

19. _____ enzymes   D. Molecules that remain at the end of a reaction.

20. _____ product   E. Reactions requiring a net input of energy

Identification [pp.82-83]

For the accompanying "energy hill" reaction diagram, identify the different amounts of energy.

21. _____

22. _____

23. _____

Identify the two reaction pathways.

24. _____

25. _____

Fill-in-the-Blanks [pp.82-83]

The specific substance, or reactant, upon which a particular enzyme acts is called its (26) _____; this substance fits into the enzyme's crevice, which is called its (27) _____. Enzyme induced changes in the substrate bring the (28) _____.

5.5. METABOLISM—ORGANIZED, ENZYME-MEDIATED REACTIONS [pp.84-85]

5.6. COFACTORS IN METABOLIC PATHWAYS [pp.86-87]

Boldfaced, Page-Referenced Terms

metabolic pathways_____

_____

feedback inhibition _____

_____

allosteric _____

_____

redox reactions _____

_____

electron transfer chain _____

_____

cofactor_____

_____

coenzymes_____

_____

antioxidant _____

_____

phosphorylation _____

_____

ATP_____

_____

ATP/ADP cycle _____

_____

Fill-in-the-Blanks [pp.84-85]

(1) _____ refers to the activity by which cells acquire and use energy. A (2) _____ is a series of enzyme-mediated reactions in which cells build, breakdown, or rearrange organic molecules. Metabolic pathways may be either (3) _____ or (4) _____. Enzymatic reactions may run in (5) _____ with products being converted back to (6) _____. The (7) _____ depends on the concentration of reactants and products. (8) _____ help cells maintain the needed concentration of thousands of different substances. These mechanisms may increase how fast (9) _____ are made. Others may activate or (10) _____ enzymes that are already made. An (11) _____ is a region of an enzyme that can bind regulatory molecules. Allosteric effects can cause activation of (12) _____.

Fill-in-the-Blanks [p.85]

Oxidation–reduction (or redox) refers to (13) _____ transfers. In terms of oxidation–reduction reactions, a molecule in the sequence that donates electrons is said to be (14) _____, while molecules accepting electrons are said to be (15) _____. Electrons that enter an (16) _____ are at a higher energy state than when they leave. If we think of the electron transport system as a staircase, excited electrons at the top of the staircase have the (17) _____ [choose one] (most, least) energy. Many (18) _____ deliver electrons to electron transfer chains. Energy released at certain steps is used to synthesize (19) _____.

Matching [p.85]

Match the lettered statements to the numbered items on the sketch.

20. _____

21. _____

22. _____

23. _____

24. _____

A. Represent the electron carrier molecules in an electron transport system

B. Electrons at their highest energy level

C. Released energy harnessed and used to produce ATP

D. Electrons at their lowest energy level

E. The separation of hydrogen atoms into protons and electrons

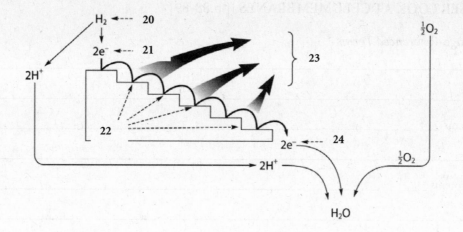

Labeling [p.87]

Identify the accompanying molecule and label its parts.   Use Figure 5.18 to guide you.

25. _____

26. _____

27. _____

28. The name of this molecule is _____

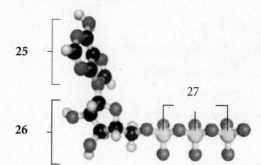

Matching [p.86]

Study the sequence of reactions indicated by arrows in the accompanying diagram. Identify the numbered components of the reactions by entering the correct letter in the appropriate blank. [pp.84-86]

29. _____

30. _____

31. _____

32. _____

33. _____

A. cofactor

B. intermediates

C. reactants

D. product

E. enzymes

5.7. A CLOSER LOOK AT CELL MEMBRANES [pp.88-89]

Boldfaced, Page-Referenced Terms

fluid mosaic _____

_____

adhesion proteins _____

_____

recognition proteins _____

_____

receptor proteins_____

_____

transport proteins _____

_____

Matching [pp.88-89]

Choose the most appropriate answer for each term.

1. _____ fluid mosaic model
2. _____ phospholipid
3. _____ active transporters
4. _____ transport proteins
5. _____ cholesterol
6. _____ integral proteins
7. _____ recognition proteins
8. _____ peripheral proteins
9. _____ receptor proteins
10. _____ lipid bilayer
11. _____ passive transporters
12. _____ adhesion proteins

A. The main steroid component of animal membranes

B. A composition of phospholipids, proteins, sterols, and glycolipids

C. The general name for proteins that are physically embedded within the cell membrane

D. The primary component of the cell membrane; consists of both hydrophobic and hydrophilic regions

E. Bind extracellular substances that trigger changes in the cell's activity

F. Help cells of the same type stick together

G. A general group of proteins positioned at the surface of the membrane

H. Proteins that move molecule through the plasma membrane by using energy

I. Allow materials to pass through the cell membrane using the interior of the protein; may or may not require energy

J. Act as molecular fingerprints to identify tissues or individuals

K. Proteins that allow molecules to move through the plasma membrane without expending energy

L. The double layer of phospholipids that forms the cell membrane

13. With reference to the fluid mosaic model of the plasma membrane, what contributes to the mosaic nature? What contributes to the fluidity? _____

_____

_____

14. Why are proteins so important to the lipid bilayer? _____

_____

_____

15. What is the basic purpose of a cell membrane? _____

_____

_____

5.8. DIFFUSION AND MEMBRANES [pp.90-91]

5.9. MEMBRANE TRANSPORT MECHANISMS [pp.92-93]

5.10. MEMBRANE TRAFFICKING [pp.94-95]

Boldfaced, Page-Referenced Terms

diffusion _____

_____

concentration _____

_____

concentration gradient _____

_____

hypotonic _____

_____

hypertonic _____

_____

isotonic _____

_____

osmosis _____

_____

turgor _____

_____

osmotic pressure _____

_____

passive transport _____

_____

active transport _____

_____

calcium pump _____

_____

Fill-in-the-Blanks [pp.90-93]

The amount of solute in a given amount of fluid is called a (1) _____. A(n) (2) _____ is a difference between the number of molecules or ions of a given substance in adjoining regions. (3) _____ is the name for the net movement of like molecules or ions down a (2); it is a factor in the movement of substances across cell membranes and through cytoplasmic fluid. The five factors that influence the rate of (3) are (4) _____, (5) _____, (6) _____, (7) _____, and (8) _____. If a membrane is (9) _____, it possesses a molecular structure that permits some substances but not others to cross it in certain ways, at certain times. For example, the lipid bilayer of biological membranes is impermeable to (10) _____ and (11) _____ molecules while allowing (12) _____ and (13) _____ to pass. (14) _____ transport uses energy to move down the molecular concentration gradient without the need for energy. (15) _____ transport uses energy to pump a solute (16) _____ its concentration gradient. In the bulk movement of substances across a membrane, the process of (17) _____ moves particles into the cell by forming a vesicle from the plasma membrane. In (18) _____, a membrane-bound vesicle inside the cell fuses with the plasma membrane, allowing particles to exit the cell.

Matching [pp.90-91]

Choose the most appropriate answer for each term.

19. _____ osmosis

20. _____ tonicity

21. _____ hypotonic solution

22. _____ hypertonic solution

23. _____ isotonic solutions

24. _____ osmotic pressure

25. _____ turgor

A. Refers to the relative solute concentrations of the fluids

B. Have the same solute concentrations

C. The amount of force that prevents further increase in a solution's volume

D. The fluid on one side of a membrane that contains more solutes than the fluid on the other side of the membrane

E. The diffusion of water in response to a water concentration gradient between two regions separated by a selectively permeable membrane

F. The fluid on one side of a membrane that contains fewer solutes than the fluid on the other side of the membrane

G. The general term for a fluid force exerted against a cell wall and/or membrane enclosing the fluid

Short Answers [pp.90-91]

Questions 26-30 refer to the following diagram, in which side A has 25 milliliters of a 3% sucrose solution and side B has 25 milliliters of a 6% sucrose solution. The membrane separating the sides is permeable to water but impermeable to sucrose.

26. In what direction will water move through the membrane? _____

_____

27. In what direction is the net movement of water? _____

_____

28. In what direction will sucrose move through the membrane? _____

_____

29. What will happen to the sucrose concentration in side A? _____

_____

30. What will happen to the fluid level in side B? _____

_____

True/False [pp.90-91]

If the statement is true, write T in the blank. If the statement is false, rewrite it to make it correct.

31. _____ A water concentration gradient is influenced by the number of solute molecules present on both sides of the membrane.

32. _____ An animal cell placed in a hypertonic solution will swell and perhaps burst.

33. _____ Multicelled organisms can counter shifts in tonicity by selectively transporting solutes across the plasma membrane.

34. _____ A plan wilts when the concentration of salt in the soil increases, causing the soil to become hypotonic.

Matching [pp.94-95]

Choose the most appropriate answer for each term.

35. _____ exocytosis

36. _____ receptor-mediated endocytosis

37. _____ bulk-phase endocytosis

38. _____ phagocytosis

39. _____ membrane cycling

A. A cell engulfs microorganisms, large edible particles, and cellular debris

B. Membrane initially used for endocytic vesicles returns receptor proteins and lipids back to the plasma membrane

C. Vesicles form around small volumes of extracellular fluid of various contents

D. A cytoplasmic vesicle moves to the cell surface; its own membrane fuses with the plasma membrane while its contents are released to the environment

E. Chemical recognition and binding of specific substances; coated pits sink into the cytoplasm and close on themselves

SELF-TEST

___ 1. An important principle of the second law of thermodynamics states that:
    a. energy can be transformed into matter, and because of this, we can get something for nothing
    b. energy can only be destroyed during nuclear reactions, such as those that occur inside the sun
    c. if energy is gained by one region of the universe, another place in the universe also must gain energy in order to maintain the balance of nature
    d. energy in a system tends to become less concentrated

___ 2. What essentially does the first law of thermodynamics state?
    a. one form of energy cannot be converted into another
    b. entropy is increasing in the universe
    c. energy cannot be created or destroyed
    d. energy cannot be converted into matter or matter into energy

___ 3. An enzyme is best described as:
    a. an acid
    b. a protein
    c. a catalyst
    d. a fat
    e. both b and c

4. Which is not true of enzyme behavior?
   a. Enzyme shape may change during catalysis.
   b. The active site of an enzyme orients its substrate molecules, thereby promoting interaction of their reactive parts.
   c. All enzymes have an active site where substrates are temporarily bound.
   d. An individual enzyme can catalyze a wide variety of different reactions.

5. When NAD+ combines with hydrogen and electrons, the NAD is _____.
   a. reduced
   b. oxidized
   c. phosphorylated
   d. denatured

6. A substance that gains electrons is _____.
   a. oxidized
   b. a catalyst
   c. reduced
   d. a substrate

7. In pathways, carbohydrates, lipids, and proteins are broken down in stepwise _____ reactions that lead to products of lower energy.
   a. intermediate
   b. biosynthetic
   c. induced
   d. degradative

8. As to major function, NAD^+, FAD, and $NADP^+$ are classified as _____.
   a. enzymes
   b. phosphate carriers
   c. cofactors
   d. end products of metabolic pathways

9. The outer phosphate bond in ATP _____.
   a. absorbs a large amount of free energy when the phosphate group is attached during hydrolysis
   b. is formed when ATP is hydrolyzed to ADP and one phosphate group
   c. is usually found in each glucose molecule; that is why glucose is chosen as the starting point for glycolysis
   d. releases usable energy when the phosphate group is split off during hydrolysis

10. An allosteric enzyme _____.
    a. has an active site where substrate molecules bind and another site that binds with intermediate or end-product molecules
    b. is an important energy-carrying nucleotide
    c. carries out either oxidation reactions or reduction reactions but not both
    d. raises the activation energy of the chemical reaction it catalyze

11. White blood cells use _____ to devour disease agents invading your body.
    a. diffusion
    b. bulk flow
    c. osmosis
    d. phagocytosis

12. _____ proteins bind extracellular substances, such as hormones, that trigger changes in cell activities.
    a. Receptor
    b. Adhesion
    c. Transport
    d. Recognition

13. In a lipid bilayer, the phospholipid tails point inward and form a(n) _____ region that excludes water.
    a. acidic
    b. basic
    c. hydrophilic
    d. hydrophobic

14. Which of the following is not an example of an active transport mechanism?
    a. Calcium pump
    b. Glucose transporter
    c. Sodium–potassium pump
    d. All of the above are examples of active transport

15. Which of the following is not a form of passive transport?
    a. Osmosis
    b. Diffusion
    c. Bulk flow
    d. Exocytosis

16. O_2, CO_2, H_2O, and other small, electrically neutral molecules usually move across the cell membrane by _____.
    a. electric gradients
    b. receptor-mediated endocytosis
    c. passive transport
    d. active transport

17. Ions such as H^+, Na^+, K^+, and Ca^{++} move across cell membranes against a concentration gradient by _____.
   a. receptor-mediated endocytosis
   b. pressure gradients
   c. passive transport
   d. active transport

18. The fluid mosaic model is used to describe _____.
   a. the process by which particles are exported from the cell
   b. the structure of the cell membrane
   c. the movement of water across a membrane
   d. the action of transport proteins

19. A cell is placed in a beaker containing a solution of 40 percent NaCl and 60 percent water. After a few minutes you notice that the cytoplasm of the cell is shrinking in size. The cell is _____ in relation to the contents of the beaker.
   a. isotonic
   b. hypertonic
   c. hypotonic
   d. saturated

20. Which of the following types of membrane proteins would be used as a molecular identification tag?
   a. Recognition proteins
   b. Transport proteins
   c. Adhesion proteins
   d. Receptor proteins

21. What is the significance of selective permeability to biological membranes?
   a. Selective permeability allows the plasma membrane to control traffic into and out of the cell it surrounds.
   b. Selective permeability prevents toxic materials from entering the cell.
   c. Selective permeability permits the selective uptake of nutrients and the elimination of wastes.
   d. All of the above are correct.

22. What do you think would happen to an animal cell that was placed into a drop of 100% pure water?
   a. The cell will be unaffected by the pure water.
   b. The cell will shrivel and shrink down
   c. The cell will swell up and eventually burst.
   d. The cell would first shrink, and then swell and burst.

23. Which of the following is NOT an example of potential energy?
   a. A reservoir of water behind a dam?
   b. Oil in an oil tank
   c. A donkey powering a grist mill
   d. A piece of candy

24. Metabolism is a constant process whether the organism needs more or less energy.
   a. True
   b. False

CHAPTER OBJECTIVES/REVIEW QUESTIONS

1. Define energy; be able to state the first and second laws of thermodynamics. [pp.78-79]
2. Explain how the world of life maintains a high degree of organization. [pp.78-79]
3. When the "energy hill" is made smaller by enzymes so that particular reactions may proceed, it may be said that the enzyme has lowered the energy. [pp.80-81]
4. Describe the various types of metabolic pathways. [pp.84-85]
5. Give the function of each of the following participants in metabolic pathways: substrates, intermediates, enzymes, cofactors, energy carriers, and end products. [pp.84-87]
6. What are enzymes? Explain their importance. [pp.82-83]
7. Explain what happens to enzymes in the presence of extreme temperatures and pH. [p.83]
8. _____ are small molecules or metal ions that assist enzymes or carry atoms or electrons from one reaction site to another. [p.86-87]
9. Explain the functioning of the ATP/ADP cycle. [p.87]
10. Identify which molecules are the most abundant component of cell membranes. [p.88]
11. Describe what is meant by the term *fluid mosaic model*. [p.88]
12. Explain why the structure of a phospholipid results in the formation of a lipid bilayer. [p.88]
13. Explain the concept of selective permeability as it applies to cell membrane function. [pp.90-91]

14. The diffusion of water molecules in response to water concentration gradients between two regions separated by a selectively permeable membrane is known as _____. [pp.90-91]
15. What is turgor pressure? [p.90]
16. Define *osmotic pressure*. [p.90]
17. List the factors that can have an influence on the rate of diffusion. [p.90]
18. Generally distinguish passive transport from active transport. [pp.92-93]
19. Explain why the glucose transporter is an example of passive transport. [p.92]
20. Describe how ATP is used in active transport and how it improves the passage of solutes. [pp.92-93]
21. Describe mechanisms involved in receptor-mediated endocytosis. [pp.94-95]
22. Explain the role of endocytosis and exocytosis in membrane recycling; cite an example. [pp.95]

INTEGRATING AND APPLYING KEY CONCEPTS

1. A piece of dry ice left sitting on a table at room temperature vaporizes. As the dry ice vaporizes into CO_2 gas, does its entropy increase or decrease? Tell why you answered as you did.
2. Why is feedback inhibition important to the well-being of organisms? What would happen if it did not exist?
3. What is bioluminescence? Why is it important to organisms?
4. Describe the various environmental factors that can affect enzyme activity.
5. If there were no such thing as active transport, how would the lives of organisms be affected?
6. Diagram a typical biological membrane including the lipid bilayer and both integral and peripheral proteins.
7. Compare and contrast active and passive transport. Why is each important? Why can some substances use both?
8. How is a freshwater protist able to function in an osmotically hypotonic environment? What challenges are faced by organisms that live in high-salt environments?

6

WHERE IT STARTS—PHOTOSYNTHESIS

INTRODUCTION

The unique ability of photoautotrophs to harness solar energy and store it in the bonds of glucose begins the flow of energy through the biosphere. This chapter details the steps of photosynthesis from capture of solar energy by photosynthetic pigments and conversion of that light energy to chemical energy in ATP, to fixation of the carbon in CO_2 into carbohydrates. The light dependent and light independent reactions are discussed in relation to the structure of the chloroplast. Relationships between autotrophs and heterotrophs are discussed as is the origin of today's atmosphere. Both the introduction and conclusion to this chapter address the challenges a technological society faces to supply itself with plant-derived fuels while continuing to protect the Earth, its organisms, and the environment.

STUDY STRATEGIES

- Read all of Chapter 6 with the goal of familiarizing yourself with the boldface terms.
- Identify the importance of sunlight for living systems and the role of pigments in harnessing certain wavelengths of light.
- Recognize the overall reaction for photosynthesis.
- Describe the role of the chloroplast and identify the specific regions of the organelle responsible for portions of the photosynthetic process.
- Explain the processes of the light-dependent and light-independent reactions.
- Explain why photorespiration is such a problem for plants.
- Describe the different types of photosynthetic adaptations plants have acquired for different climates.

FOCAL POINTS

- Figure 6.2 [p.102] shows the relationship between color and wavelength in the electromagnetic spectrum.
- Figure 6.3 [p.103] summarizes various photopigments, their reflective colors, and sources.
- Figure 6.4 [p.104] (animated) compares the absorption spectra of several photopigments and illustrates Engelmann's original experiment on visible spectrum light and photosynthesis.
- Figure 6.5 [p.105] (animated) summarizes the interrelationship of chloroplast structure and function.
- Figure 6.8 [p.107] (animated) is an excellent illustration of the intricate processes involved in the noncyclic pathway of photosynthesis (light dependent reactions).
- Figure 6.9 [p.108] (animated) compares cyclic and noncyclic photophosphorylation.
- Figure 6.10 [p.109] (animated) illustrates the molecular changes and energy usage in the Calvin-Benson Cycle (light independent reactions).
- Figures 6.11-6.13 [pp.110-111] compare differences and similarities between C3, C4, and CAM plants.

INTERACTIVE EXERCISES

6.1. BIOFUELS [p.101]

6.2. SUNLIGHT AS AN ENERGY SOURCE [pp.102-103]

6.3. EXPLORING THE RAINBOW [p.104]

Boldfaced, Page-Referenced Terms

autotrophs _____

_____

photosynthesis _____

_____

heterotrophs _____

_____

wavelength _____

_____

pigment _____

_____

chlorophyll a _____

_____

Dichotomous Choice [p.102]

Select the correct answer from the terms between parentheses in the following questions.

1. _____ Violet light has the (highest/lowest) energy value of all visible light.

2. _____ Infrared radiation has a (longer/shorter) wavelength than ultraviolet light.

3. _____ Gamma rays have the (shortest/longest) wavelengths of energy on the electromagnetic spectrum.

Matching [pp.101-104]

Choose the most appropriate answer.

4. _____ chlorophyll *a*

5. _____ accessory pigments

6. _____ carotenoids

7. _____ UV light

8. _____ photons

9. _____ absorption spectrum

10. _____ visible light

11. _____ pigment

12. _____ biofuels

13. _____ fossil fuels

14. _____ prism

A. Packets of electromagnetic energy that have an undulating motion through space

B. A graph that shows how efficiently different wavelengths of light are absorbed by a pigment

C. An organic molecule that absorbs specific wavelengths of light

D. Appear red, orange, and yellow; absorb violet and blue wavelengths

E. Wavelengths shorter than 380 nm; can damage DNA and other biological molecules

F. Nonrenewable organic remains of ancient plants

G. Absorbs violet, blue, and red wavelengths; the reason leaves appear green

H. Renewable source of energy; made from living plants

I. Wavelengths of light 380nm to 750 nm; drives photosynthesis

J. Absorb wavelengths of light that chlorophyll cannot absorb

K. Separates white light into its component colors by bending light waves

Short Answer [pp.104-105]

15. Explain how Theodor Engelmann discovered which wavelengths of light were used most for photosynthesis.

_____

_____

_____

6.4. OVERVIEW OF PHOTOSYNTHESIS [p.105]

Boldfaced, Page-Referenced Terms

chloroplast _____

_____

stroma _____

_____

thylakoid membrane _____

_____

light-dependent reactions _____

_____

light-independent reactions _____

_____

Completion [p.105]

Complete the following equation that summarizes the metabolic pathway of photosynthesis.

$$6H_2O + 6(1) \longrightarrow 6C(2) + (3)$$

1. _____

2. _____

3. _____

Fill-in-the-Blanks [p.105]

Supply the appropriate information to state the equation (above) for photosynthesis in words.

(4)_____ molecules of water plus six molecules of (5)_____ (in the presence of enzymes and sunlight)

yield six molecules of (6) _____ plus one molecule of (7) _____.

Fill-in-the-Blanks [p.105]

The two major sets of reactions of photosynthesis are the (8) _____ reactions and the (9) _____

reactions. The internal membranes and channels of the chloroplast form the (10) _____ membranes and are

organized into stacks. These membranes contain light-harvesting (11) _____ and (12) _____ which

convert light energy into chemical energy in the form of the molecule (13) _____. The semifluid interior area

surrounding the thylakoid membranes is known as the (14) _____ and is the area where the products of

photosynthesis are assembled.

Labeling [p.105]

For the figure below, identify the portions of a chloroplast. Use Figure 6.5 as a reference.

15. _____

16. _____

17. _____

18. _____

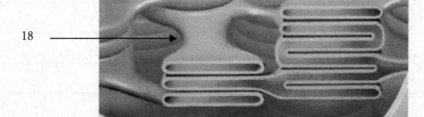

Labeling [p.105]

For the figure below, identify which of the following reactants and products are present in each of the reactions.

19. _____   A. ADP   F. ATP

20. _____   B. O_2 G. NADPH

21. _____   C. energy   H. glucose

22. _____   D. $NADP^+$ I. CO_2

                           E. H_2O

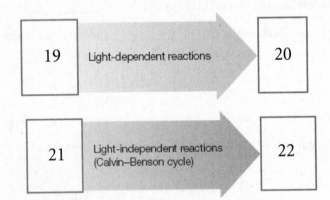

6.5. LIGHT-DEPENDENT REACTIONS [pp.106-107]

6.6. ENERGY FLOW IN PHOTOSYNTHESIS [p.108]

Boldfaced, Page-Referenced Terms

photosystems _____

_____

photolysis _____

_____

electron transfer _____

_____

phosphorylation _____

_____

Fill-in-the-Blanks [pp.106-108]

There are two sets of light-dependent reactions, a (1) _____ pathway and a (2) _____ pathway. Both convert light energy to chemical energy in the form of (3) _____. The (4) _____ pathway is the main one in organisms containing chloroplasts. It yields (5) _____ and (6) _____ in addition to ATP. The noncyclic pathway includes the molecules of both (7) _____ and (8) _____, while the cyclic pathway relies on the activity of only (9) _____. Photosystem I absorbs light at a wavelength of (10) _____ nm; Photosystem II absorbs light at a wavelength of (11) _____ nm. Absorbed light causes (12) _____ to pop off chlorophyll *a* and enter an (13)_____. Photosystem II can replace lost electrons by stripping electrons from a molecule of water. This causes the water to split into (14) _____ ions and (15) _____ in a process known as (16) _____. Oxygen is reactive and is released from the cell as a by-product of photosynthesis. The energy captured from the flow of electrons through the electron transfer chain pulls in hydrogen ions from the stroma creating a (17) _____ across the thylakoid membrane. These ions flow from the thylakoid compartment back to the (18) _____ through membrane transport proteins called (19) _____ causing a phosphate group to be attached to ADP turning it into (20) _____. In the noncyclic pathway, electrons are passed from Photosystem II to (21) _____ and enter another electron transfer chain. At the end, electrons and (22) _____ are accepted by $NADP^+$ and (23) _____ is formed. If too much NADPH builds up, the noncyclic pathway stops running and cells continue to make ATP through the (24) _____ pathway.

Choice [pp.106-108]

Indicate the pathway in which each of the following occurs.

    a. cyclic pathway   b. noncyclic pathway   c. both pathways

25. _____   Uses H_2O as a reactant
26. _____   Includes Photosystem I (p700)
27. _____   Includes Photosystem II (p680)
28. _____   ATP produced
29. _____   NADPH produced
30. _____   Does not consume water nor produce NADPH and oxygen
31. _____   Causes H^+ to be shunted into the thylakoid compartments from the stroma
32. _____   Produces O_2 as a product
33. _____   Produces H^+ by breaking apart H_2O
34. _____   Uses ADP and P_i as reactants

Ordering [p.107]

Using numbers 1-8, put the following statements in the correct order for the noncyclic pathway of photosynthesis.

35. _____ The electrons move through a second electron transfer chain, then combine with NADP+ and H+, so NADPH forms.

36. _____ Hydrogen ions in the thylakoid compartment are propelled through the interior of ATP synthases by their gradient across the thylakoid membrane.

37. _____ Hydrogen ion flow causes ATP synthases to attach phosphate to ADP, so ATP forms in the stroma.

38. _____ Energy lost by the electrons as they move through the electron transfer chain is used to pump hydrogen ions from the stroma into the thylakoid compartment. A hydrogen ion gradient forms across the thylakoid membrane.

39. _____ The photosystem pulls replacement electrons from water molecules, which break apart into oxygen and hydrogen ions. The oxygen leaves the cell as O2.

40. _____ Light energy ejects electrons from a photosystem II.

41. _____ Light energy ejects electrons from a photosystem I. Replacement electrons come from an electron transfer chain in the thylakoid membrane.

42. _____ The electrons enter an electron transfer chain in the thylakoid membrane.

6.7. LIGHT-INDEPENDENT REACTIONS: THE SUGAR FACTORY [p.109]

6.8. ADAPTATIONS: DIFFERENT CARBON-FIXING PATHWAYS [pp.110-111]

Boldfaced, Page-Referenced Terms

Calvin–Benson cycle _____

_____

carbon fixation _____

_____

rubisco _____

_____

Matching [pp.109-111]

Supply the full names for the following abbreviations and match it to its lettered description.

1. _____ RuBP
2. _____ PGAL
3. _____ PGA
4. _____ C3 Plants

5. _____ C4 Plants
6. _____ CAM Plants

A. Plants that fix carbon twice, first in mesophyll cell, then in bundle sheath cells

B. A 5-carbon sugar precursor that fixes inorganic carbon from carbon dioxide

C. Two 3-carbon molecules formed from an unstable 6-carbon molecule

D. Water-conserving desert plants that use an alternate carbon-fixing system at night

E. A 3-carbon sugar that has been modified by the addition of a phosphate from ATP and hydrogen and electrons from NADPH.

F. Plants that use only the Calvin-Benson cycle in mesophyll cells to fix carbon

Fill-in-the-Blanks [pp.109-111]

The light-independent reactions of the (7) _____ cycle build sugars in the (8) _____ of chloroplasts. They run on (9) _____ and (10) _____, molecules made in the light-dependent reactions. During carbon fixation, the enzyme (11) _____ is used to attach CO_2 to a five-carbon molecule (12) _____. The resulting unstable six-carbon molecule splits into two three-carbon (13) _____. A phosphate group from (14) _____ and hydrogen and electrons from (15) _____ transform each (20) _____ into a molecule of (16) _____. Two (23) combine to form (17) _____. The ten remaining (23) combine to regenerate six (18) _____. Most of the (24) made by plants is converted to (19) _____ or (20) _____. Plants that use only the Calvin-Benson cycle to fix carbon are called (21) _____. On hot, dry days, plants close their (22) _____ to prevent excess water loss. This also prevents (23) _____ gas from entering the leaf and (24) _____ from exiting the leaf. The buildup of oxygen in the leaf causes (25) _____ to attach carbon dioxide to RuBP in a process known as (26) _____. As a result, the cell loses (27) _____ instead of fixing it. To compensate for the inefficiency caused by photorespiration, some plants called (28) _____ evolved a secondary carbon-fixation process. These plants fix carbon first in (29) _____ cells by an enzyme that does not use oxygen as rubisco does under stress conditions. The fixed carbon is transported to (30) _____ cells where it is converted to carbon dioxide and fixed for a second time in the (31) _____ cycle. A third carbon-fixing pathway is found in many desert plants such as succulents and cacti. These plants are called (32) _____ plants. These plants only open their stomata at (33) _____ when temperatures are cooler. Carbon dioxide enters the plant then and is fixed in the (34) _____ cycle into a four-carbon acid. The next day, the four-carbon acid is converted to (35) _____ that enters the Calvin-Benson cycle.

Biofuels [p.112-113]

Boldfaced, Page-Referenced Terms

chemoautotrophs _____

_____

photoautotrophs _____

_____

Fill-in-the-Blanks [pp.112-113]

"Self-nourishing" organisms that make food from inorganic chemicals are called (1) _____.

Because plants use solar energy they are a kind of (2) _____. In the process of (3) _____, they make

their own food by using energy from the sun and carbon from inorganic molecules and produce huge amounts of

(4) _____ and (5) _____. (6) _____, "nourished by others," get their energy and carbon by

consuming organic molecules that have been made by other organisms including plants. The first cells on earth were

(7) _____ that obtained energy and carbon from molecules like (8) _____ and (9) _____ gases

that were plentiful in the earth's early noxious atmosphere. Chemoautotrophs flourished for a (10) _____

years until the evolution of (11) _____ in the first photoautotrophs. Shortly afterward, (12) _____

evolved. The (13) _____ gas released by the splitting of water in this process began to accumulate in the

atmosphere. This gas is reactive and produced toxic (14) _____ that caused many organisms to become (15)

_____. Only a few survived in anaerobic environments. In time, oxygen-detoxifying pathways evolved. One

of these was the ATP-forming reactions of (16) _____. Oxygen molecules that rose high in the ancient

atmosphere combined to form (17) _____, molecules capable of absorbing (18) _____ radiation in

sunlight. Accumulation of this gas formed the (19) _____ that shielded, and still shields, life from damaging

radiation. New (20) _____ emerged, flourished, and diversified across the face of the earth. Photosynthesis

removes carbon dioxide from the atmosphere and fixes it in organic molecules, while aerobic respiration releases

organic carbon back into the atmosphere as carbon dioxide in a balanced cycle known as the (21) _____.

Since the (22) _____ in the mid 1800's, the levels of (23) _____ in the atmosphere have been rising

largely due to the burning of (24) _____. We are adding more carbon dioxide than photoautotrophs can

remove. Accumulating carbon dioxide acts like an atmospheric blanket contributing to (25) _____.

SELF-TEST

____ 1. _____ and _____ light are best at driving photosynthesis.
   a. orange; red
   b. red; green
   c. violet; red
   d. green; yellow

____ 2. The cyclic pathway of the light-dependent reactions functions mainly to _____.
   a. fix CO_2
   b. make ATP
   c. produce PGAL
   d. regenerate ribulose bisphosphate

____ 3. Chemoautotrophs obtain energy by oxidizing such inorganic substances as _____.
   a. PGA
   b. PGAL
   c. hydrogen sulfide
   d. water

____ 4. The ultimate electron and hydrogen acceptor in noncyclic photophosphorylation is _____.
   a. $NADP^+$
   b. ADP
   c. O_2
   d. H_2O

____ 5. C4 plants have an advantage in hot, dry conditions because _____.
   a. their leaves are covered with thicker wax layers than those of C3 plants
   b. their stomata open wider than those of C3 plants, thus cooling their surfaces
   c. they have a two-step CO_2 fixation that reduces photorespiration
   d. they are also capable of carrying on photorespiration

____ 6. Light-harvesting pigments are found _____.
   a. on the outer chloroplast membrane
   b. in the cytoplasm of the cell
   c. in the stroma
   d. in the thylakoid membrane system

____ 7. In both C4 and CAM plants, the first product of carbon fixation is _____.
   a. malate
   b. PGA
   c. oxaloacetate
   d. glycolate

____ 8. O_2 released during photosynthesis comes from _____.
   a. splitting CO_2
   b. splitting H_2O
   c. RuBP
   d. degradation of sugar molecules

____ 9. Plants need _____ and _____ to carry on photosynthesis.
   a. oxygen; water
   b. oxygen; CO_2
   c. CO_2; H_2O
   d. sugar; water

____ 10. The two products of the light-dependent reactions that are required for the light-independent chemistry are _____ and _____.
   a. CO_2; H_2O
   b. O_2; NADPH
   c. O_2; ATP
   d. ATP; NADPH

____ 11. Global climate change is due in a large part to the accumulation of _____ in the atmosphere.
   a. water
   b. oxygen
   c. carbon dioxide
   d. hydrogen sulfide

____ 12. O_2 and NADPH are produced only by _____.
   a. cyclic photophosphorylation
   b. noncyclic photophosphorylation
   c. the Calvin-Benson cycle
   d. Photosystem I

____ 13. A pigment is an organic molecule that selectively _____ light of a specific wavelength.
   a. reflects
   b. refracts
   c. transmits
   d. absorbs

____ 14. The most common photosynthetic pigments in plants are _____.
   a. anthocyanins and retinal
   b. phycobilins and anthocyanins
   c. chlorophylls and phycobilins
   d. chlorophylls and carotenoids

15. Photosynthesis uses sunlight to convert water and carbond dioxide into _____.
    a. oxygen.
    b. high-energy sugars.
    c. ATP and oxygen.
    d. oxygen and high-energy sugars.

16. Plants gather the sun's energy with light-absorbing molecules called
    a. pigments.
    b. thylakoids.
    c. chloroplasts.
    d. glucose.

17. Where do the light-dependent reactions take place?
    a. in the stroma
    b. outside the chloroplasts
    c. in the thylakoid membranes
    d. only in chlorophyll molecules

18. The Calvin-Benson cycle takes place in the
    a. stroma
    b. photosystems
    c. thylakoid membranes
    d. chlorophyll molecules

19. Rubisco
    a. begins the light-dependent reactions.
    b. acts as an enzyme.
    c. is a 6-carbon molecule.
    d. breaks into two equal parts.

20. In non-cyclic electron flow, electrons that leave the chlorophyll
    a. return to the chlorophyll.
    b. are used to turn NADP+ into NADPH.
    c. produce only ATP.
    d. are channeled to break apart water.

21. Carbon dioxide enters the leaf from the atmosphere via
    a. leaf hairs.
    b. xylem and phloem.
    c. stomata.
    d. enzymes.

22. _____ plants utilize a day/night photosynthesis cycle to conserve water in hot, dry climates.
    a. C3
    b. C4
    c. CAM

23. When humans cut down and burn trees and burn fossil fuels, excess _____ is released into the atmosphere, contributing greatly to climate change.
    a. CO_2
    b. O_2
    c. ATP
    d. chlorophyll

CHAPTER OBJECTIVES/REVIEW QUESTIONS

1. Distinguish between organisms known as autotrophs and those known as heterotrophs. [p.101]
2. The most common photosynthetic pigment is _____. [p.102]. It forms the reactive center of both _____ [pp.105-106] by absorbing light in wavelengths of 680-700 nm.
3. Name the two major stages of photosynthesis, state in what part of the chloroplast they occur, and give an overview of the reactions they involve. [p.105]
4. The flattened channels and disk-like compartments inside the chloroplast are organized into interconnected stacks. This is the _____ membrane system that is surrounded by a semifluid interior, the _____ [p.105]
5. Describe how the pigments found on thylakoid membranes are organized into photosystems and how they relate to photon light energy. [pp.105-106]
6. Contrast the components and functioning of the cyclic and noncyclic pathways of the light-dependent reactions. [pp.105-106]
7. Two energy-carrying molecules produced in the noncyclic pathways are _____ and _____; explain why these molecules are necessary for the light-independent reactions. [pp.105-106]
8. Describe the Calvin–Benson cycle in terms of its reactants and products. [p.109]
9. State the various ways that glucose produced by the Calvin-Benson cycle is used by photoautotrophs. [p.109]
10. Describe the variations on carbon fixation seen in C3, C4, and CAM plants. Include locations and timing of processes and structural and chemical differences. How do these variations adapt plants to survive climatic stresses? [pp.110-111]
11. Following evolution of the noncyclic pathway, _____ accumulated in the atmosphere and made aerobic respiration possible. [pp.112-113]

INTEGRATING AND APPLYING KEY CONCEPTS

1. Suppose oxygen-producing pathways of photosynthesis had never evolved. What would be the effect on the lifestyle of organisms on Earth?
2. Suppose that humans acquired all the enzymes needed to carry out photosynthesis. Speculate about the attendant changes in human anatomy, physiology, and behavior that would be necessary for those enzymes to actually carry out photosynthetic reactions.
3. Consider the consequences of the development of a photosynthetic cell symbiont for human epidermal tissues.
4. Human activities, such as deforestation and polluting of waterways, have the capacity to greatly reduce the numbers of photosynthetic organisms on earth. On a chemical level, consider how the resulting reduction in consumption of photosynthetic reactants and evolution of photosynthetic products would affect the Earth and it inhabitants.

7

HOW CELLS RELEASE CHEMICAL ENERGY

INTRODUCTION

Chapter 7 looks at the various ways that cells can extract energy from food. Both aerobic and anaerobic mechanisms are covered, but a major emphasis of the chapter is aerobic respiration.

STUDY STRATEGIES

- Read all of Chapter 7 with the goal of familiarizing yourself with the boldface terms.
- Identify the importance of the mitochondria organelle and what the result of a mitochondrial disorder may be.
- Describe the general process of a heterotroph using chemical food to derive energy.
- Explain the steps required in glycolysis, the intermediate reaction, and the Kreb's cycle and identify the reactants and products of these metabolic pathways.
- Describe the series of steps in the electron transfer pathway, including the roles of coenzymes and oxygen in the process.
- Explain the role of ion gradients in generating ATP.
- Compare and contrast aerobic and anaerobic respiration.
- Identify the role fermentation plays in nature and the two fermentation pathways.
- Recognize other forms of food energy besides carbohydrates and describe how they are broken down differently.

FOCAL POINTS

- The introductory section [p.117] looks at the problems that arise when mitochondria do not function properly.
- Figure 7.3 [p.119] provides an overview of aerobic respiration.
- The reaction pathway of glycolysis is presented in a detailed flow chart [p.121].
- Figure 7.6 [p.122] provides an overview of aerobic respiration in the mitochondria.
- Figure 7.7 [p.123] looks at the second stage of aerobic respiration, acetyl–CoA formation, and the Krebs cycle.
- Figure 7.8 [p.124] details the electron transfer pathway.
- Figure 7.9 [p.125] summarizes aerobic respiration.
- Figures 7.10-7.11 [pp.126-127] looks at two of the many fermentation pathways seen in living organisms.
- Figure 7.12 [p.128] shows how various foods in your diet can enter the various stages of aerobic respiration.

INTERACTIVE EXERCISES

7.1. MIGHTY MITOCHONDRIA [p.117]

7.2. OVERVIEW OF CARBOHYDRATE BREAKDOWN PATHWAYS [pp.118-119]

Boldfaced, Page-Referenced Terms

anaerobic _____

_____

aerobic _____

_____

aerobic respiration _____

_____

glycolysis _____

_____

pyruvate _____

_____

fermentation _____

_____

Short Answer [p.117-118]

1. Why are mitochondria so crucial to normal life? _____

_____

_____

2. Compare and contrast aerobic respiration and anaerobic fermentation. _____

_____

_____

_____

3. What types of problems arise when mitochondria are mutated or are malfunctioning? _____

_____

_____

_____

Fill-in-the-Blanks [pp.118-119]

Most organisms make (3) _____ by (4) _____ which breaks down organic molecules such as carbohydrates. The (5) _____ pathway uses oxygen while the (6) _____ pathway does not. Both of these pathways begin with the same reaction, (7) _____. This reaction converts one molecule of (8) _____ to two molecules of (9) _____. Aerobic respiration ends in the (10) _____, while anaerobic fermentation ends in the (11) _____. Aerobic respiration yields about (12) _____ molecules of ATP per glucose, while fermentation ends with a net yield of (13) _____ molecules of ATP. During aerobic respiration, glycolysis is followed by the (14) _____. In order to begin the (14), pyruvate must be converted to (15) _____. Hydrogen ions and electrons released by these reactions are picked up by (16) _____ and (17) _____. The hydrogen ions and electrons are then delivered to the (18) _____, the third stage of the aerobic pathway. Operation of the electron transfer chain allows for the formation of many molecules of (19) _____.

True or False [pp.118-119]

If the statement is true, write a T in the blank. If the statement is false, make it correct by changing the underlined word(s) and writing the correct word(s) in the answer blank.

20. _____ Energy flows from photosynthesis to <u>anaerobic respiration</u> in a one-way direction.

21. _____ The <u>Kreb's Cycle</u> is the second step in aerobic respiration.

22. _____ As many as <u>18</u> ATP may be generated by aerobic respiration.

23. _____ Fermentation begins with <u>glycolysis</u>.

Identification [p.119]

24. For each of the three stages of aerobic respiration in the figure below, name each and list all of molecular inputs and outputs for each reaction. Refer to Figure 7.3 in your text.

Stage	Name	Inputs	Outputs
A			
B			
C			

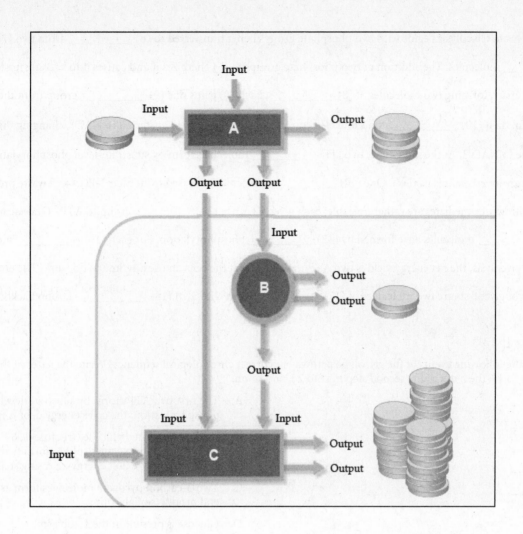

7.3. GLYCOLYSIS—GLUCOSE BREAKDOWN STARTS [pp.120-121]

Boldfaced, Page-Referenced Term

substrate-level phosphorylation _____

_____

Fill-in-the-Blanks [pp.120-121]

(1) [Recall Ch.6] _____ organisms can synthesize and stockpile energy-rich carbohydrates and other

food molecules from inorganic raw materials for later use. When used as an energy source, (2) _____ is

partially broken down by the glycolytic pathway; at the end of this process some of its stored energy remains in two

(3) _____ molecules. Some of the energy of glucose is released during the breakdown reactions and used in

forming the energy carrier (4) _____ and the reduced coenzyme (5) _____. These reactions take place in the cytoplasm. Glycolysis begins with two phosphate groups being transferred to (6) _____ from two (7)

_____ molecules. The addition of two phosphate groups to (6) energizes it and causes it to become unstable

and split apart, forming two molecules of (8) _____. Each (8) gains one (9) _____ group from the

cytoplasm, then (10) _____ atoms and electrons from each PGAL are transferred to NAD$^+$, changing this

coenzyme to NADH. At the same time, two (11) _____ molecules form by substrate-level phosphorylation; the

cell's energy investment is paid off. One (12) _____ molecule is released from each 2-PGA as a waste product.

The resulting intermediates are rather unstable; each gives up a(n) (13) _____ group to ATP. Once again, two

(14) _____ molecules have formed by (15) _____ phosphorylation. For each (16) _____ molecule

entering glycolysis, the net energy yield is two ATP molecules that the cell can use anytime to do work. The end

products of glycolysis are two molecules of (17) _____, each with a(n) (18) _____ -carbon backbone.

Sequence [p.121]

Arrange the following events of the glycolysis pathway in correct chronological sequence. Write the letter of the first
step next to 19, the letter of the second step next to 20, and so on.

19. _____

20. _____

21. _____

22. _____

23. _____

24. _____

25. _____

26. _____

A. The first two ATPs form by substrate-level phosphorylation; the cell's energy debt is paid off

B. Diphosphorylated glucose (fructose-1, 6-bisphosphate) molecules split to form two PGAL; this is the first energy-releasing step

C. Two 3-carbon pyruvate molecules form as the end products of glycolysis

D. Glucose is present in the cytoplasm

E. Two more ATPs form by substrate-level phosphorylation, the cell gains ATP; net yield of ATP from glycolysis is two ATPs

F. The cell invests two ATPs; one phosphate group is attached to each end of the glucose molecule (fructose-1, 6-bisphosphate)

G. Two PGAL gain two phosphate groups from the cytoplasm

H. Hydrogen atoms and electrons from each PGAL are transferred to NAD$^+$, reducing this carrier to NADH

7.4. SECOND STAGE OF AEROBIC RESPIRATION [pp.122-123]

7.5. AEROBIC RESPIRATION'S BIG ENERGY PAYOFF [pp.124-125]

Fill-in-the-Blanks [pp.122-123]

If sufficient oxygen is present, the end product of glycolysis enters a preparatory step, (1) _____

formation. This step converts pyruvate into (1), the molecule that enters the (2) _____ cycle. This is followed

by (3) _____ phosphorylation. During these three processes, a total of (4) _____ and (5) _____

(energy-carrier molecules) are typically generated. In the preparatory conversions prior to the Krebs cycle and within

the Krebs cycle, the food molecule fragments are further broken down into molecules of (6) _____. During

these reactions, hydrogen atoms (7) (with their _____) are stripped from the fragments and transferred to the

coenzymes (8) _____ and (9) _____.

Labeling [p.122-125]

In exercises 10-14, identify the structure or location in the top diagram; in exercises 15-18, identify the chemical
substance involved in the lower diagram. In exercise 19, name the metabolic pathway depicted.

10. _____ of mitochondrion

11. _____ of mitochondrion

12. _____ of mitochondrion

13. _____ of mitochondrion

14. _____

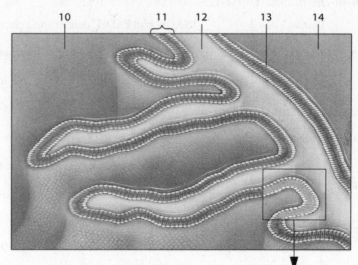

15. _____

16. _____

17. _____

18. _____

19. _____

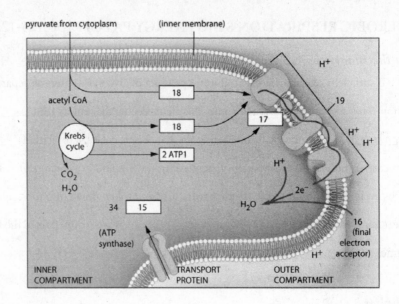

Fill-in-the-Blanks [pp.124-125]

During (20) _____ NADH and $FADH_2$ deliver their electrons into an electron transfer chain. The harnessed energy from the electrons is used to (21) _____ hydrogen ions across the (22) _____ membrane of the mitochondria. A hydrogen ion gradient forms as the hydrogen ions move from the (23) _____ to the (24) _____ compartment. The hydrogen ions move from the (23) _____ through a channel protein called (25) _____. The flow of the hydrogen ions results in the production of (26) _____ molecules of ATP. Once the hydrogen ions have moved back into the inner compartment and the electrons have moved through the electron transfer chain, they are picked up by (27) _____ to form (28) _____.

7.6. ANAEROBIC ENERGY-RELEASING PATHWAYS [pp.126-127]

Boldfaced, Page-Referenced Terms

alcoholic fermentation _____

_____

lactate fermentation _____

_____

Yeast and animal cells can use both (1) _____ and (2) _____ pathways. (3) _____ is the first step in fermentation, just like in aerobic respiration. However, in fermentation, the (4) _____ is converted to other molecules and not broken down to (5) _____ and (6) _____. Fermentation results in the formation of a net (7) _____, no more. In (8) _____ pyruvate is converted to ethanol. Bakers use (9) _____, a yeast, to make dough rise. As the yeast ferments the glucose, (10) _____ is released, thus the dough expands. Yeast is also used to form (11) _____ beverages. In (12) _____ pyruvate is converted to lactic acid. The bacteria (13) _____ are used to ferment dairy products. Skeletal muscles may be classified as (14) _____ or (15) _____ depending on how they make ATP. (14) muscle fibers have (16) _____ mitochondria and produce ATP via (17) _____. (15) muscle fibers have (18) _____ mitochondria and produce ATP via (19) _____.

7.8. ALTERNATIVE ENERGY SOURCES IN FOOD [pp.128-129]

Fill-in-the-Blanks[pp. 128-129]

After eating, increased glucose in the blood causes the pancreas to secrete (1) _____, which favors the uptake of glucose by the liver and muscles for conversion to (2) _____ for storage. Between meals, when glucose levels fall, the pancreas secretes (3) _____, which causes the (4) _____ to break down (2) and release glucose. Maintaining the proper glucose concentration is very important to the (5) _____, which uses 2/3 of the circulating glucose as its only energy source.

Labeling [p.128]

Identify the process or substance indicated in the illustration.

6. _____

7. _____

8. _____

9. _____

10. _____

11. _____

12. _____

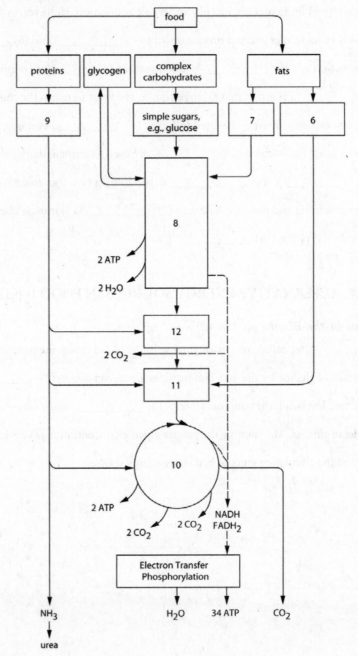

Choice [pp.128-129]

For questions 13-22, refer to the text and Figure 7.12; choose from the following:

a. glucose   b. glucose-6-phosphate   c. glycogen   d. fatty acids   e. triglycerides   f. PGAL
g. acetyl–CoA h. amino acids   i. glycerol   j. proteins

13. _____ Fats that are broken down between meals or during exercise as alternatives to glucose

14. _____ Used between meals when free glucose supply dwindles; enters glycolysis after conversion

15. _____ Its breakdown yields much more ATP than does glucose breakdown

16. _____ Absorbed in large amounts immediately following a meal

17. _____ Following removal of amino groups, the carbon backbones may be converted to fats or carbohydrates or they may enter the Krebs cycle

18. _____ Between meals liver cells can convert it back to free glucose and release it

19. _____ Amino groups undergo conversions that produce urea, a nitrogen-containing waste product excreted in urine

20. _____ Converted in the liver to PGAL, a key intermediate of glycolysis

21. _____ A storage polysaccharide produced from glucose-6-phosphate following food intake that exceeds cellular energy demand (and increases ATP production to inhibit glycolysis)

22. _____ A product resulting from enzymes cleaving circulating fatty acids; enters the Krebs cycle

Short Answer [p.128-129]

1. Why are other forms of chemical energy, besides carbohydrates, used by the body? _____

_____

_____

_____

SELF-TEST

___ 1. Glycolysis would quickly halt if the process ran out of _____, which serves as the hydrogen and electron acceptor.
    a. NADP$^+$
    b. ADP
    c. NAD$^+$
    d. H$_2$O

___ 2. The ultimate electron acceptor in aerobic respiration is _____.
    a. NADH
    b. carbon dioxide (CO_2)
    c. oxygen (O_2)
    d. ATP

___ 3. When glucose is used as an energy source, the largest amount of ATP is generated by the portion _____ of the entire respiratory process.
    a. glycolytic pathway
    b. acetyl–CoA formation
    c. Krebs cycle
    d. electron transfer phosphorylation

___ 4. The process by which about 10 percent of the energy stored in a sugar molecule is released as it is converted into two small organic-acid molecules is _____.
    a. photolysis
    b. glycolysis
    c. fermentation

5. During which of the following phases of respiration is ATP produced directly by substrate-level phosphorylation?
   a. Glycolysis
   b. Krebs cycle
   c. Both a and b

6. ATP production by electron transfer phosphorylation involves _____.
   a. H^+ concentration and electron gradients across a membrane
   b. ATP synthases
   c. both a and b
   d. neither a nor b

7. What is the name of the process by which reduced NADH transfers electrons along a chain of acceptors to oxygen so as to form water and in which the energy released along the way is used to generate ATP?
   a. Glycolysis
   b. Acetyl–CoA formation
   c. The Krebs cycle
   d. Electron transfer phosphorylation

8. Pyruvate can be regarded as the end product of _____.
   a. glycolysis
   b. acetyl–CoA formation
   c. fermentation
   d. the Krebs cycle

9. Which of the following is not one of three macromolecules from which heterotrophs derive energy?
   a. carbohydrates
   b. lipids
   c. oxygen
   d. proteins

10. When oxygen is reduced by the electrons in the electron transport chain, which of the following molecules is reduced?
    a. Hydrogen
    b. Water
    c. NADH
    d. $FADH_2$

11. One round of the Kreb's cycle generates _____ molecules of ATP.
    a. 2
    b. 4
    c. 12
    d. 24

12. Cellular respiration that occurs without oxygen is classified as _____.
    a. chemotrophic
    b. phototrophic
    c. aerobic
    d. anaerobic

13. How many carbon atoms make up one molecule of glucose?
    a. 1
    b. 3
    c. 6
    d. 9

14. This metabolic process within cellular respiration generates the most ATP.
    a. Glycolysis
    b. The citric acid cycle
    c. The electron transport chain
    d. Oxidative phosphorylation

15. All of the following events take place in the mitochondria EXCEPT _____.
    a. glycolysis
    b. the Kreb's cycle
    c. the electron transport chain
    d. oxidative phosphorylation

16. Which of the following produces ethanol as a by-product?
    a. The citric acid cycle
    b. Homolactic fermentation
    c. Oxidative phosphorylation
    d. Alcoholic fermentation

17. Which of the following is not a structural component of the mitochondria?
    a. Matrix
    b. Cytosol
    c. Inner membrane
    d. Intermembrane space

18. Which of the following molecules contains three phosphate groups?
    a. AMP
    b. ADP
    c. APP
    d. ATP

CHAPTER OBJECTIVES/REVIEW QUESTIONS

1. No matter what the source of energy may be, organisms must convert it to _____, a form of chemical energy that can drive metabolic reactions. [p.118]
2. The main energy-releasing pathway is respiration. [p.118]
3. Which stage of aerobic respiration has the highest yield of ATP? [pp.119]
4. By the end of the second stage of aerobic respiration, which includes the _____ cycle, _____ has been completely degraded to carbon dioxide and water. [pp.119, 122-123]
5. Explain the purpose served by the cell investing two ATP molecules into the chemistry of glycolysis. [pp.120-121]
6. Consult Figure 7.6 in the main text. Relate the events that happen during acetyl–coenzyme A formation and explain how the process of acetyl–CoA formation relates glycolysis to the Krebs cycle. [p.122]
7. Explain, in general terms, the role of oxygen in aerobic respiration. [pp.124-125]
8. Explain how the flow of hydrogen ions through the mitochondrion membrane accounts for the production of ATP molecules. [pp.124-125]
9. Briefly describe the process of electron transfer phosphorylation by stating what reactants are needed and what the products are. State how many ATP molecules are produced through operation of the transport system. [pp.124-125]
10. List some environments where very little oxygen is present and where anaerobic organisms might be found. [pp.126-127]
11. List the main anaerobic energy-releasing pathways and the examples of organisms that use them. [pp.126-127]
12. Describe what happens to pyruvate in anaerobic organisms. Then explain the necessity for pyruvate to be converted to a fermentation product. [pp.126-127]
13. List some sources of energy (other than glucose) that can be fed into the respiratory pathways. [pp.128-129]

INTEGRATING AND APPLYING KEY CONCEPTS

1. What problems might humans and other organisms experience if their mitochondria were defective?
2. Human skeletal muscle has both slow-twitch and fast-twitch fibers. Where in the body would you expect each type to predominate?
3. How is the "oxygen debt" experienced by runners and sprinters related to aerobic respiration and fermentation in humans?
4. Predict what your body would do if you switched to a diet of 100 percent protein.

8

DNA STRUCTURE AND FUNCTION

INTRODUCTION

This chapter looks at the details of DNA structure. It also links the structure of DNA to its replication and function. In addition, the topic of cloning and how it impacts our lives is discussed.

STUDY STRATEGIES

- Read all of Chapter 8 with the goal of familiarizing yourself with the boldface terms.
- Identify what a clone is and what techniques are used in making them.
- Define the following terms: *chromosome, chromatid, centromere, diploid, autosome,* and *sex chromosome.*
- Explain the role of the following scientists in the discovery of the structure and function of DNA: Miescher, Griffith, Avery and McCarty, Hershey and Chase, Chargaff, Watson and Crick, Franklin.
- Describe the structure of DNA and the process of DNA replication.
- Recognize the causes of genetic mutations and how DNA can repair itself.
- Discuss the processes of therapeutic and reproductive cloning, including some of the issues surrounding them.

FOCAL POINTS

- Figure 8.2 [p.134] details the chromosome structure.
- Figure 8.3 [p.135] illustrates a karyotype.
- Figures 8.5-8.6 [pp.136-137] show the classic experiments that led scientists to the understanding that DNA is the hereditary material.
- Figures 8.7-8.8 [pp.138-139] details the structure of DNA.
- Figures 8.9-8.10 [pp.140-141] illustrates the steps of DNA replication.
- Figure 8.14 [p.144] details the procedure used to transfer cell nuclei in cloning cattle.

INTERACTIVE EXERCISES

8.1. A HERO DOG'S GOLDEN CLONES [p.133]

8.2. EUKARYOTIC CHROMOSOMES [pp.134-135]

Boldfaced, Page-Referenced Terms

clone_____

_____

chromosomes _____

_____

sister chromatids _____

_____

centromere _____

_____

histones _____

_____

nucleosome _____

_____

chromosome number _____

_____

diploid _____

_____

karyotype _____

_____

autosomes _____

_____

sex chromosomes _____

_____

Short Answer [pp.133-135]

1. Explain the relationship between "chromosome", "sister chromatids", "centromere", "histones", and "nucleosome".

_____

_____

_____

_____

_____

2. What does it mean to be a "diploid" organism?

_____

_____

3. Explain the difference between autosomes and sex chromosomes.

_____

_____

_____

8.3. THE DISCOVERY OF DNA'S FUNCTION [pp.136-137]

8.4. THE DISCOVERY OF DNA'S STRUCTURE [pp.138-139]

Boldfaced, Page-Referenced Terms

bacteriophage_____

_____

DNA sequence _____

_____

Short Answer [pp.138-139]

1. List the three parts of a DNA nucleotide.

_____

_____

Labeling [p.138-139]

Four nucleotides are illustrated below. In the blank, label each nitrogen-containing base correctly as guanine, thymine, cytosine, or adenine. In the parentheses following each blank, indicate whether that nucleotide base is a purine (pu) or a pyrimidine (py).

2._____ ( )   3._____ ( )   4._____ ( )   5._____ ( )

Labeling and Matching [pp.138-139]

Identify each indicated part of the accompanying DNA illustration. Use Figure 8.8 for reference. Choose from these answers: phosphate group, purine, pyrimidine, nucleotide, deoxyribose

Complete the exercise by matching and entering the letter of the proper structure description in the parentheses following each label. The following memory devices may be helpful: Use *pyrCUT* to remember that the pyrimidines are cytosine, uracil (in RNA), and thymine. Use *purAG* to remember that the purines are adenine and guanine; to help recall the number of hydrogen bonds between the DNA bases, remember that AT = 2 and CG = 3.

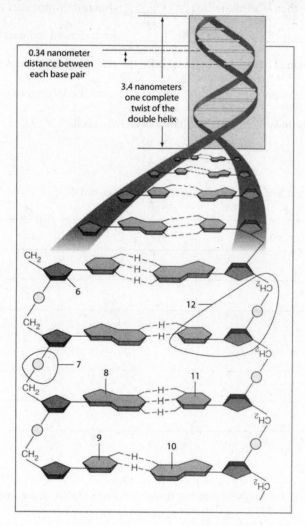

6. _____ ( )

7. _____ ( )

8. _____ ( )

9. _____ ( )

10. _____ ( )

11. _____ ( )

12. A complete _____ ( )

A. Thymine

B. Five-carbon sugar

C. Guanine

D. Cytosine

E. Adenine

F. Composed of three smaller molecules: a phosphate group, deoxyribose, and a nitrogenous base

G. A chemical group that joins two sugars in the DNA ladder

DNA Structure and Function   **105**

Fill-in-the-Blanks [pp.136-137]

A bacteriophage is a kind of (13) _____ that can infect (14) _____ cells. These infectious

particles carry (15) _____ information about how to make new viruses. Once injected into a host cell, the

bacteriophage injects its (16) _____ material into bacteria and incorporates that new material into the host

DNA. In Hershey and Chase's experiments, the (17) _____ (not the DNA) was labeled with a radioisotope of

sulfur, ^{35}S. Hershey and Chase then labeled the (18) _____ (but not the protein) with the radioisotope of

phosphorus known as (19) _____. When (20) _____ were allowed to infect bacterial cells, the (21)

_____ radioisotope remained outside the bacterial cells; the (22) _____ was part of the material injected

into the bacterial cells. Through many experiments, researchers accumulated strong evidence that (23) _____,

not (24) _____, serves as the molecule of inheritance in all living cells.

Short Answer [pp.136-137]

What are the contributions of each of the following to our knowledge of DNA?

25. Miescher _____

_____

26. Griffith _____

_____

27. Avery (et al.) _____

_____

28. Hershey and Chase _____

_____

Short Answer [pp.138-139]

29. What was Rosalind Franklin's contribution to our understanding of DNA structure?

_____

_____

30. Why didn't Rosalind Franklin share in the 1962 Nobel Prize that went to Watson, Crick, and Wilkins for the
discovery of the structure of DNA?

_____

_____

8.5. DNA REPLICATION [pp.140-141]

Boldfaced, Page-Referenced Terms

DNA replication _____

_____

DNA polymerase _____

_____

primer _____

_____

DNA ligase _____

_____

semiconservative replication _____

_____

Short Answer [pp.140-141]

1. Discuss the necessity for Okazaki fragments in DNA synthesis.

_____

_____

2. Explain how errors are corrected in DNA.

_____

_____

Choice [pp.140-141]

For questions 3-10, choose from the following:

      a. DNA polymerase   b. primer   c. DNA ligase   d. initiators   e. helicase

3. ____ Short, single strand of DNA or RNA complimentary to a target DNA sequence.

4. ____ Seals any gaps in the newly formed DNA strand

5. ____ Large numbers of proteins that bind to certain sequences within the DNA

6. ____ Enzyme central to the process of DNA replication.

7. ____ Serve as attachment points for DNA polmerases

8. ____ An enzyme that breaks the bonds of hydrogen holding the two strands of DNA together.

9. ____ Recruit the enzymes that will break the DNA apart.

10. ____ Attaches free nucleotides to the 3' end of a DNA strand.

11. The term *semiconservative replication* refers to the fact that each new DNA molecule resulting from the replication process is "half old, half new." The following sequences represent DNA in the midst of replication. Complete the replication required in the middle of the molecule by adding the required letters for the missing nucleotide bases. Note that ATP energy and the appropriate enzymes are actually required in order to complete this process.

T– _____   _____ –A

G– _____   _____ –C

A– _____   _____ --T

C– _____   _____ –G

C– _____   _____ –G

C– _____   _____ –G

Old New   New Old

True/False [pp.140-141]

If a statement below is true, write T in the blank. If it is false, explain why by changing one or more of the underlined words.

12. _____ The hydrogen bonding of adenine to <u>guanine</u> is an example of complementary base pairing.

13. _____ The replication of DNA is considered a <u>conservative process</u> because each new molecule is really half new and half old.

14. _____ Each <u>parent single strand remains intact</u> during replication, and a new companion strand is assembled on each of those parent strands.

15. _____ DNA <u>ligases</u> govern the assembly of nucleotides on a parent strand.

16. _____ <u>DNA polymerases, DNA ligases, and other enzymes</u> engage in DNA replication.

8.6. MUTATIONS: CAUSE AND EFFECT [pp.142-143]

8.7. ANIMAL CLONING [pp.144-145]

Boldfaced, Page-Referenced Terms

mutation_____

_____

reproductive cloning _____

_____

somatic cell nuclear transfer _____

_____

therapeutic cloning _____

_____

Sequence [pp.144-145]

Correctly order the following events in somatic cell nuclear transfer (SCNT).

1. _____ Egg is treated with an electric current.

2. _____ A micropipette is used to deliver a cell from the donor animal.

3. _____ The receiving egg is held in place by suction through a micropipette.

4. _____ The embryo is transplanted into a surrogate mother.

5. _____ The polar body and chromosomes are removed from the receiving egg.

Matching [pp.144-145]

Match the vocabulary term to its definition.

6. _____ Production of embryos for research

7. _____ Animal that will carry the cloned embryo to term

8. _____ Cell that has not been committed to developing into a certain kind of cell

9. _____ Natural process where an embryo forms identical twins

10. _____ Involves removal of the nucleus from an unfertilized egg

11. _____ Results from a single embryo splitting to form two separate individuals

12. _____ Cell that has been committed to specialize as a certain kind of cell

13. _____ Production of genetically cloned individuals

A. surrogate mother

B. differentiated cell

C. undifferentiated cell

D. identical twins

E. embryo splitting

F. therapeutic cloning

G. reproductive cloning

H. somatic cell nuclear transfer (SCNT)

Short Answer [p.145]

14. How can therapeutic cloning be used to help individuals with spinal cord injuries or who need organ transplants?

_____

_____

_____

15. How is reproductive cloning used in animal husbandry?

_____

_____

_____

DNA Structure and Function **109**

16. What are some scientific problems that have arisen out of the cloning process?

_____

_____

_____

_____

SELF-TEST

____ 1. Each DNA strand has a backbone that consists of alternating _____.
   a. purines and pyrimidines
   b. nitrogen-containing bases
   c. hydrogen bonds
   d. sugar and phosphate molecules

____ 2. In DNA, complementary base-pairing occurs between _____.
   a. cytosine and uracil
   b. adenine and guanine
   c. adenine and uracil
   d. adenine and thymine

____ 3. Adenine and guanine are _____.
   a. double-ringed purines
   b. single-ringed purines
   c. double-ringed pyrimidines
   d. single-ringed pyrimidines

____ 4. Franklin used the technique known as _____ to determine many of the physical characteristics of DNA.
   a. transformation
   b. cloning
   c. density-gradient centrifugation
   d. x-ray diffraction

____ 5. The significance of Griffith's experiment in which he used two strains of pneumonia-causing bacteria is that _____.
   a. DNA is conservative in its replication
   b. it demonstrated that pathogenic cells had become permanently transformed into harmless cells through a change in the bacteriophage protein coat
   c. it established that pure DNA extracted from disease-causing bacteria and injected into harmless strains transformed them into "pathogenic strains"
   d. it demonstrated that radioactively labeled bacteriophages transfer their DNA but not their protein coats to their host bacteria

____ 6. The significance of the experiments in which ^{32}P and ^{35}S were used is that _____.
   a. the semiconservative nature of DNA replication was finally demonstrated
   b. it demonstrated that harmless cells had become permanently transformed through a change in the bacterial hereditary system
   c. it established that pure DNA extracted from disease-causing bacteria transformed harmless strains into "killer strains"
   d. it demonstrated that radioactively labeled bacteriophages transfer their DNA but not their protein coats to their host bacteria

____ 7. Franklin's research contribution was essential in _____.
   a. establishing the principle of base pairing
   b. establishing most of the principal structural features of DNA
   c. both a and b
   d. neither a nor b

____ 8. Chargaff's requirement that A binds to T and G binds to C suggested that _____.
   a. cytosine molecules pair up with guanine molecules, and thymine molecules pair up with adenine molecules
   b. the two strands in DNA run in opposite directions (are antiparallel)
   c. the number of adenine molecules in DNA relative to the number of guanine molecules differs from one species to the next
   d. the replication process must necessarily be semiconservative

____ 9. A single strand of DNA with the base-pairing sequence C–G–A–T–T–G is compatible only with what sequence?
   a. C–G–A–T–T–G
   b. G–C–T–A–A–G
   c. T–A–G–C–C–T
   d. G–C–T–A–A–C

____ 10. Rosalind Franklin's data indicated that the DNA molecule had to be long and thin with a width (diameter) that is 2 nanometers along its length. Double-ringed nucleotides are wider than single-ringed ones, so the uniform width of the DNA molecule results from _____ (as Watson and Crick declared).
  a. complementary base-pairing processes that match purine with pyrmidine
  b. semiconservative replication processes
  c. hydrogen bonding of the sugar–phosphate backbones
  d. the antiparallel nature of DNA

____ 11. Which of these enzymes adds new base pairs to the growing DNA strand?
  a. DNA polymerase
  b. ligase
  c. helicase
  d. toposiomerase

____ 12. The total number of human chromosomes is
  a. 2
  b. 23
  c. 46
  d. 208

____ 13. This includes all but one pair of chromosomes in an individual:
  a. sister chromatids
  b. autosomes
  c. sex chomosomes
  d. nucleosomes

____ 14. A map of an individual's chomosomes lined up in order is a(n) _____.
  a. diploid
  b. chromosome number
  c. pattern
  d. karyotype

____ 15. All of the following can cause DNA mutations EXCEPT:
  a. food
  b. pollutants
  c. UV
  d. some natural chemicals

____ 16. DNA replication occurs with very few mistakes because of
  a. mistakes always stay in the DNA
  b. DNA polymerase proofreading
  c. a map that DNA follows
  d. no mistakes are ever made

____ 17. A technique that artificially leads to identical twinning is.
  a. standard breeding
  b. therapeutic cloning
  c. reproductive cloning

____ 18. The "rungs" of the ladder of DNA are.
  a. sugars
  b. paired nitrogenous bases
  c. phosphate groups

____ 19. In DNA there are equal parts of adenine and thymine versus guanine and cytosine. If a strand of DNA is 27% cytosine, what is its percentage of adenine?
  a. 23%
  b. 27%
  c. 46%
  d. 54%

____ 20. In the semiconservative replication of DNA, progeny DNA molecules consist of:?
  a. Half of the strands would be all parental DNA and half of the strands would be all new DNA
  b. Segments of parental and new DNA would be in all strands.
  c. Each strand would be half parental and half new DNA
  d. Each strand would be considered new DNA.

DNA Structure and Function **111**

CHAPTER OBJECTIVES/REVIEW QUESTIONS

1. Summarize the research carried out by Miescher, Griffith, Avery, and Hershey and Chase; state the specific advances made by each in the understanding of genetics. [pp.136-139]
2. DNA is composed of double-ring nucleotides known as _____ and single-ring nucleotides known as _____; the two purines are _____ and _____, whereas the two pyrimidines are _____ and _____. [p.138]
3. Assume that the two parent strands of DNA have been separated and that the base sequence on one parent strand is A–T–T–C–G–C; the base sequence that will complement that parent strand is _____ . [p.139]
4. The two scientists who assembled the clues to DNA structure and produced the first model were _____ and _____. [pp.138-139]
5. Explain what is meant by the pairing of nitrogen-containing bases (base-pairing), and explain the mechanism that causes bases of one DNA strand to join with bases of the other strand. [pp.140-141]
6. Generally describe how double-stranded DNA replicates from stockpiles of nucleotides. [pp.140-141]
7. Draw the basic shape of a deoxyribose molecule, and show how a phosphate group is joined to it when forming a nucleotide. [p.141]
8. List the pieces of information about DNA structure that Rosalind Franklin discovered through her x-ray diffraction research. [p.143]
9. Describe the process of making a genetically identical copy of yourself. [pp.144]

INTEGRATING AND APPLYING KEY CONCEPTS

1. Review the stages of mitosis and meiosis, as well as the process of fertilization. Relate what was learned in the chapter about DNA replication and the relationship of DNA to a chromosome. As you pass through fertilization and the stages of both types of cell division, use a diploid number of 2n = 4. Show the proper number of DNA threads in each cell at each of the stages of mitosis and meiosis. Include the cells that represent the end products of mitosis and meiosis.
2. How might somatic cell nuclear transfer be used to alleviate health problems in humans? What are the ethical considerations with using SCNT?
3. How do the results of cloning and sexual reproduction differ? How might each interact with evolution?

9

FROM DNA TO PROTEIN

INTRODUCTION

This chapter details two processes: transcription, to make RNA copies of DNA, and translation, to convert the information in the sequence of a nucleic acid into the amino acid sequence of a protein. In addition, the effects of mutations on protein formation are outlined.

STUDY STRATEGIES

- Read all of Chapter 9 with the goal of familiarizing yourself with the boldface terms.
- Identify the importance of ribosomes and how RIPs affect an organism.
- Compare and contrast DNA and RNA.
- Explain the process of transcription.
- Describe post-transcriptional processing.
- Identify the importance of mRNA, rRNA, and tRNA to translation.
- Explain the process of translation.
- Discuss the possible mutations of DNA and their effects on protein function.

FOCAL POINTS

- Figure 9.3 [p.151] compares DNA to RNA.
- Figures 9.4, and 9.5 [pp.152-153] outline the steps of transcription.
- Figure 9.6 [p.153] details the modification of mRNA before it leaves the nucleus.
- Figures9.7 and 9.8 [pp.154-155] explain the mRNA genetic code.
- Figures 9.11 and 9.12 [pp.156-157] outline the steps of translation.

INTERACTIVE EXERCISES

9.1. THE APTLY ACRONYMED RIPS [p.149]

9.2. DNA, RNA, AND GENE EXPRESSION [pp.150-151]

Boldfaced, Page-Referenced Terms

genes_____

_____

transcription _____

_____

ribosomal RNA (rRNA) _____

_____

transfer RNA (tRNA) _____

_____

messenger RNA (mRNA) _____

_____

translation _____

_____

gene expression _____

_____

Fill-in-the-Blanks [pp.150-151]

The (1) _____ of four bases contains the genetic information in DNA. The two steps from genes to

proteins are called (2) _____ and (3) _____. In (4) _____, single-stranded molecules of RNA are

assembled on DNA templates in the nucleus. In (5) _____, the RNA molecules are shipped from the nucleus

into the cytoplasm, where they are used as templates for assembling the sequence of (6) _____ of a protein. (7)

_____ is the multistep process by which genetic information encoded by a gene is converted into a structural or

functional part of a cell or body. Proteins – (8) _____ – assemble lipids and complex carbohydrates from

simple building blocks, replicate DNA, and make RNA.

Complete the Table [p.151]

9. Three types of RNA are transcribed from DNA in the nucleus (two are from genes that code only for RNA).
Complete the following table, which summarizes information about these molecules.

RNA Molecule	Abbreviation	Description/Function
a. Ribosomal RNA		
b. Messenger RNA		
c. Transfer RNA		

10. List three ways in which a molecule of RNA differs structurally from a molecule of DNA.

_____

_____

_____

11. Explain, in a general sense, how DNA codes genetic information and how this is utilized by the cell?

_____

_____

_____

9.3. TRANSCRIPTION: DNA TO RNA [pp.152-153]

Boldfaced, Page-Referenced Terms

RNA polymerase _____

_____

promoter _____

_____

introns _____

_____

exons _____

_____

alternative splicing _____

_____

Short Answer [pp.152-153]

1. Cite two similarities between DNA replication and transcription.

_____

_____

2. What are the three key ways in which transcription differs from DNA replication?

_____

_____

3. In eukaryotes, what modifications to "pre-mRNA" need to be made before it can be used in translation?

_____

_____

Labeling and Matching [Figure 9.6, p.153]

Newly transcribed mRNA contains more genetic information than is necessary to code for a chain of amino acids. Before the mRNA leaves the nucleus for its ribosome destination, an editing process occurs as certain portions of nonessential information are snipped out. Identify each indicated part of the following illustration; use abbreviations for the nucleic acids. Complete the exercise by entering the letter of the description in the parentheses following each label.

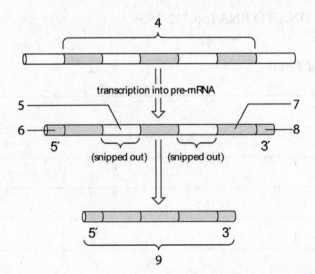

4. _____ ( )	A. An actual coding portion of pre-mRNA
5. _____ ( )	B. A noncoding portion of the newly transcribed mRNA
6. _____ ( )	C. Mature mRNA transcript
7. _____ ( )	D. Poly-A tail
8. _____ ( )	E. The region of the DNA template strand to be copied
9. _____ ( )	F. Cap on the 5' end of mRNA (the first synthesized)

9.4. RNA AND THE GENETIC CODE [pp.154-155]

Boldfaced, Page-Referenced Terms

codon _____

_____

genetic code _____

_____

anticodon _____

_____

Completion [pp.154-155]

1.   Given the following DNA sequence, deduce the composition of the mRNA transcript.

     TAC   AAG   ATA   ACA   TTA   TTT   CCT   ACC   GTC   ATC

     ____   ____   ____   ____   ____   ____   ____   ____   ____   ____

                                    (mRNA transcript)

2.   From the mRNA transcript in question 2, use Figure 9.7 of the text to deduce the composition of the amino acids of the polypeptide sequence.

     ____   ____   ____   ____   ____   ____   ____   ____   ____   ____

                                    (amino acids)

Matching [pp.154-155]

Match the following codons with the amino acid it codes for. Use Figure 9.7 as a guide.

3.   ____   AUG   A.   proline

4.   ____   GGC   B.   glycine

5.   ____   UAG   C.   aspartic acid

6.   ____   AAU   D.   asparagine

7.   ____   CCG   E.   methionine

8.   ____   GAU   F.   stop

9.5. TRANSLATION: RNA TO PROTEIN [pp.156-157]

Fill-in-the-Blanks [pp.156-157]

(1) _____ occurs in the cytoplasm of all cells and consists of three stages: (2) _____, (3) _____, and (4) _____. A small ribosomal subunit binds to (5) _____ RNA and a special initiator (6) _____ RNA base-pairs with the first (7) _____ codon. In (8) _____, the ribosome moves along the mRNA and builds a (9) _____ chain. During the last stage of translation, (10) _____, a STOP codon in the mRNA moves onto the platform, and no tRNA has a corresponding anticodon. Now proteins called (11) _____ factors bind to the ribosome. They trigger (12) _____ activity that detaches the mRNA and the polypeptide chain from the ribosome. When many ribosomes simultaneously translate the same mRNA they are called (13) _____. The energy required for translation is mainly in the form of (14) _____ transfers from the RNA nucleotide (15) _____.

Short Answer [pp.156-157]

16. What is the role of mRNA, rRNA, and tRNA in the process of translation?

_____

_____

_____

17. Explain how the various types of RNA know where to start and stop translating a polypeptide.

_____

_____

_____

9.6. MUTATED GENES AND THEIR PROTEIN PRODUCTS [pp.158-159]

Boldfaced, Page-Referenced Terms

base-pair substitution _____

_____

deletion_____

_____

insertion _____

_____

frame-shift _____

_____

transposable elements _____

_____

Matching [pp.158-159]

From the following original codes, match the change with the type of mutation demonstrated.

AUG-GGC-AAU-CCC-UCG

1. ____ AUG-GGG-CAA-UCC-CUC
2. ____ AUG-GGC-AUC-CCU-CGA   A. insertion
3. ____ AUG-CGC-AAU-CCC-UCG   B. substitution
4. ____ AUG-CAA-UCC-CUC-GAA   C. deletion
5. ____ AUG-GGC-UAU-UCC-UCG
6. ____ AUG-GGC-AAA-UCC-CUC-

Short Answer [pp.158-159]

Use Figure 9.13 to answer the following questions.

7. Which is usually the more damaging mutation: base substitution or deletion? Why?

_____

_____

8. Would a base substitution mutation in the third position of a codon be more or less likely to cause a serious mutation? Explain. _____

_____

_____

9. Would an insertion or deletion mutation be more damaging at the beginning of a mRNA strand or at the end? Why?

_____

_____

_____

10. What is a "silent" mutation?

_____

_____

Labeling and Matching [p. 160]

A summary of the flow of genetic information in protein synthesis is useful as an overview. Identify the numbered parts of the accompanying illustration by filling in the blanks with the names of the appropriate structures or functions. Choose from the following:

DNA, mRNA, rRNA, tRNA, amino acids, anticodon, intron, exon, polypeptide, ribosomal subunits, transcription, translation

Complete the exercise by matching and entering the letter of the description in the parentheses following each label.

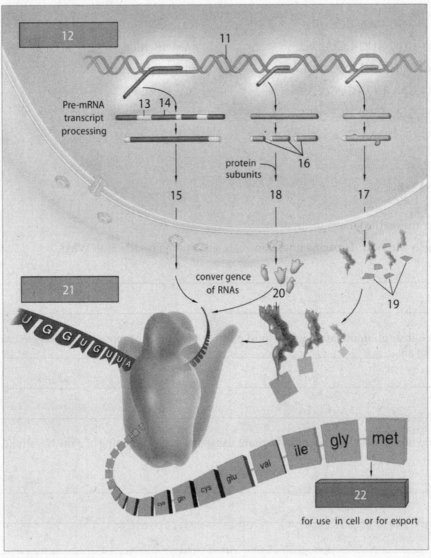

11. _____ ( )   13. _____ ( )

12. _____ ( ) (process)   14. _____ ( )

15. _____ ( )

16. _____ ( )

17. _____ ( )

18. _____ ( )

19. _____ ( )

20. _____ ( )

21. _____ ( ) (process)

22. _____ ( )

A. Pre-mRNA coding portion (will translate into proteins)

B. Carries a form of the genetic code from DNA in the nucleus to the cytoplasm

C. Transports amino acids to the ribosome and mRNA

D. The building blocks of polypeptides

E. Noncoding portion of pre-mRNA

F. Combines with proteins to form the ribosomal subunits

G. Join during the initiation step of protein synthesis

H. Holds the genetic code for protein production

I. Amino acids are joined together

J. RNA synthesized on a DNA template

K. A sequence of three bases that can pair with a specific mRNA codon

L. May serve as a functional protein (enzyme) or structural protein

SELF-TEST

___ 1. Transcription _____.
a. occurs on the surface of a ribosome
b. is the final process in the assembly of protein DNA template
c. occurs during the synthesis of any type of RNA by use of a DNA template
d. is catalyzed by DNA polymerase

___ 2. _____carries the actual instructions for a protein's sequence to the ribosome.
a. DNA
b. mRNA
c. rRNA
d. tRNA

___ 3. _____ carries amino acids to ribosomes, where amino acids are linked into the primary structure of a polypeptide.
a. mRNA
b. tRNA
c. An intron
d. rRNA

___ 4. Transfer RNA differs from other types of RNA because it _____.
a. transfers genetic instructions from cell nucleus to cytoplasm
b. specifies the amino acid sequence of a particular protein
c. carries an amino acid at one end
d. contains codons

___ 5. _____ catalyzes the process of transcription.
a. RNA polymerase
b. DNA polymerase
c. Phenylketonuria
d. Transfer RNA

___ 6. _____ is found in RNA but not in DNA.
a. Deoxyribose
b. Uracil
c. Phosphate
d. Thymine

___ 7. Each codon in the mRNA language consists of _____ letters.
a. three
b. four
c. five
d. more than five

___ 8. The genetic code is used to align _____ amino acids.
a. four
b. sixteen
c. twenty
d. sixty-four

___ 9. The genetic code is composed of
_____ codons.
   a. three
   b. twenty
   c. sixteen
   d. sixty-four

___10. The cause of sickle-cell anemia has been
traced to _____.
   a. a mosquito-transmitted virus
   b. extra bases inserted into DNA
   c. ionizing radiation
   d. a DNA mutation in a hemoglobin
      chain

___11. The process of converting the mRNA
message into a protein is _____.
   a. splicing
   b. DNA replication
   c. transcription
   d. translation

___12. Transcription takes place in the
   a. ER
   b. nucleus
   c. cytoplasm
   d. Golgi apparatus

___13. Translation takes place on the
   a. ER
   b. nucleus
   c. cytoplasm
   d. Golgi apparatus

___14. During transcription, a DNA strand with
following bases GGCTAC will produce an
mRNA strand with bases:
   a. CCGAUG
   b. TTAGCA
   c. GGCTAC
   d. CATCGG

___ 15. What type of mutation would be least
detrimental?
   a. substitution
   b. single-base deletion at the start of the
      sequence
   c. single-base deletion in the stop code
      sequence
   d. single-base insertion

___ 16. Which of the following are the non-coding
units of mRNA are removed during
processing?
   a. poly-A tail
   b. introns
   c. exons
   d. 5' cap

___ 17. The following triplet is considered the
initiator codon.
   a. AUG
   b. GUA
   c. UAC
   d. GGG

___ 18. Ribosomes preform what function in
protein synthesis?
   a. They carry the amino acids.
   b. They have positions for the attachment
      of tRNA to mRNA via
      codons/anticodons
   c. They guide the movement of other
      organelles
   d. They have no function in protein
      synthesis

CHAPTER OBJECTIVES/REVIEW QUESTIONS

1. State how RNA differs from DNA in structure and function, and indicate what features RNA has in common with DNA. [p.150]
2. RNA combines with certain proteins to form the ribosome; RNA carries genetic information for protein construction from the nucleus to the cytoplasm; RNA picks up specific amino acids and moves them to the area of mRNA and the ribosome. Identify which type of RNA does each of these jobs. [p.150]
3. Describe the process of transcription, and indicate three ways in which it differs from replication [pp.152-153]
4. Transcription starts at a(n) _____, a specific sequence of bases on one of the two DNA strands that signals the start of a gene. [p.152]
5. What RNA code would be formed from the following DNA code: TAC–CTC–GTT–CCC–GAA? [pp.152-153]
6. Distinguish introns from exons. [p.153]
7. Describe how the three types of RNA participate in the process of translation. [pp.154-155]
8. State the relationship between the DNA genetic code and the order of amino acids in a protein chain. [pp.154-157]

9. Explain how the DNA message TAC–CTC–GTT–CCC–GAA would be used to code for a segment of protein, and state what its amino acid sequence would be. [pp.154-157]
10. What is the advantage of having several different codons encode for most amino acids? [p.158]
11. Cite an example of a change in one DNA base-pair that has profound effects on the human phenotype. [pp.158-159]
12. Briefly describe the spontaneous DNA mutations known as *base-pair substitution, frameshift mutation,* and *transposons.* [pp.158-159]

INTEGRATING AND APPLYING KEY CONCEPTS

1. Genes code for specific polypeptide sequences. Not every substance in living cells is a polypeptide. Explain how genes might be involved in the production of a storage starch (such as glycogen) that is constructed from simple sugars.
2. Why is the genetic code almost universal? How did the few alterations occur?
3. In words/diagrams describe the three stages of translation.
4. Why are mutations in germ cells usually more of a problem than mutations in somatic cells?

10

GENE CONTROL

INTRODUCTION

This chapter looks at the mechanisms of controlling protein synthesis in prokaryotic and eukaryotic organisms, and why such controls are needed. Epigenetics are also discussed as it relates to survivability of a species.

STUDY STRATEGIES

- Read all of Chapter 10 with the goal of familiarizing yourself with the boldface terms.
- Explain the various points of control in protein synthesis for eukaryotes.
- Describe the role of homeotic genes in development.
- Identify the role of Barr bodies in females.
- Recognize some examples of control in eukaryotes.
- Discuss and give examples of protein synthesis controls in prokaryotes.
- Explain current research in to epigenetics and what it tells scientists about organisms' responses to the environment.

FOCAL POINTS

- Section 10.2 [pp.164-165] looks at controls in eukaryotic systems.
- Figure 10.10 [pp.170-171] outlines prokaryotic control.

INTERACTIVE EXERCISES

10.1. BETWEEN YOU AND ETERNITY[p.163]

10.2. SWITCHING GENES ON AND OFF[pp.164-165]

Boldfaced, Page-Referenced Terms

cancer _____

_____

differentiation _____

_____

transcription factors_____

_____

repressor _____

_____

activator _____

_____

enhancers _____

_____

Matching [p.164]

Match the situation to the point of control over eukaryotic gene expression.

1. _____ Alternative splicing occurs to modify the mRNA before it leaves the nucleus

2. _____ Enzyme-mediated modifications occur to activate the newly-made protein

3. _____ mRNA stability determines how long it's translated

4. _____ Binding factors must locate special sequences on the DNA

5. _____ Proteins bind to RNA to allow transport through the nuclear pore

A. transcription

B. RNA processing

C. RNA transport

D. translation

E. protein processing

Short Answer [pp.164-165]

6. Although a complex organism such as a human being arises from a single cell, the zygote, differentiation occurs in development. Define *differentiation*. _____

_____

7. How does a cell know which genes to express? _____

_____

8. How do polytene genes affect how the final protein product is made? _____

_____

9. How does a poly-A tail affect how long an mRNA strand is translated? _____

_____

_____

Gene Control   **125**

10. What types of translational control affect the outcome of protein synthesis? _____

_____

_____

Labeling and Matching [pp.164-165]

11. Using the figure below, place the number of the step in eukaryotic gene control next to the appropriate point of control.

a. _____ Translation

b. _____ mRNA transport

c. _____ mRNA processing

d. _____ Protein processing

e. _____ Transcription

10.3. MASTER GENES [pp.166-167]

10.4. EXAMPLES OF GENE CONTROL IN EUKARYOTES [pp.168-169]

Boldfaced, Page-Referenced Terms

master genes_____

_____

pattern formation_____

_____

homeotic genes_____

_____

knockout _____

_____

X chromosome inactivation_____

_____

dosage compensation_____

_____

Matching [pp.166-167]

Match the gene to what it develops.

1. ____ mutated genes that block development   A. dunce
2. ____ learning and memory   B. groucho
3 ____ heart   C. homeotic
4. ____ bristles above the eyes   D. knockout
5. ____ named for the German word for "cool"   E. PAX6
6. ____ homologue of the eyeless gene   F. tinman
7. ____ master genes that control development   G. toll

Fill-in-the-Blanks [pp.168-169]

In mammalian females, the majority of one X chromosome's genes are shut down in a process called (8)

_____. The inactive X is condensed and afterwards called a (9) _____. Since the shutdown is random

in the embryonic cells formed by the time inactivation occurs, all mature mammalian females are a (10)

_____ for the expression of X chromosome genes. As an example, human females with the X-linked disorder

(11) _____ have patches of skin with lighter and darker pigmentation. The inactivation of one X in the

female is thought to balance gene expression between males and females, a theory called (12) _____. The

shutdown of the X is accomplished by the (13) _____ gene. This gene's product, a large (14) _____,

sticks to the chromosome and causes it to condense into a Barr body.

15. What determines male sexual development in humans?

_____

_____

16. What controls activation of the ABC master genes in flower formation?

_____

_____

17. What is a homeodomain and how does it relate to homeotic genes?

_____

_____

18. What role does X chromosome inactivation play in the development of a female human?

_____

_____

19. How are knockout experiments used to study gene expression?

_____

_____

10.5. EXAMPLES OF GENE CONTROL IN PROKARYOTES [pp.170-171]

Boldfaced, Page-Referenced Terms

operator _____

_____

operon _____

_____

True/False [pp.170-171]

Determine if each statement is true or false.

1. _____ Prokaryotes use master genes much like eukaryotes.

2. _____ *E. coli* prefers to use glucose, but can also use other sugars like lactose.

3. _____ cAMP binds to a protein and acts as an activator for transcription of lac operon genes.

4. _____ Allolactose is a conversion product of lactose.

5. _____ Lactose is absorbed directly by the intestine.

6. _____ RNA polymerase transcribes the operon genes when lactose is present.

7. _____ Operators are DNA regions that are binding sites for a promoter.

8. _____ Production of the enzyme lactase increases at about the age of five years.

9. _____ Prokaryotes control their gene expression by adjusting the rate of translation.

10. _____ To some extent, everyone is lactose intolerant.

11. _____ The lac operon is in the *E. coli* chromosome.

12. _____ RNA polymerases cannot bind to twisted promoters and therefore cannot transcribe operon genes.

Labeling and Matching [pp.170-171]

Escherichia coli, a bacterial cell living in mammalian digestive tracts, uses a negative type of gene control over lactose metabolism. Use the numbered blanks to identify each part of the accompanying diagram of the genes involved. Choose from the following:

lactose, lactose enzyme genes, regulatory gene, repressor–operator complex, promoter, mRNA, lactose operon, repressor protein, RNA polymerase, operator

Complete the exercise by matching and entering the letter of the proper function description in the parentheses following each label.

13. _____ (   )

14. _____ (   )

15. _____ (   )

16. _____ (   )

17. _____ (   )

18. _____ (   )

19. _____ (   )

20. _____ (   )

21. _____ (   )

22. _____ (   )

A. Includes promoter, operator, and the genes coding for lactose-metabolizing enzymes

B. Short DNA base sequences on both sides of the promoter

C. A nutrient molecule; a form of it binds to the repressor to allow operation of the lactose operon

D. Major enzyme that catalyzes transcription

E. Capable of preventing RNA polymerases from binding with DNA

F. Genes that code for lactose-metabolizing enzymes

G. Binds to operator and prevents RNA polymerase from binding to DNA and initiating transcription

H. Specific base sequence that signals the beginning of a gene, binding site for RNA polymerase

I. Gene that contains coding for production of repressor protein

J. Carries genetic instructions to ribosomes for production of lactose enzymes

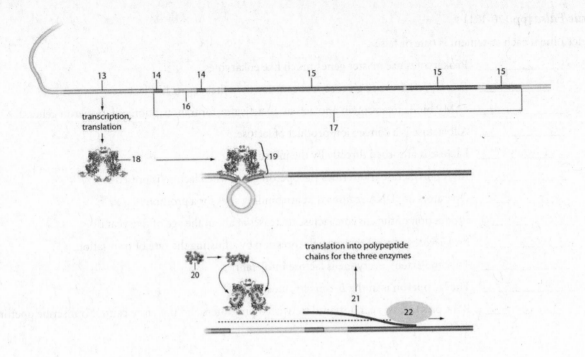

10.6. EXAMPLES OF GENE CONTROL IN PROKARYOTES [pp.172-173]

Boldfaced, Page-Referenced Terms

epigenetic _____

_____

Short Answer [pp.172-173]

1. What is the role of methylation in DNA transcription and how does it relate to cancer?

_____

_____

_____

2. How do environmental events, such as a famine, affect the children conceived during the event?

_____

_____

_____

3. What is epigenetics and what might its role be in survivability of a species?

_____

_____

_____

_____

SELF-TEST

___ 1. Mutated tumor suppressor genes associated with breast and ovarian cancers include _____.
   a. BRCA 1 and 2 genes
   b. PAX6 genes
   c. lac operon genes
   d. tinman genes

___ 2. _____ controls govern the rates at which mRNA transcripts that reach the cytoplasm will be translated into polypeptide chains at the ribosomes.
   a. Transport
   b. Transcript processing
   c. Translational
   d. Transcriptional

___ 3. _____ refers to the processes by which cells with identical genotypes become structurally and functionally distinct from one another according to the genetically controlled developmental program of the species.
   a. Metamorphosis
   b. Metastasis
   c. Cleavage
   d. Differentiation

___ 4. In multicelled eukaryotes, cell differentiation occurs as a result of _____.
   a. growth
   b. expression of different subsets of genes
   c. repressor molecules
   d. the death of certain cells

___ 5. One type of gene control discovered in female mammals is _____.
   a. a conflict in maternal and paternal alleles
   b. slow embryo development
   c. X chromosome inactivation
   d. operon

___ 6. Due to inactivation of either the paternal or maternal X chromosome, human females with incontinentia pigmenti _____.
   a. have mosaic tissues of lighter and darker patches of skin.
   b. develop benign growths
   c. have mosaic patches of skin that lack sweat glands
   d. develop malignant growths

___ 7. Genes that control the information about the basic body plan of eukaryotes are called _____.
   a. homeotic genes
   b. knockout genes
   c. repressor genes
   d. toll genes

___ 8. A(n) _____ binds to operator whenever lactose concentrations are low.
   a. operon
   b. repressor
   c. promoter
   d. operator

___ 9. Any gene or group of genes together with its promoter and operator sequence is a(n)_____ .
   a. repressor
   b. operator
   c. promoter
   d. operon

___ 10. The operon model explains the regulation of _____ in prokaryotes.
   a. replication
   b. transcription
   c. induction
   d. Lyonization

11. A(n) _____ is a transcription factor that speeds up the transcription process.
    a. enhancer
    b. repressor
    c. activator
    d. operon

12. A(n) _____ is a transcription factor that slows or stops the transcription process.
    a. enhancer
    b. repressor
    c. activator
    d. operon

13. A polytene chromosome ensures many copies of the same DNA molecule because _____.
    a. the mRNA always starts from locations on polytene chromosomes.
    b. there are many copies of the molecule on polytene chromosomes that can be transcribed simultaneously.
    c. they have a promoter that directs more transcription at that location than any other.
    d. none of the above.

14. Each of the following is a process of gene control in eukaryotes EXCEPT _____.
    a. transcriptional
    b. replication
    c. translational
    d. mRNA processing

15. Post-translational control in eukaryotic cells takes place in the _____.
    a. nucleus
    b. mitochondria
    c. cytoplasm
    d. cell membrane

16. Which statement is NOT true about the *lac* operon?
    a. It is normally turned off if glucose is present.
    b. It regulates a series of eight genes.
    c. Allolactose binds to the repressor and inactivates it.
    d. The genes make products for lactose metabolism.

17. Which of the following is the primary process of gene control in prokaryotes?
    a. transcription
    b. replication
    c. translation
    d. mRNA processing

18. Which of the following is the most permanent method of inactivating DNA?
    a. methylation
    b. histones
    c. repressors

19. X chromosome inactivation occurs _____.
    a. at birth
    b. by 3 months
    c. at conception
    d. during the embryonic stage

20. The SRY gene on the Y chromosome is important for what purpose?
    a. It is the master development gene for all humans.
    b. Turning off female development.
    c. Development of male gonads.

CHAPTER OBJECTIVES/REVIEW QUESTIONS

1. List and define the levels of gene control in eukaryotes. [pp.164-165]
2. How do genes relate to cell differentiation in multicelled eukaryotes? [pp.166-167]
3. Genes that control the basic body plans of organisms are called _____. [pp.166-167]
4. The condensed X chromosome seen on the edge of each nucleus of female mammals is known as the _____ body. [p.168]
5. Explain how X chromosome inactivation provides evidence for selective gene expression; use the example of incontinentia pigmenti. [p.168]
6. Some control agents bind to _____, which are short regulatory genes on either side of a promoter. [pp.170-171]
7. A(n) _____ speeds up transcription when it binds to a promoter. [pp.170-171]
8. Explain why newly-synthesized polypeptides must be modified before they become functional. [pp.170-171]

INTEGRATING AND APPLYING KEY CONCEPTS

1. Suppose you have been restricting yourself to a completely vegetarian diet for the past six months. Quite unexpectedly, you find yourself in a social situation that requires you to eat a half-pound sirloin steak. Would you expect to digest the steak as easily as you digest soybean burgers? Explain your yes or no answer in terms of transcriptional controls or feedback inhibition.
2. Describe the sequence of events that occurs on the chromosome of *E. coli* after you drink a glass of milk.
3. What problems would face bacteria that had lost their transcriptional control mechanisms?

11

HOW CELLS REPRODUCE

INTRODUCTION

We all began life as a fertilized egg. That single cell then reproduced trillions of times; resulting in the complex humans we are today. In this chapter you will examine the principles of cell division, specifically how eukaryotic cells make exact copies of themselves by the process of mitosis. In the next chapter, you will build on this knowledge to examine the principles of meiosis, a more complex process for the production of reproductive cells. In the last section of this chapter you will examine how scientists continue to research the mechanisms of cell division, specifically the control of the cell cycle, since unregulated cell growth can lead to cancer.

STUDY STRATEGIES

- Read all of Chapter 11 with the goal of familiarizing yourself with the boldface terms.
- Explain the cell cycle by identifying the major stages and describing the role of checkpoints.
- Discuss the phases of mitosis.
- Identify the differences in the final stage of mitosis between plant and animal cells.
- Recognize the importance of telomeres.
- Describe the role of oncogenes, tumor-suppressor genes, and mutations in the development of metastatic cancers.

FOCAL POINTS

- Figure 11.2[p.178] illustrates how cell division fits into the "life" cycle of the cell.
- Page 181 provides an overview of the stages of mitosis. These stages will be revisited in Chapter 10.
- Figure 11.11[p.185] illustrates the differences between benign and malignant neoplasms.

INTERACTIVE EXERCISES

11.1. HENRIETTA'S IMMORTAL CELLS [p.177]

11.2. MULTIPLICATION BY DIVISION [pp.178-179]

Boldfaced, Page-Referenced Terms

cell cycle _____

_____

interphase _____

_____

mitosis _____

_____

asexual reproduction _____

_____

homologous chromosomes _____

_____

Matching [pp.178-179]

Choose the most appropriate answer for each term.

1. _____ checkpoints
2. _____ cell cycle
3. _____ interphase
4. _____ end of mitosis
5. _____ sister chromatids
6. _____ mitosis
7. _____ G₁ interval
8. _____ S interval
9. _____ G₂ interval

A. The two attached DNA molecules of a duplicated chromosome

B. Cell division responsible for growth, tissue repair, and replacement of cells

C. Synthesis, or duplication of the DNA and proteins

D. Period after duplication of DNA during which the cell prepares for division

E. Period of cell growth before DNA duplication; a "gap" of interphase

F. Period of cytoplasmic division

G. Period that includes G1, S, G2

H. Points in the cell cycle that determine progression to the next phase

I. Period of time that encompasses the production of a new cell through cell division

Fill-in-the-Blanks [p.178-179]

During mitosis, a(n) (10) _____ parent cell can produce two diploid (11) _____ cells. This does

not mean each merely gets forty-six (12) _____. If only the total mattered, then one cell may get two pairs of

chromosome 22 and no pairs whatsoever of chromosome 9. But neither cell could function like its parent without

(13) _____ of each type of chromosome. A bipolar (14) _____ forms during nuclear division. This

dynamic network consists of (15) _____, some of which attached to duplicated (16) _____.

Microtubules separate sister (17) _____ and move them to (18) _____ ends of the cell.

Labeling [p.178] For each number, identify the component of the cell cycle.

19. _____

20. _____

21. _____

22. _____

23. _____

24. _____

25. _____

26. _____

27. _____

28. _____

Labeling [p.178-179]

29. In the preceding diagram, first shade in those stages of the cell cycle that are responsible for the process of mitosis. Second, shade in the stage in which the sister chromatids are formed.

True/False [pp.178-179]

If a statement below is true, write T in the blank. If it is false, explain why by changing one or more of the underlined words.

30. _____ DNA is <u>replicated</u> during S phase.

31. _____ Interphase includes G_1, S, and <u>mitosis</u> phases.

32. _____ <u>G_2</u> is the stage of interphase directly before DNA replication.

33. _____ Homologous chromosomes <u>are those with same length, shape, and genes.</u>

11.3. A CLOSER LOOK AT MITOSIS [pp.180-181]

Boldfaced, Page-Referenced Terms

prophase_____

_____

spindle _____

_____

metaphase _____

_____

anaphase_____

_____

telophase _____

_____

Labeling and Matching [pp.180-181]

Identify each of the mitotic stages shown by entering the correct stage in the blank beneath the sketch. Select from the following:

early prophase, late prophase, transition to metaphase, cell at interphase, metaphase, anaphase, telophase, interphase—daughter cells

Complete the exercise by matching and entering the letter of the correct phase description in the parentheses following each label.

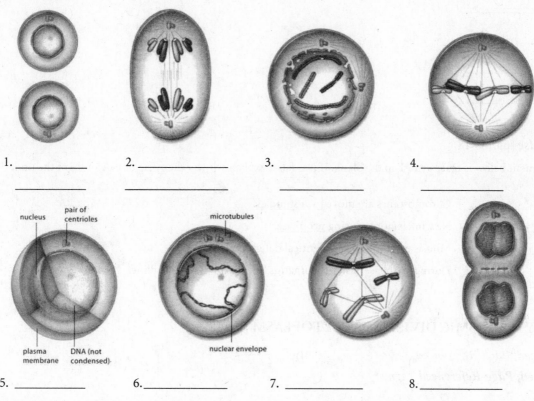

1. _____   2. _____   3. _____   4. _____

_____   _____   _____   _____

5. _____   6. _____   7. _____   8. _____

_____   _____   _____   _____

A. Attachments between two sister chromatids of each chromosome break; the two are now separate chromosomes that move to opposite spindle poles.

B. Microtubules penetrate the nuclear region and collectively form the spindle apparatus; microtubules become attached to the two sister chromatids of each chromosome.

C. The DNA and its associated proteins have started to condense.

D. All the chromosomes are now fully condensed and lined up at the equator of the spindle.

E. DNA is duplicated and the cell prepares for nuclear division.

F. Two daughter cells have formed, each diploid with two of each type of chromosome, just like the parent cell's nucleus.

G. Chromosomes continue to condense. New microtubules are assembled, and they move one of two centriole pairs toward the opposite end of the cell. The nuclear envelope begins to break up.

H. Patches of new membrane fuse to form a new nuclear envelope around the decondensing chromosomes.

Chronological Order [pp.180-181]

Using the illustration from the previous exercise, place the phases of mitosis in their correct chronological order. Write the correct name of the stage in the answer blank, and then place the number corresponding to the mitosis diagram.

9. _____ ( )

10. _____ ( )

11. _____ ( )

12. _____ ( )

13. _____ ( )

14. _____ ( )

15. _____ ( )

16. _____ ( )

True/False [pp.180-181]

If a statement below is true, write T in the blank. If it is false, explain why by changing one or more of the underlined words.

17. _____ Chromosomes are moved via <u>spindles</u>.

18. _____ New nuclei form during <u>prophase</u>.

19. _____ Mitosis ends with <u>three</u> identical daughter cells.

20. _____ During <u>anaphase</u>, sister chromatids line up on the cell midline.

11.4. CYTOPLASMIC DIVISION OF CYTOPLASM [pp.182]

Boldfaced, Page-Referenced Terms

cytokinesis _____

_____

cleavage _____

_____

furrow _____

_____

cell plate_____

_____

Matching [pp.182]

Choose the most appropriate answer for each term.

1. _____ Cytoplasmic division
2. _____ Forms from the cell plate
3. _____ Filaments that contract to divide cytoplasm in animal cells
4. _____ Band of actin and myosin that divides cytoplasm in animal cells
5. _____ Indentation in the cell membrane between the cell's poles
6. _____ Formed from vesicles and their wall-building contents in plant cells

A. Contractile ring
B. Cell plate
C. Cleavage furrow
D. Cytokinesis
E. Actin
F. Primary cell wall

Choice [pp.182]

For exercises 1-9 on cytoplasmic division, choose from the following forms of division.

   a. plant cell division   b. animal cell division   c.   both plant and animal cell cytoplasmic division

7. _____ A cell plate forms
8. _____ Actin filaments contract
9. _____ Involves cells with rigid cell walls
10. _____ A cleavage furrow forms via the use of ATP
11. _____ The end result is the formation of two daughter cells
12. _____ Microtubules orient cellulose fibers prior to cytoplasmic division

Sequence [pp.182]

Correctly order the following events in animal cytoplasmic division by placing the letter of the diagram corresponding to the step in the appropriate blank and so on.

13. _____ Contractile ring pulls cell surface inwards while shrinking.
14. _____ Spindles begin to disassemble.
15. _____ Cell is pinched in two.
16. _____ Actin and myosin filaments begin to contract along the cell midline.

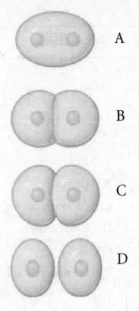

A

B

C

D

Sequence [pp.182]

Correctly order the following events in plant cytoplasmic division by placing a "1" in front of the first step and so on.

17. _____ A cell plate forms

18. _____ The cytoplasm is divided when the cell plate joins the plasma membrane

19. _____ Microtubules orient cellulose fibers prior to cytoplasmic division

20. _____ The end result is the formation of two daughter cells

11.5. MARKING TIME WITH TELOMERES [pp.182]

Boldfaced, Page-Referenced Terms

telomeres _____

_____

Short Answer [pp.182]

1. Explain the importance of telomeres.

_____

_____

2. Why do telomeres shorten with each cell division and what normally happens when they get too short?

_____

_____

3. What relationship do telomeres and stem cells have?

_____

_____

4. How does aging relate to telomeres?

_____

_____

11.6. WHEN MITOSIS BECOMES PATHOLOGICAL [pp.184-185]

Boldfaced, Page-Referenced Terms

neoplasm _____

_____

tumor _____

_____

oncogene _____

_____

proto-oncogene _____

_____

growth factors _____

_____

metastasis _____

_____

Matching [pp.184-185]

Match each of the following definitions with the correct term.

1. _____ Cell masses that may be classified as
either benign or malignant

2. _____ Checkpoint genes that stimulate mitosis

3. _____ Abnormally growing and dividing cells
of a malignant neoplasm

4. _____ Form of skin cancer

5. _____ Promote tumor formation

6. _____ Checkpoint genes that inhibit mitosis

A. growth factors
B. proto-oncogenes
C. basal cell carcinoma

D. tumor suppressors
E. cancers
F. neoplasms

SELF-TEST

___ 1. The replication of DNA and formation of the sister chromatids occurs _____.
   a. during a growth phase of interphase
   b. immediately before prophase of mitosis
   c. during prophase of mitosis
   d. during telophase of mitosis

___ 2. Which of the following inhibits the process of mitosis?
   a. proto-oncogenes
   b. actin
   c. growth factors
   d. tumor suppressors

___ 3. Diploid refers to _____.
   a. having two chromosomes of each type in somatic cells
   b. triple the parental chromosome number
   c. half the parental chromosome number
   d. having one chromosome of each type in somatic cells

___ 3. Cells need to have ____ copy of each chromosome to function properly
   a. one
   b. two
   c. four
   d. eight

___ 5. If a parent cell has sixteen chromosomes and undergoes mitosis, the resulting cells will have _____ chromosomes.
   a. sixty-four
   b. thirty-two
   c. sixteen
   d. eight
   e. four

___ 6. Microtubules bind to the _____, a constriction where sister chromatids attach.
   a. cell membrane
   b. nucleosome
   c. centromere
   d. mitotic spindle
   e. histone

___ 7. The correct order of the stages of mitosis is _____.
   a. prophase, metaphase, telophase, anaphase
   b. telophase, anaphase, metaphase, prophase
   c. telophase, prophase, metaphase, anaphase
   d. anaphase, prophase, telophase, metaphase
   e. prophase, metaphase, anaphase, telophase

___ 8. What is the stage of mitosis that is characterized by the alignment of the chromosomes along a central line in the cell called?
   a. prophase
   b. metaphase
   c. prometaphase
   d. anaphase
   e. telophase

___ 9. During _____, sister chromatids of each chromosome are separated from each other, and those former partners, now chromosomes, move to opposite poles.
   a. prophase
   b. metaphase
   c. anaphase
   d. telophase

___ 10. Asexual reproduction is also known as
   a. replication
   b. mitosis
   c. meiosis
   d. interphase

___ 11. Animal and plant cytoplasm division differs in that plants produce _____.
   a. extra DNA
   b. more cytoplasm
   c. a cell plate
   d. less plasma membrane

12. Which of the following statements about telomeres is INCORRECT?
   a. They are repeating sequences.
   b. They are located at the ends of chromosomes.
   c. The cell cycle does not stop even if the telomeres are too short.
   d. They are an evolutionary protection against runaway cell division.

13. Mice that have had their telomeres reduced in length _____.
   a. age prematurely
   b. always develop cancer
   c. are normal
   d. have an overgrowth of tissues

14. _____ are molecules that stimulate cells to divide and differentiate.
   a. Oncogenes
   b. Growth factors
   c. Tumors
   d. Checkpoint genes

15. Metastasis is the process of
   a. abnormal growth and division of cells.
   b. development of a neoplasm.
   c. cancer cells forming a mass.
   d. cancer cells moving into other tissues and organs of the body.

16. Each year, cancer causes _____ percent of all human deaths in the developed world.
   a. 0-5
   b. 5-10
   c. 15-20
   d. 50

17. It is possible to reduce cancer risk through all of the following EXCEPT:
   a. avoiding fresh fruits and vegetables
   b. avoiding excessive sun exposure
   c. avoiding toxic chemicals and radiation
   d. quitting smoking

18. HeLa cells have taught science about all of the following EXCEPT:
   a. protein synthesis
   b. cancer
   c. viral growth
   d. all of the above have been researched through HeLa cells

CHAPTER OBJECTIVES/REVIEW QUESTIONS

1. Discuss how HeLa cells can be used to understand cancers today. [p.177]
2. Explain the steps of the cell cycle. [pp.178-179]
3. List and describe, in order, the various activities occurring in the eukaryotic cell life cycle. [pp.178-179]
4. Understand why a diploid parent cell produces two diploid daughter cells by mitosis. [pp.178-179]
5. Explain how the terms centromere, and sister chromatid relate to chromosome structure. [pp.180-181]
6. Describe the structure and function of the bipolar mitotic spindle. [pp.180-181]
7. Explain the purpose of the centrosome in animal cell division. [pp.180-181]
8. Give a detailed description of the cellular events occurring in the prophase, metaphase, anaphase, and telophase of mitosis. [pp.180-181]
9. Compare and contrast cytokinesis as it occurs in plant and animal cell division; use the following concepts: cleavage furrow, microfilaments at the cell's midsection, and cell plate formation. [p.182]
10. Explain the function of telomeres [p.183]
11. Describe the role of tumor suppressors and proto-oncogenes in the regulation of the cell cycle. [pp.184-185]
12. Understand the process by which cancers form. [pp.184-185]
13. Describe the process of metastasis [p.185]
14. Describe the three major forms of skin cancer. [p.185]

INTEGRATING AND APPLYING KEY CONCEPTS

1. Runaway cell division is characteristic of cancer. Imagine the various points of the mitotic process that might be sabotaged in cancerous cells in order to halt their multiplication. How might one discriminate between cancerous and normal cells in order to guide those methods of sabotage most effective in combating cancer?
2. As a cancer researcher, how might you use your knowledge of cell cycle regulation to inhibit the division of cancer cells?
3. Why can the process of mitosis also be called "cellular cloning"?
4. The rate at which the cell cycle progresses depends on the kind of cell reproducing. How might a researcher treat damaged cells that have stopped the cell cycle by understanding the behavior of fast cell cycle cells?

12

MEIOSIS AND SEXUAL REPRODUCTION

INTRODUCTION

One of the first things you will notice in this chapter is that meiosis is very similar to mitosis. However, while mitosis is a form of cloning, meiosis represents a mechanism of introducing variation. In order to understand meiosis, you must first recognize why variation is important for a species. Once you have mastered the "why," proceed to the explanation of "how" cells accomplish reduction division. Study the diagrams carefully; the ability to visualize the process is the key to comprehending meiosis.

STUDY STRATEGIES

- Read all of Chapter 12 with the goal of familiarizing yourself with the boldface terms.
- Discuss the importance of variation in the genetic structure of a species and how sexual reproduction maintains this.
- Explain the process and function of meiosis.
- Describe the ways that variation can be achieved in a species.
- Compare and contrast spermatogenesis and oogenesis.
- Compare and contrast the strategies and life stages of plants and animals.
- Discuss the possible ancient connections between mitosis and meiosis.

FOCAL POINTS

- Figure 12.5 [pp.192-193] is the core of this chapter (see also the related animation on the CD).
- Figure 12.11 [pp.198-199] compares mitosis and meiosis, providing an excellent review of Chapters 11 and 12.

INTERACTIVE EXERCISES

12.1 WHY SEX? [p.189]

12.2. MEIOSIS HALVES THE CHROMOSOME NUMBER [pp. 190-191]

Boldfaced, Page-Referenced Terms

sexual reproduction _____

_____

meiosis _____

_____

somatic _____

_____

alleles _____

_____

gametes_____

_____

germ cells _____

_____

haploid_____

_____

fertilization_____

_____

zygote_____

_____

Fill-in-the-Blank [pp.190-191]

The first cell of a sexually reproducing eukaryotic organism has (1) _____ genes on (2) _____ chromosomes, where one chromosome is (3) _____ and one is (4) _____. Information in the gene pairs is not identical or sexual reproduction would produce (5) _____. Variations in gene pairs are caused by (6) _____ that are permanently incorporated into the information they carry. Offspring of sexual reproducers inherit new (7) _____ of (8) _____, which is the basis of new combinations of traits.

Matching [pp.190-191]

Match each of the following terms with the appropriate statement.

9. _____ chromosome number
10. _____ germ cells
11. _____ diploid
12. _____ haploid
13. _____ gametes
14. _____ meiosis
15. _____ egg
16. _____ sperm cell
17. _____ homologous chromosomes
18. _____ zygote

A. A fertilized egg

B. Mature reproductive structures in animals

C. Male gamete

D. Nuclear division in sexually reproducing eukaryotic species

E. Female gamete

F. A cell that contains one of each type of chromosome

G. A cell that contains two of each type of chromosomes

H. Sum of the chromosomes in a cell

I. Chromosomes with the same length, shape, and collection of genes

J. Immature reproductive cells

True/False [pp.190-191]

Use Figure 12.4 to complete the following statements.

19. _____ The chromosome pairs pictured are homologous pairs.

20. _____ The chromosomes pictured are from a human male.

21. _____ The chromosomes pictured illustrate a haploid cell.

22. _____ One chromosome from each pair comes from the maternal parent.

23. _____ The chromosomes pictured might result from fusion of an egg and sperm cell.

24. _____ Genes on chromosome pair 21 are identical to genes on chromosome pair 22.

25. _____ Genes on one chromosome in pair 21 are identical to genes on the other chromosome in pair 21.

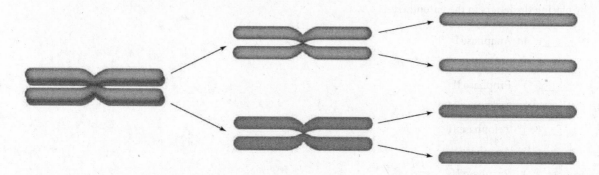

Labeling and Matching [pp.190-191]

Identify each of the meiotic organs in the following diagram matching the letter from the diagram with the description.

25. _____ Ovary (where sexual spores that give rise to eggs are located)

26. _____ Anther (where sexual spores that give rise to sperm cells form)

27. _____ Testis (where sperm originate)

28. _____ Ovary (where eggs develop)

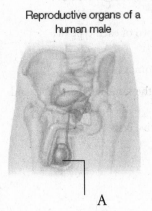

Reproductive organs of a human male

A

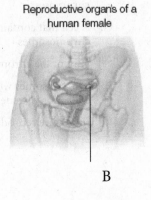

Reproductive organs of a human female

B

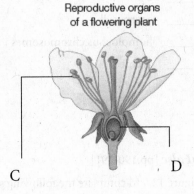

Reproductive organs of a flowering plant

C   D

12.3. VISUAL TOUR OF MEIOSIS [pp.192-193]

Choice and Ordering [pp.192-193]

1. Recall that in diploid cells the sister chromatids are attached; in haploid cells the sister chromatids have been separated. Place the steps of meiosis in chronological order starting with "1" for the first stage and so on. Then, use Figure 12.5 to determine whether the individual cells in each phase of meiosis are haploid (H) or diploid (D) by placing the letters in the parentheses.

a. _____ ( ) Anaphase I

b. _____ ( ) Metaphase II

c. _____ ( ) Prophase II

d. _____ ( ) Telophase I

e. _____ ( ) Telophase II

f. _____ ( ) Anaphase II

g. _____ ( ) Prophase I

h. _____ ( ) Metaphase I

Labeling and Matching [pp.192-193]

Identify each of the meiotic stages in the following diagram by entering the correct stage of either meiosis I or meiosis II in the blank beneath the sketch. Choose from these stages:

prophase I, metaphase I, anaphase I, telophase I, prophase II,

metaphase II, anaphase II, telophase II

Complete the exercise by matching and entering the letter of the correct stage description in the parentheses following each label.

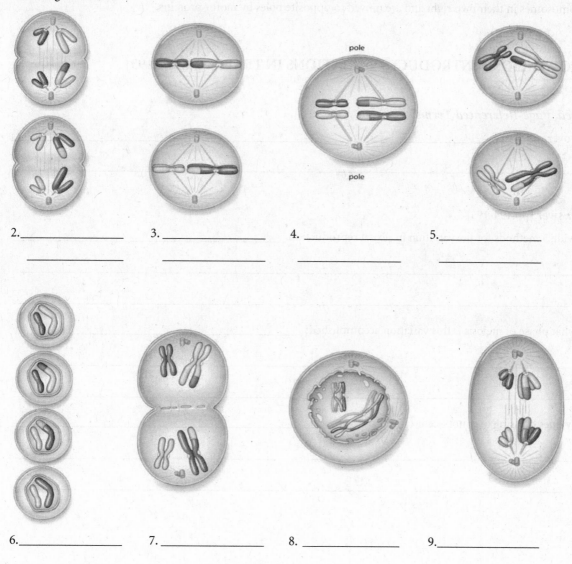

2._____   3. _____   4. _____   5._____

6._____   7. _____   8. _____   9._____

A.   The spindle is now fully formed; all chromosomes are positioned midway between the poles of one cell.

B.   In each of two daughter cells, microtubules attach to the kinetochores of chromosomes, and motor proteins drive the movement of chromosomes toward the spindle's equator.

C.   Four daughter nuclei form; when the cytoplasm divides, each new cell has a haploid chromosome number, all in the unduplicated state. The cells may develop into gametes in animals or spores in plants.

D.   In one cell, each duplicated chromosome is pulled away from its homologous partner; the partners are moved to opposite spindle poles.

E. Duplicated chromosomes condense, each chromosome pairs with its homologous partner, and crossing over and genetic recombination (swapping of gene segments) occurs. Each chromosome becomes attached to some microtubules of a newly forming spindle.

F. Motor proteins and spindle microtubule interactions have moved all the duplicated chromosomes so that they are positioned at the spindle equator, midway between the poles.

G. Two haploid cells form, each having one of each type of chromosome that was present in the parent cell; the chromosomes are still in the duplicated state.

H. Attachment between the two chromatids of each chromosome breaks; former "sister chromatids" are now chromosomes in their own right and are moved to opposite poles by motor proteins.

12.4. HOW MEIOSIS INTRODUCES VARIATIONS IN TRAITS [pp.194-195]

Boldfaced, Page-Referenced Terms

crossing over _____

_____

Short Answer [pp.194-195]

1. Provide 3 methods for the variation in sexual reproduction._____

_____

_____

2. In what phase of meiosis is this variation accomplished?_____

_____

_____

3. Why does crossing over not occur in mitosis? _____

_____

_____

Matching [pp.194-195]

Choose the most appropriate answer for each term.

4. _____ telophase II

5. _____ 2²³

6. _____ paternal chromosomes

7. _____ prophase I

8. _____ nonsister chromatids

9. _____ crossing over

10. _____ metaphase I

11. _____ maternal chromosomes

A. The stage in the cell cycle when the maternal and paternal chromosomes randomly position at the spindle equator

B. The stage in the cell cycle when the chromosomes break at the same places along their length and then exchange corresponding segments

C. The twenty-three chromosomes inherited from your mother

D. The twenty-three chromosomes inherited from your father

E. The number of possible combinations of maternal and paternal chromosomes at metaphase I

F. The molecular exchange of material between nonsister chromatids

G. Chromatids from homologous chromosomes; not linked by a centromere

H. Results in four haploid nuclei

12.5. FROM GAMETES TO OFFSPRING [pp.196-197]

Boldfaced, Page-Referenced Terms

sporophytes _____

_____

gametophyte _____

_____

sperm _____

_____

egg _____

_____

Choice [pp.196-197]

For questions 1–14, choose from the following:

    a. animal life cycle   b. plant life cycle   c. both animal and plant life cycles

1. _____ Meiosis results in the production of haploid spores

2. _____ A zygote divides by mitosis

3. _____ Meiosis results in the production of haploid gametes

4. _____ Haploid gametes fuse in fertilization to form a diploid zygote

5. _____ A zygote divides by mitosis to form a diploid sporophyte

6. _____ During meiosis, one egg and three polar bodies form

7. _____ A spore divides by mitosis to produce a haploid gametophyte

8. _____ Gametes form by oogenesis and spermatogenesis

9. _____ A haploid gametophyte divides by mitosis to produce haploid gametes

10. _____ A secondary oocyte gets nearly all the cytoplasm

11. _____ A haploid spore divides by mitosis to produce a gametophyte

12. _____ A diploid body forms from mitosis of a zygote

13. _____ A gamete-producing body and a spore-producing body develop during the life cycle

14. _____ One daughter cell of the secondary oocyte develops into a second polar body

Sequence [pp.196-197]

Refer to Figure 12.9 and 12.10 to arrange the following entities in correct order of development, entering a 1 by the stage that appears first and a 4 by the stage that completes the process of spermatogenesis. Then correctly order the stages of development of oogenesis. Complete the exercise by indicating, in the parentheses next to each blank, if each cell is *n* or *2n*.

Spermatogenesis

15. _ ( ) primary spermatocyte

16. _ ( ) sperm

17. _ ( ) spermatid

18. _ ( ) secondary spermatocyte

Oogenesis

19. _ ( ) ovum

20. _ ( ) oogonium

21. _ ( ) secondary oocyte

22. _ ( ) primary oocyte

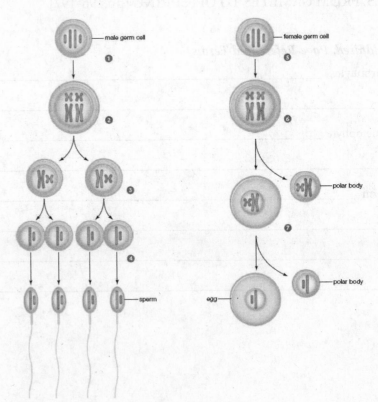

23. Use Figure 12.9 and 12.10 to compare the sperm and egg formed by meiosis. How are the resulting gametes similar?

_____

_____

24. How are they different?_____

_____

12.6. MITOSIS AND MEIOSIS—AN ANCESTRAL CONNECTION? [pp.198-199]

Matching [pp.198-199]

The cell model used in this exercise has two pairs of homologous chromosomes, one long pair and one short pair. Match the descriptions to the letters beneath each picture in the following sketches.

1. ____ One cell at the beginning of meiosis II

2. ____ A daughter cell at the end of meiosis II

3. ____ Metaphase I of meiosis

4. ____ Metaphase of mitosis

5. ____ G1 in a daughter cell after mitosis

6. ____ Prophase of mitosis

A   B   C   D   E   F

Short Answer [pp.198-199]

The following thought questions refer to the sketches in the preceding exercise; enter answers in the blanks following each question.

7. How many chromosomes are present in cell E? _____

8. How many chromatids are present in cell E? _____

9. How many chromatids are present in cell C? _____

10. How many chromatids are present in cell D? _____

11. How many chromosomes are present in cell F? _____

SELF-TEST

___ 1. Which of the following does not occur in prophase I of meiosis?
   a. A cytoplasmic division
   b. A cluster of four chromatids
   c. Homologues pairing tightly
   d. Crossing over

___ 2. Crossing over is one of the most important events in meiosis because _____.
   a. it establishes new genetic combinations that were not present in the parents
   b. homologous chromosomes must be separated into different daughter cells
   c. the number of chromosomes allotted to each daughter cell must be halved
   d. it sorts the chromatids into gametes for fertilization

___ 3. Which of the following is the most correct sequence of events in plant life cycles?
   a. Fertilization, sporophyte, zygote, meiosis, spores, gametophytes, gametes
   b. Fertilization, zygote, sporophyte, meiosis, spores, gametophytes, gametes
   c. Fertilization, zygote, sporophyte, meiosis, gametes, gametophyte, spores
   d. Fertilization, zygote, gametophyte, meiosis, gametes, sporophyte, spores

___ 4. While observing a given cell under a microscope, you notice that pairs of condensed homologous chromosomes appear to be aligning in the center of the cell. The most likely stage of meiosis for this cell is _____.
   a. anaphase II
   b. metaphase II
   c. metaphase I
   d. prophase II
   e. telophase I

___ 5. The end result of meiosis in human females is _____.
   a. two diploid clones of the parent cell
   b. four haploid ovum
   c. one haploid ovum and three diploid polar bodies
   d. one haploid ovum and three haploid polar bodies

___ 6. The end result of meiosis in human males is _____.
   a. one haploid sperm and three haploid polar bodies
   b. one diploid sperm and three haploid polar bodies
   c. four haploid sperm
   d. four diploid sperm

___ 7. If an organism has a chromosome number of 5, how many different combinations of homologous chromosomes are possible at metaphase I?
   a. 2^5
   b. 5^2
   c. 5
   d. 10

___ 8. Which of the following does not increase genetic variation?
   a. Crossing over
   b. Random fertilization
   c. Prophase of mitosis
   d. Random homologue alignments at metaphase I

___ 9. Which of the following is the most correct sequence of events in animal life cycles?
   a. Meiosis, fertilization, gametes, diploid organism
   b. Diploid organism, meiosis, gametes, fertilization
   c. Fertilization, gametes, diploid organism, meiosis
   d. Diploid organism, fertilization, meiosis, gametes

___ 10. In sexually reproducing organisms, the zygote is _____.
   a. an exact genetic copy of the female parent
   b. an exact genetic copy of the male parent
   c. completely unlike either parent genetically
   d. a genetic mixture of male parent and female parents

___ 11. A cells with a diploid number of 24 chromosomes undergoes meiosis. How many chromosomes are in each daughter cell?
   a. 4
   b. 12
   c. 24
   d. 48

___ 12. An advantage of sexual reproduction over asexual reproduction is that it allows _____.
   a. protection of the offspring
   b. more offspring
   c. more genetic variation in the offspring
   d. there are no advantages

13. What divides unequally during cytokinesis
for the female?
a. the ovary
b. the cytoplasm
c. the DNA
d. the plasma membrane

14. During alternation of generations, cells
reproduce by _____.
a. mitosis
b. meiosis
c. both mitosis and meiosis
d. none of the above

15. The process in which sperm and egg cells
join is called _____.
a. fertilization
b. combination
c. mitosis
d. asexual reproduction

16. Which of the following does not provide
new genetic combinations?
a. crossing-over
b. random fertilization
c. independent assortment
d. cytokenesis

17. What is different about interphase before
and after meiosis I?
a. DNA replication occurs at half the rate
before meiosis I
b. DNA replication occurs at double the
rate before meiosis I
c. There is DNA replication before, but
not after, meiosis I
d. There is DNA replication after, but not
before, meiosis I

18. What is the state of daughter cells following
meiosis I?
a. They are diploid (2n)
b. They are haploid (n)
c. They are duplicated
d. They are larger than at the start of
meiosis.

CHAPTER OBJECTIVES/REVIEW QUESTIONS

1. Give two advantages of sexual reproduction. [p.189]
2. Explain why variation is important to a species. [p.189]
3. Distinguish between sexual and asexual reproduction. [p.189]
4. Explain what is meant by diploid and haploid chromosome number. [pp.190-191]
5. Explain the terms gene and allele. [p.190]
6. Describe the relationships between homologous chromosomes and sister chromatids. [pp.190-191]
7. Explain why anaphase I results in haploid cells. [pp.192-193]
8. Explain the importance of two consecutive cell divisions in meiosis. [pp.192-193]
9. Sketch cell models of the various stages of meiosis I and meiosis II. Using color pencils or crayons to distinguish
two pairs of homologous chromosomes (a long pair and a short pair) is helpful. [pp.192-193]
10. Explain how crossing over, the distribution of random mixes of homologous chromosomes into gametes, and
fertilization all contribute to variation in the traits of offspring. [pp.194-195]
11. Explain the importance and mechanism of crossing over. [pp.194-195]
12. Discuss the differences between plant and animal life cycles. [pp.196-197]
13. Describe the differences in gamete production in male and female animals. [pp.196-197]

INTEGRATING AND APPLYING KEY CONCEPTS

1. The process of cloning organisms has been in the news for over a decade. It may be possible in the near future to clone humans. Based upon your knowledge of mitosis and meiosis, what are the advantages of cloning to individuals? What are the disadvantages? What about the species as a whole?
2. Imagine a pond where 95% of the fish are green. If a virus that kills green fish infected the pond, what would happen to the population? How does this example illustrate the importance of variation in sexually reproducing organisms?
3. Based upon your knowledge of meiosis, what would provide more variation to a species—a few large chromosomes, or lots of small chromosomes? Explain your answer.

13

OBSERVING PATTERNS IN INHERITED TRAITS

INTRODUCTION

This chapter introduces you to some of the principles of genetics, starting with the work of Gregor Mendel. Advances in the study of genetics are frequently in the news, and most people have an interest in understanding more about inheritance. The key to understanding genetics is the ability to solve problems using probability. In this chapter you will be exposed to many different forms of inheritance, but all use probability to predict the outcome. You need to not only understand the terminology in the chapter but also work the problems if you are going to gain an understanding of genetics.

STUDY STRATEGIES

- Read all of Chapter 13 with the goal of familiarizing yourself with the boldface terms.
- Identify the role that Mendel played in the history of genetics and be able to define the terms used in genetics.
- Summarize the Law of Segregation and the Law of Independent Assortment.
- Complete a testcross, monohybrid cross, and a dihybrid cross.
- Describe the patterns of complete dominance, codominance, incomplete dominance, and epigenetics.
- Describe the role of the environment in the expression of genes.
- Explain the role of polygenetics and environment in continuous variation.

FOCAL POINTS

- The diagram in Figure 13.4 [p.205] and the terms introduced on this page are the basic "language" of geneticists. You need to understand these terms before proceeding to the remainder of the chapter.
- Figure 13.5 [p.206] illustrates the Punnett square, a useful tool for predicting outcomes in introductory genetics.
- Figure 13.7 [p.208] shows how the principles of independent assortment are executed during meiosis.
- Figure 13.8 [p.209] outlines how to construct a Punnett square for a dihybrid experiment.
- Figure 13.9, 13.10, and 13.11 illustrate some of the more complex aspects of genetics such as codominance, incomplete dominance, and epistasis.

INTERACTIVE EXERCISES

13.1. MENACING MUCUS [p.203]

13.2. MENDEL, PEA PLANTS, AND INHERITANCE PATTERNS [pp. 204-205]

Boldfaced, Page-Referenced Terms

locus _____

_____

homozygous _____

_____

genotype _____

_____

phenotype _____

_____

hybrids _____

_____

heterozygous _____

_____

dominant _____

_____

recessive _____

_____

Matching [p. 204-205]

Choose the correct definition for each term.

1. _____ genotype
2. _____ alleles
3. _____ heterozygous
4. _____ dominant
5. _____ phenotype
6. _____ genes
7. _____ homozygous recessive
8. _____ recessive
9. _____ hybrids
10. _____ diploid organism
11. _____ gene locus
12. _____ homozygous dominant
13. _____ homologous chromosomes

A. All the different molecular forms of the same gene

B. Particular location of a gene on a chromosome

C. Describes an individual having a pair of nonidentical alleles

D. An individual with a pair of recessive alleles, such as *aa*

E. Allele whose effect is masked by the effect of the dominant allele

F. Offspring of a genetic cross that inherit a pair of nonidentical alleles for a trait

G. Refers to an individual's observable traits

H. The effect of an allele on a trait masks that of any recessive allele paired with it

I. An individual with a pair of dominant alleles, such as *AA*

J. Units of information about specific traits; passed from parents to offspring

K. A pair of similar chromosomes, one obtained from the father and the other obtained from the mother

M. Having pairs of genes on homologous chromosomes

L. The particular alleles that an individual carries for a trait

True /False [pp.170-171]

If a statement below is true, write T in the blank. If it is false, explain why by changing one or more of the underlined words.

14. _____ A genotype refers to the alleles carried by an individual.

15. _____ A dominant allele will be hidden by the presence of a recessive allele.

16. _____ A cell that is diploid has three copies of each chromosome.

17. _____ A gene's locus is its location on a chromosome.

Short Answer [pp.170-171]

18. Was Mendel's choice of *Pisum sativum*, the garden pea plant, a good choice to use to study how traits are passed from parent to offspring?

_____

_____

_____

_____

13.3. MENDEL'S LAW OF SEGREGATION [pp.206-207]

13.4. MENDEL'S LAW OF INDEPENDENT ASSORTMENT [pp.208-209]

Boldfaced, Page-Referenced Terms

testcross _____

_____

monohybrid experiments_____

_____

probability _____

_____

Punnett squares _____

_____

segregation _____

_____

dihybrid cross _____

_____

law of independent assortment _____

_____

linkage group _____

_____

Short Answer

1. Why is the predicted ratio likely closer to the observed ratio when a large number of offspring are sampled?

    _____

    _____

2. Describe Mendel's Law of Segregation.

    _____

    _____

3. Describe Mendel's Law of Independent Assortment.

    _____

    _____

True/False [pp.206-207]

If a statement below is true, write T in the blank. If it is false, explain why by changing one or more of the underlined words.

4. _____ A testcross may be used to indicate if an individual is heterozygous or homozygous for a dominant trait.

5. _____ Punnett squares are used to determine the observed ratios of testcrosses.

6. _____ During meiosis, gene pairs are separated into different gametes. This is Mendel's Law of Segregation.

7. _____ Mendel's experiments did not exactly match his predicted ratios because of errors he made in counting.

8. _____ The larger the sample size, the more closely the predicted ratios will come to the observed ratios.

9. _____ The F_1 generation are offspring of the F_2 generation.

10. _____ Punnett squares calculate the probability of both genotypes and phenotypes.

Matching [pp.208-209]

Choose the correct definition for each term.

11. ____ independent assortment

12. ____ possible genetic outcomes

13. ____ Punnett-square method

14. ____ testcross

15. ____ dihybrid experiments

16. ____ segregation

17. ____ P generation

18. ____ F₁ generation

19. ____ F₂ generation

A. Used to determine the genotype of an individual when it displays the dominant phenotype

B. The mathematical chance that a given event will occur

C. The separation of traits during a genetic cross

D. Original individuals involved in a testcross

E. The process by which each pair of homologous chromosomes is sorted out into gametes

F. Genetic crosses that examine the inheritance of two traits

G. A graphic means of representing the distribution of gametes and possible zygotes in a genetic cross

H. 3rd generation offspring

I. Offspring of the parent generation

Problems [pp.206-209]

20. In garden pea plants, tall *(T)* is dominant over dwarf *(t)*. In the cross *Tt x tt*, the *Tt* parent would produce a gamete carrying *T* (tall) and a gamete carrying *t* (dwarf) through segregation; the *tt* parent could only produce gametes carrying the *t* (dwarf) gene. Use the Punnett-square method (refer to Figure 13.6 in the text) to determine the following results of a *Tt x tt* cross.

a. The genotypic probabilities of the offspring: _____

_____

b. The phenotypic probabilities of the offspring: _____

_____

21. In fruit flies, the trait vestigial wings *(a)* is recessive to normal wings *(A)*. You wish to determine whether a fly with a dominant phenotype is homozygous dominant or heterozygous. Using a testcross, you mate the fly with a homozygous recessive individual. Following are two possibilities for the F₁ generation. For each, state if the results indicate that the genotype of the parent is *homozygous dominant* or *heterozygous*.

a. 78 normal-winged offspring:_____

_____

b. 37 normal-winged and 41 vestigial-winged offspring: _____

_____

22. Albinos cannot form the pigments that normally produce skin, hair and eye color, so albinos exhibit white hair and pink eyes and skin (because the blood shows through). An albino is homozygous recessive *(aa)* for the pair of genes that code for the key enzyme in pigment production. Suppose a woman of normal pigmentation *(Aa)* with an albino mother marries an albino man. State the possible kinds of pigmentation for this couple's children, and specify the ratio of each kind of child the couple is likely to have.

_____

_____

Observing Patterns in Inherited Traits   **161**

23. In horses, black coat color is influenced by the dominant allele *(B)* and chestnut coat color is influenced by the recessive allele *(b)*; trotting gait is due to a dominant gene *(T)*, pacing gait to the recessive allele *(t)*. A homozygous black trotter is crossed to a chestnut pacer.

a. What is the predicted appearance and gait of the F_1 and F_2 generations?_____

_____

b. Which phenotype will likely be most common?_____

_____

c. Which genotype will likely be most common? _____

_____

d. Which of the potential offspring will be certain to breed true? _____

_____

24. In pea plants, green pod color (G) is dominant to yellow pod color (g). Use the Punnett-square method (refer to Figure 13.6 in the text) to determine the following results of a *Gg x Gg* cross.

a. The genotypic probabilities of the offspring: _____

_____

b. The phenotypic probabilities of the offspring: _____

_____

25. In guinea pigs, black hair (B) is dominant to brown hair (b), short hair (S) is dominant to long hair (s). Use the Punnett-square method (refer to Figure 13.6 in the text) to determine the following results of a homozygous black long-haired male guinea pig with a homozygous brown short-haired female.

a. The genotypic probabilities of the offspring: _____

_____

b. The phenotypic probabilities of the offspring: _____

_____

Fill-in-the-Blanks [p.208-209]

We now know that there are many (26) _____ on each type of autosome and sex chromosome. All of the genes on one (27) _____ are called a(n) (28) _____ group. If genes on the same chromosome stayed together through (29) _____, then there would be no surprising mixes of parental (30) _____. You could expect parental phenotypes among the F2 offspring of dihybrid experiments to show up in a predictable (31) _____. Genes that have a relatively (32) _____ distance between them are said to be tightly linked.

13.5. BEYOND SIMPLE DOMINANCE [pp.210-211]

Boldfaced, Page-Referenced Terms

codominance _____

_____

multiple allele systems _____

_____

incomplete dominance_____

_____

epistasis _____

_____

pleiotropic gene _____

_____

Matching [pp.210-211]

Match the term to its definition.

1. ____ Occurs when one allele of a pair is not fully dominant over the other

2. ____ Occurs when one gene influences multiple traits

3. ____ Effect that occurs when a trait is affected by interactions among different gene products

4. ____ Marfan Syndrome is caused by a defect in these molecules

5. ____ Pigment molecules that denote color variations in different individuals

6. ____ Occurs when three or more alleles of a gene persist among individuals of a population

7. ____ Nonidentical alleles of a gene that are both fully expressed in heterozygotes so neither is dominant or recessive

A. codominance

B. epistasis

C. fibrillin

D. incomplete dominance

E. melanin

F. multiple allele systems

G. pleiotropy

Labelling [pp.210-211]

Match the phenotype (bloodtype) with the letter in the figure.

8. _____ Type AB

9. _____ Type O

10. _____ Type A

11. _____ Type B

Genotypes: AA or AO | AB | BB or BO | OO

Phenotypes (blood type): A | B | C | D

Short Answer [p.211]

12. List the phenotypes in ABO blood typing. Next to each phenotype list the possible genotypes.

_____

_____

13. Which phenotype in ABO blood typing is homozygous recessive?

_____

_____

Choice [pp.210-211]

Choose the effect that applies to each of the following situations.

a. epistasis   b. pleiotropy   c. multiple allele systems (codominance)   d. incomplete dominance

14. _____ ABO blood typing

15. _____ Pink offspring from a cross between red and white parent plants

16. _____ Multiple health problems from some conditions such as mucus duct clogs and low male fertility

17. _____ Defects in fibrillin molecules that cause multiple phenotypic effects

18. _____ Coat color in Labrador retrievers

19. In four o'clock plants, red flower color is determined by gene R and white flower color by R', while the heterozygous condition, RR', is pink. For each of the following crosses, give the phenotypic ratios of the offspring.

    a. RR x R^1R^1 _____

    _____

    b. RR' x RR' _____

    _____

20. Using Figure 13.9 as a reference, indicate the possible blood types that may be present in the offspring of the following crosses.

    a. AO x AB _____

    _____

    b. BO x AO _____

    _____

    c. AA x OO _____

    _____

    d. OO x OO _____

    _____

    e. AB x AB _____

    _____

21. In sweet peas, there is an epistatic interaction between two genes involved in flower color. Genes C and P are necessary for colored flowers. In the absence of either (_ _*pp* or *cc*_ _), or both (*ccpp*), the flowers are white. What color and proportions will the offspring of the following offspring exhibit?

    a. *CcPp* x *ccpp* _____

    _____

    b. *CcPP* x *Ccpp* _____

    _____

22. In the inheritance of the coat (fur) color of Labrador retrievers, allele *B* specifies that black is dominant over allele *b*, brown (chocolate). Allele *E* permits full deposition of color pigment, but two recessive alleles, *ee*, reduce deposition, and a yellow coat results. Predict the phenotypes of the coat color and their proportions resulting from the cross *BbEe* x *Bbee*.

_____

_____

_____

23. In the inheritance of the coat (fur) color of Labrador retrievers, allele *B* specifies that black is dominant over allele *b*, brown (chocolate). Allele *E* permits full deposition of color pigment, but two recessive alleles, *ee*, reduce deposition, and a yellow coat results. Predict the phenotypes of the coat color and their proportions resulting from the cross *BBEe* x *bbee*.

_____

_____

_____

13.6. NATURE & NURTURE [pp.212-213]

Short Answer

1. How does altitude affect the phenotype of yarrow plants?

_____

_____

2. How does temperature determine the coat color of snowshoe hares?

_____

_____

3. What causes a pointed head feature in Daphnia pulex?

_____

_____

4. What substance is released when humans are dealing with stress?

_____

_____

13.7. COMPLEX VARIATION IN TRAITS [pp.214-215]

Boldfaced, Page-Referenced Terms

continuous variation _____

_____

bell curve _____

_____

Choice [pp.214-215]

For items 1–5, choose the primary contributing factor from the options below.

    a. environment   b. polygenic inheritance (epistasis)

1. ____ Distribution of height in human populations
2. ____ Continuous variation in a trait
3. ____ Plant height in *Achillea millefolium*
4. ____ The range of eye colors in the human population
5. ____ Heat-sensitive version of one of the enzymes required for melanin production in Himalayan rabbits

---

SELF-TEST

____ 1. Mendel's theory of independent assortment is best stated as
    a. One allele is always dominant to another.
    b. Hereditary units from the male and female parents are blended in the offspring.
    c. The two hereditary units that influence a certain trait separate during gamete formation.
    d. Genes on each pair of chromosomes are sorted into gametes independent of genes on other chromosomes.

____ 2. All the different molecular forms of the same gene are called
    a. hybrids
    b. alleles
    c. autosomes
    d. loci

____ 3. If two heterozygous individuals are crossed in a monohybrid experiment involving complete dominance, the expected phenotypic ratio is
    a. 3:1
    b. 1:1:1:1
    c. 1:2:1

    d. 1:1
    e. 9:3:3:1

____ 4. The phenotypic ratio of a cross between a pink-flowered snapdragon and a white-flowered snapdragon is
    a. 75% red, 25% white
    b. 100% red
    c. 50% pink, 50% white
    d. 100% pink

____ 5. In a testcross, an individual with a dominant phenotype, but unknown genotype, is crossed to an individual known to be homozygous recessive for that trait. If the F1 offspring are all heterozygous, what can you say about the genotype of the dominant parent?
    a. The dominant parent is heterozygous.
    b. The dominant parent is homozygous dominant.
    c. The dominant parent is homozygous recessive.
    d. The dominant parent has no effect on the genotype of the offspring.

6. The tendency for dogs to bark while trailing is determined by a dominant gene, *S*, whereas silent trailing is due to the recessive gene, *s*. In addition, erect ears, *D*, is dominant over drooping ears, *d*. What combination of offspring would be expected from a cross between two erect-eared barkers who are heterozygous for both genes?
   a. 1/4 erect barkers, 1/4 drooping barkers, 1/4 erect silent, 1/4 drooping silent
   b. 9/16 erect barkers, 3/16 drooping barkers, 3/16 erect silent, 1/16 drooping silent
   c. 1/2 erect barkers, 1/2 drooping barkers
   d. 9/16 drooping barkers, 3/16 erect barkers, 3/16 drooping silent, 1/16 erect silent

7. If a mother has type O blood and the father has type AB blood, which of the following blood types could not be present in their children?
   a. Type A
   b. Type B
   c. Type O
   d. Type AB
   e. More than once answer is possible

8. A single gene that affects several seemingly unrelated aspects of an individual's phenotype is said to be
   a. pleiotropic
   b. epistatic
   c. allelic
   d. continuous

9. A bell curve of phenotypes is typical of what form of inheritance?
   a. pleiotropic
   b. polygenic
   c. environmental
   d. multiple allele systems

10. What decreases the probability that two genes on the same chromosome will separate during crossing over?
   a. The dominance of the alleles
   b. The age of the chromosome
   c. A short distance between the genes
   d. Epistatic interactions
   e. None of the above

11. What is the phenotypic outcome of a cross between two heterozygous purple-flowered pea plants?
   a. 25% purple: 75% white
   b. 50% purple: 50% white
   c. 75% purple: 25% white
   d. 100% purple

12. Albinism in humans is caused by a recessive allele. A normal couple have six children. Five are normal and one is albino. What must the genotype of the parents be?
   a. Both parents are heterozygous
   b. One parent is homozygous dominant while the other is homozygous recessive
   c. One parent is homozygous dominant while the other is heterozygous
   d. It is impossible to determine given the information.

13. With reference to question #12, what is the probability that a next child will be albino?
   a. 25%
   b. 50%
   c. 75%
   d. 100%

14. The best definition of an allele is
   a. a gene
   b. all of the various forms of a particular gene
   c. a genotype
   d. a phenotype

15. Gregor Mendel was best known from his work on which of the following plants?
   a. Wheat
   b. Corn
   c. Pea
   d. Rice

16. Punnett squares are helpful in that they
   a. help predict genotypic ratios
   b. help predict phenotypic ratios
   c. indicate all possible gametic combinations
   d. all of the above are true

17. If a chesnut horse breeds with a white horse, a palamino (chesnut and white spots) will result. What aspect of genetics can explain this?
   a. Codominance
   b. Incomplete dominance
   c. Complete dominance
   d. Epistasis

18. If the P gene and the R gene are not linked, the genotype PpRr can produce how many different gametic combinations?
   a. 1
   b. 2
   c. 3
   d. 4

___ 19. Marfan syndrome is pleiotropic and is characterized by all of the following physical symptoms EXCEPT
a. thinness
b. tall height
c. loose jointed
d. hair loss

___ 20. Seasonal changes in coat color in snowshoe hares is best described by the science of _____.
a. codominance
b. complete dominance
c. epigenetics
d. epistasis

CHAPTER OBJECTIVES/REVIEW QUESTIONS

1. Describe the concept of blending inheritance. [p.204]
2. Explain why the pea plant was an ideal model organism for Mendel's studies. [pp.204-205]
3. Distinguish between genes and alleles. [p.205]
4. Explain the terms genotype and phenotype, and relate them to the terms homozygous dominant, homozygous recessive, and heterozygous. [p.205]
5. Explain the purpose and structure of a testcross. [p.206]
6. Determine the genotypic and phenotypic ratios of a monohybrid experiment. [pp.206-207]
7. Explain the significance of Mendel's theory of segregation. [p.207]
8. Determine the genotypic and phenotypic ratios of a dihybrid experiment. [pp.208-209]
9. Explain the significance of Mendel's theory of independent assortment. [pp.208-209]
10. Explain crossing over and the factors that determine the probability of crossing over between two genes on the same chromosome. [p.209]
11. Define *epistatic* and *pleiotropic* inheritance. [p.211]
12. Distinguish among complete dominance, incomplete dominance, and codominance. [pp.210-211]
13. Explain how the environment can influence the expression of a trait. [p.212-213]
14. Explain what is meant by continuous variation and polygenic inheritance. [pp.214-215]

INTEGRATING AND APPLYING KEY CONCEPTS

1. Solve the following genetics problem: In garden peas, one pair of alleles controls the height of the plant and a second pair of alleles controls flower color. The allele for tall (D) is dominant to the allele for dwarf (d), and the allele for purple (P) is dominant to the allele for white (p). A tall plant with purple flowers crossed with a tall plant with white flowers produces offspring that are 3/8 tall purple, 3/8 tall white, 1/8 dwarf purple, and 1/8 dwarf white. What are the genotypes of the parents?
2. As a breeder of exotic plants, you advertise that you have produced a true-breeding yellow flowered line. Assuming that yellow flower color (Y) is dominant over the less-desirable (w) color, how do you test to ensure that your line is not heterozygous for the trait?
3. Height and eye color in humans are frequently given as examples of continuous variation. What are some other human traits that may display continuous variation? What about in other species?
4. Sickle-cell anemia is a genetic disease in which children that are homozygous for a defective gene ($Hb^S Hb^S$) produce defective hemoglobin. The genotypes of normal persons are $Hb^A Hb^A$. If the level of blood oxygen drops below a certain level in a person with the $Hb^S Hb^S$ genotype, the hemoglobin chains stiffen and cause the red blood cells to form sickle, or crescent, shapes. These cells clog and rupture capillaries, which results in oxygen-deficient tissues where metabolic wastes collect. Several body functions are badly damaged. Severe anemia and other symptoms develop, and death nearly always occurs before adulthood. The sickle-cell gene is considered *pleiotropic*. Persons that are heterozygous ($Hb^A Hb^S$) are said to possess *sickle-cell trait*. They are able to produce enough normal hemoglobin molecules to appear normal, but their red blood cells will sickle if they encounter oxygen tension (such as at high altitudes). A man whose sister died of sickle-cell anemia married a woman whose blood is found to be normal. What advice would you give this couple about the inheritance of this disease as they plan their family? If a man and a woman, each with sickle-cell trait, planned to marry, what information could you provide for them regarding the genotypes and phenotypes of their future children?
5. Human genetics has reached a point at which it is possible to screen an individual's genome for specific alleles related to disease. What are the positive and negative possibilities of this ability? Are there ethical aspects that should be taken into consideration?

Observing Patterns in Inherited Traits **169**

14

CHROMOSOMES AND HUMAN INHERITANCE

INTRODUCTION

This chapter takes the genetics you learned in previous chapters and applies that knowledge to humans. You will also learn about tools that are used to study patterns in human genetics.

STUDY STRATEGIES

- Read all of Chapter 14 with the goal of familiarizing yourself with the boldface terms.
- Discuss the usefulness of pedigrees in understanding genetics and traits.
- Explain the difference between genetic abnormalities and genetic disorders.
- Compare and contrast autosomal dominant and autosomal recessive inheritance patterns.
- Cite examples of autosomal dominant and autosomal recessive inheritance patterns.
- Describe and cite examples of the X-linked recessive inheritance pattern.
- Compare and contrast the effects of duplication, deletion, inversion, and translocation on chromosomes.
- Explain the process of nondisjunction and its possible effects.
- Compare and contrast the various prenatal diagnostic toos including the procedure, results, and risks.

FOCAL POINTS

- Figure 14.2 [p.220] illustrates a pedigree by looking at polydactyly in an Amish family.
- Table 14.1 [p.221] describes many genetic disorders and abnormalities that occur in humans.
- Figure 14.7 [p.225] illustrates how a pedigree is used to follow hemophilia inherited by the descendants of Queen Victoria of England.
- Figure 14.9 [p.227] illustrates the evolution of the male Y chromosome.
- Figure 14.11 [p.228] illustrates nondisjunction.
- Figure 14.12 [p.229] shows a karyotype of an individual with Downs Syndrome, illustrating the extra chromosome that leads to this disorder.
- Figures 14.13-14.15 [pp.230-231] describe prenatal diagnostic tools.

INTERACTIVE EXERCISES

14.1 SHADES OF SKIN [p.219]

14.2. HUMAN CHROMOSOMES [pp.220-221]

Boldfaced, Page-Referenced Terms

pedigrees _____

_____

Fill-in-the-Blanks [pp.220-221]

Human inheritance patterns are usually studied through a (1) _____, which tracks observable (2)

_____ over multiple generations. These charts can indicate whether a (3) _____ is associated with a (4)

_____ or (5) _____ allele. It can also help determine whether the allele is carried on an (6) _____

or on a (7) _____. Finally, a (1) _____ can ascertain whether or not this (3) _____ will recur in

future (8) _____. Alleles that cause severe (9) _____ usually are found in very small quantities in a

population because they compromise the (10) _____ and (11) _____ ability of the individuals that carry

them. These alleles probably continue to show up in the population due to (12) _____. In addition,

individuals that are (13) _____ for a gene will carry that allele. They may possibly pass that (3) _____

onto the next generation.

Matching [p.220]

Using the symbols in Figure 14.2 B, match each of the following individuals to the description. For each generation
row, count across from left to right for each individual.

14. _____ II, 1 & 2

15. _____ I, 2

16. _____ IV, 2

17. _____ III, 1

18. _____ V, 2

19. _____ IV, 4, 5, 6, 7, 8

20. _____ III, 5

A. Male showing the trait being tracked

B. Male not showing the trait being tracked

C. Female showing the trait being tracked

D. Female not showing the trait being tracked

E. Married individuals

F. Siblings

G. Individual whose sex is not specified

Short Answer [p.219-221]

21. Why is human skin pigmented and what causes the pigmentation?

_____

_____

22. What is the difference between a genetic abnormality and a genetic disorder?

_____

_____

Chromosomes and Human Inheritance   **171**

23. Why is studying human inheritance patterns difficult for researchers and how do they work around these problems?

_____

_____

14.3. EXAMPLES OF AUTOSOMAL INHERITANCE PATTERNS [pp.222-223]

Matching [pp.222-223]

Match the disorder to its description.

1. _____ An autosomal dominant allele causes this disorder that alters a protein necessary for brain development

2. _____ An autosomal dominant allele causes this disorder of abnormally short arms and legs

3. _____ This autosomal recessive disorder is due to muations in the genes for lysosome enzymes. It is carried in 1:300 people.

4. _____ An autosomal dominant allele causes this disorder that leads to extremely accelerated aging.

5. _____ Mutations in the genes for production of melanin leads to this autosomal recessive disorder.

A. Achondroplasia

B. Huntington's disease

C. Hutchinson-Gilford Progera

D. Albanism

E. Tay-Sachs

Problems [pp.222-223]

6. The autosomal allele that causes albanism(a) is recessive to the allele for normal melanin pigment(A). A normal woman whose father was albino marries an albino who had normal parents. They have three children, two normal and one with albanism. List the genotypes for each person involved.

_____

_____

_____

7. Huntington disease is a rare form of autosomal dominant inheritance, H; the normal gene is h. The disease causes progressive degeneration of the nervous system with onset exhibited near middle age. An apparently normal man in his early twenties learns that his father has recently been diagnosed as having Huntington disorder. What are the chances that the son will develop this disorder?

_____

_____

_____

8. Hutchinson-Gilford Progeria is a rare form of accelerated aging via an autosomal dominant trait (P). The normal gene is p. The disease causes death in an individual's early teens. What are the changes of this trait being inherited from two homozygous recessive parents? Does this trait run in families? Why or why not?

_____

_____

_____

9. Describe the basic characteristics of progeria.

_____

_____

_____

10. What protein is affected in persons with progeria?

_____

_____

_____

Choice [pp.222-223]

For questions 11-21, choose from the following patterns of inheritance. Some items may require more than one letter.

     a. autosomal recessive   b. autosomal dominant

11. _____ Heterozygotes can remain undetected

12. _____ The trait usually appears in each generation

13. _____ Both parents may be heterozygous normal

14. _____ If one parent is heterozygous and the other homozygous recessive, there is a 50 percent chance that any child of theirs will be affected

15. _____ The allele is usually expressed, even in heterozygotes

16. _____ Heterozygous normal parents can expect that one-fourth of their children will be affected by the disorder

17. _____ The trait is expressed in heterozygotes of either sex

18. _____ Individuals displaying this type of disorder will always be homozygous for the trait

19. _____ The trait is expressed in both the homozygote and the heterozygote

20. _____ Heterozygous unaffected parents have two children affected with the disorder

21. _____ Heterozygous affected parents have a child that is unaffected

14.4. EXAMPLES OF X-LINKED INHERITANCE PATTERNS [pp.224.225]

14.5. HERITABLE CHANGES IN CHROMOSOME STRUCTURE [pp.226-227]

Boldfaced, Page-Referenced Terms

duplications _____

_____

inversion_____

_____

translocation _____

_____

Choice [pp.224-225]

Answer the following questions with one of these choices:

    a. sons   b. daughters   c. sons and daughters   d. fathers   e. mothers   f. fathers and mothers

1. ____ Male humans transmit their Y chromosome to their _____

2. ____ Male humans receive their X chromosome from their _____

3. ____ Human mothers and fathers each provide an X chromosome for their _____

4. ____ Female humans pass their X chromosomes to their _____

Labeling and Matching [p.226]

On rare occasions, chromosome structure becomes abnormally rearranged. Such changes may have profound effects on the phenotype of an organism. Label the following diagrams of abnormal chromosome structure as a deletion, a duplication, an inversion, or a translocation. Complete the exercise by matching and entering the letter of the proper description in the parentheses following each label.

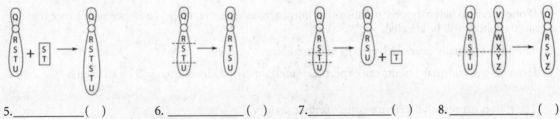

5._____( )   6. _____( )   7._____( ) 8._____( )

A. The loss of a chromosome segment; an example is the cri-du-chat disorder

B. A gene sequence in excess of its normal amount in a chromosome; this alters the position and order of the chromosome's genes; possibly promoted human evolution

C. A chromosome segment that separated from the chromosome and then was inserted at the same place, but in reverse; may cause mispairing in meiosis

D. The transfer of part of one chromosome to a nonhomologous chromosome; an example is when chromosome 14 ends up with a segment of chromosome 8; the Philadelphia chromosome is also an example

9. Explain the evolution of the male Y chromosome.

_____

_____

_____

10. Describe how speciation can occur through large changes in chromosomes.

_____

_____

_____

14.6. HERITABLE CHANGES IN THE CHROMOSOME NUMBER [pp.228-229]

Boldfaced, Page-Referenced Terms

polyploidy _____

_____

nondisjunction_____

_____

aneuploidy _____

_____

Short Answer [p.228]

1. If a nondisjunction occurs at anaphase I of the first meiotic division, what will be the proportion of abnormal gametes (for the chromosomes involved in the nondisjunction)?

_____

_____

_____

2. If a nondisjunction occurs at anaphase II of the second meiotic division, what will be the proportion of abnormal gametes (for the chromosomes involved in the nondisjunction)?

_____

_____

_____

3. Define the following terms: *polyploid, trisomic,* and *monosomic.*

_____

_____

_____

_____

Choice [pp.228-229]

For questions 4–13, choose from the following:

a. Down syndrome   b. Turner syndrome   c. Klinefelter syndrome   d. XYY condition

4. _____ XXY male

5. _____ Females whose ovaries are nonfunctional and who fail to develop secondary sexual traits at puberty

6. _____ Testes smaller than normal, sparse body hair, and some breast enlargement

7. _____ Could only be caused by a nondisjunction in males

8. _____ As a group, they tend to be cheerful and affectionate; many have heart problems

9. _____ XO female; most abort early; distorted female phenotype

10. _____ Males that tend to be taller than average; some mildly retarded but most are phenotypically normal

11. _____ Injections of testosterone reverse feminized traits but not the fertility

12. _____ Trisomy 21; muscles and muscle reflexes weaker than normal

13. _____ At one time these males were thought to be genetically predisposed to become criminals

14.7. GENETIC SCREENING [pp.230-231]

Matching [pp.230-231]

Match the term to its definition.

1. _____ High quality images taken from inside the uterus.

2. _____ Describes a human up to 8 weeks of development in utero.

3. _____ Identifying possible harmful genetic traits carried by parents that may be passed onto offspring.

4. _____ Testing the cells of the fetus that have been shed in the uterus for genetic testing.

5. _____ A technique of producing images of a fetus's limbs and organs via ultrasound waves.

6. _____ Describes a human past 8 weeks of development in utero.

7. _____ Checks for abnormalities (physical and genetic) before birth.

8. _____ Testing of the chorion for genetic disorders.

A. electric sonography

B. fetus

C. prenatal diagnosis

D. fetoscopy

E. chorionic villus

F. genetic screening

G. amniocentesis

H. embryo

Fill-in-the-Blanks [pp.230-231]

The effects of genetic disorders can often be minimized by (9) _____ treatments such as surgery,

prescription drugs, or dietary modification. In one disorder, (10) _____, reduction of the amount of

phenylalanine in the diet can slow or even stop deterioration of the brain. Sometimes, (11) _____ diagnosis is

used to diagnose disorders before birth. In (12) _____, a syringe is used to draw fluid from the amniotic sac.

This fluid can then be analyzed for many genetic disorders. A diagnostic tool that can be used even earlier is (13)

_____, in which cells are removed from one of the placental membranes for further analysis. Once a diagnosis

is made, parents often face the decision whether or not to have an induced (14) _____.

SELF-TEST

____ 1. A _____ is assembled using
chromosomes stopped in metaphase of
mitosis.
   a. karyotype
   b. bridging cross
   c. wild-type allele
   d. linkage group

____ 2. Chromosomes other than those involved in
sex determination are known as
_____.
   a. nucleosomes
   b. heterosomes
   c. alleles
   d. autosomes

____ 3. Karyotype analysis is _____.
   a. a means of detecting and reducing
      mutagenic agents
   b. a surgical technique separating
      chromosomes that have failed to
      segregate properly during meiosis
      II
   c. used to reveal abnormalities in
      chromosome structure and number
   d. a process that substitutes defective
      alleles with normal ones

____ 4. A man with achondroplasia marries a
normal woman. What is the chance that
their first child will be achondroplastic?
   a. 0%
   b. 25%
   c. 50%
   d. 75%

____ 5. Red–green color blindness is a sex-linked
recessive trait in humans. A color-blind
woman and a man with normal vision have
a son. What are the chances that the son is
color blind? If the parents ever have a
daughter, what is the chance that a
daughter will be color blind? (For the latter
question, consider only the female
offspring.)
   a. 100 percent, 0 percent
   b. 50 percent, 0 percent
   c. 100 percent, 100 percent
   d. 50 percent, 100 percent
   e. none of the above

____ 6. Suppose that a hemophiliac male (X-linked
recessive allele) and a female carrier for the
hemophiliac trait have a non-hemophiliac
daughter with Turner syndrome.
Nondisjunction could have occurred in.
   a. either parent
   b. neither parent
   c. the father only
   d. the mother only
   e. the non-hemophiliac daughter

____ 7. Nondisjunction involving the X
chromosome occurs during oogenesis and
produces two kinds of eggs, XX and O (no
X chromosome). If normal Y sperm
fertilize the two types, which genotypes are
possible?
   a. XX and XY
   b. XXY and YO
   c. XYY and XO
   d. XYY and YO
   e. YY and XO

___ 8. Of all phenotypically normal males in prisons, the type once thought to be genetically predisposed to becoming criminals was the group with _____.
   a. XXY disorder
   b. XYY disorder
   c. Turner syndrome
   d. Down syndrome
   e. Klinefelter syndrome

___ 9. Amniocentesis is _____.
   a. a surgical means of repairing deformities
   b. a form of chemotherapy that modifies or inhibits gene expression or the function of gene products
   c. used in prenatal diagnosis; a small sample of amniotic fluid is drawn to detect chromosomal mutations and metabolic disorders in embryos
   d. a form of gene-replacement therapy
   e. a diagnostic procedure; cells for analysis are withdrawn from the chorion

___ 10. John Nash, Pablo Picasso, Virginia Woolf, and Abraham Lincoln were all thought to have _____.
   a. phenylketonuria
   b. progeria
   c. hemophilia
   d. achondroplasia
   e. neurobiological disorder

___ 11. In a genetic disorder that has autosomal recessive inheritance,
   a. If both parents are homozygous recessive, the children have a 25% chance of inheriting the disorder.
   b. If both parents are homozygous recessive, the children have a 50% chance of inheriting the disorder.
   c. If both parents are heterozygous, the children have a 25% chance of inheriting the disorder.
   d. If both parents are heterozygous, the children have a 50% chance of inheriting the disorder.

___ 12. Which of the following are examples of genetic disorders with autosomal dominant inheritance?
   a. Huntington disorder
   b. achondroplasia
   c. hemophilia
   d. Tay-Sachs

___ 13. Which of the following are examples of genetic disorders with autosomal dominant inheritance?
   a. Huntington disorder
   b. achondroplasia
   c. hemophilia
   d. Tay-Sachs

___ 14. From a pedigree, how would you guess that inheritance of a genetic disorder were X-linked recessive?
   a. Males exhibit phenotype more often than females
   b. Females exhibit phenotype more often than males
   c. Sons have not inherited the recessive allele from the father.
   d. Daughters have not inherited the recessive alleles from the father.

___ 15. Even though humans are not good for studying inheritance through experimentation, scientists can study human inheritance through
   a. molecular studies
   b. genetic studies
   c. pedigrees
   d. family history

___ 16. A karyotype shows a portion of chromosome #5 turned upside down. This is most likely due to a _____.
   a. translocation
   b. inversion
   c. deletion
   d. duplication

___ 17. Which of the following is an X-linked disorder?
   a. Down's syndrome
   b. Tay-Sachs
   c. hemophilia
   d. progeria

___ 18. Which of the following is a non-disjunction disorder?
   a. Down's syndrome
   b. Tay-Sachs
   c. hemophilia
   d. progeria

___ 19. Which of the following can be done in utero to test a fetus for genetic abnormalities?
   a. chorionic villus
   b. amniocentesis
   c. fetoscopy
   d. sonography
   e. all of the above

___ 20. Three or more full sets of chromosomes is called _____.
   a. duplication
   b. monosomy
   c. nondisjunction
   d. polyploidy

CHAPTER OBJECTIVES/REVIEW QUESTIONS

1. A(n) _____ chart or diagram is used to study genetic connections between individuals. [p.220]
2. A genetic is an uncommon but not unhealthy version of a trait, whereas an inherited genetic causes mild to severe medical problems. [pp.220-221]
3. Describe the Hutchinson–Gilford progeria syndrome and its probable mode of inheritance. [p.222]
4. Carefully characterize patterns of autosomal recessive inheritance, autosomal dominant inheritance, and X-linked recessive inheritance. [pp.222-225]
5. A(n) _____ is a loss of a chromosome segment; a(n) _____ is a gene sequence separated from a chromosome and then inserted at the same place, but in reverse; a(n) _____ is a repeat of several gene sequences on the same chromosome; a(n) _____ is the transfer of part of one chromosome to a non-homologous chromosome. [pp.226-227]
6. _____ is the failure of the chromosomes to separate in either meiosis or mitosis. [p.228]
7. Trisomy 21 is known as _____ syndrome; Turner syndrome has the chromosome constitution _____; XXY chromosome constitution is syndrome; taller-than-average males with sometimes slightly lower-than-average IQs have the condition _____. [pp.228-229]
8. When gametes or cells of an affected individual end up with one extra or one less chromosome than the parental number, the condition is known as _____; relate this concept to monosomy and trisomy.[p228]
9. List some benefits to a society of genetic screening and genetic counseling. [pp.230-231]
10. Explain the procedures and purpose of three types of prenatal diagnosis: amniocentesis, chorionic villi analysis, and fetoscopy; compare the risks. [pp.230-231]

INTEGRATING AND APPLYING KEY CONCEPTS

1. The parents of a young boy bring him to their doctor. They explain that the boy does not seem to be going through the same vocal developmental stages as his older brother. The doctor orders a common cytogenetics test to be done, and it reveals that the young boy's cells contain two X chromosomes and one Y chromosome. Describe the test that the doctor ordered and explain how and when such a genetic result, XXY, most logically occurred.
2. Solve the following genetics problem. Show rationale, genotypes, and phenotypes. A husband sues his wife for divorce, arguing that she has been unfaithful. His wife gave birth to a girl with a fissure in the iris of her eye, an X-linked recessive trait. Both parents have normal eye structure. Can the genetic facts be used to argue for the husband's suit? Explain your answer.
3. A test is available to identify heterozygotes for a common, incurable, recessive genetic defect. A man and woman, both of whom have family members with the defect, are trying to decide whether to have the test. What are some of the ethical issues involved?
4. Figure 14.10 in the text shows the relationship of chromosome 2 in humans with two separate chromosomes in apes. How might this relate to evolution? Explain how this change could have occurred.

15

STUDYING AND MANIPULATING GENOMES

INTRODUCTION

This chapter covers the many modern techniques used in genetics including production of recombinant DNA, using DNA fragments, and producing transgenic organisms.

STUDY STRATEGIES

- Read all of Chapter 15 with the goal of familiarizing yourself with the boldface terms.
- Discuss the process of DNA cloning with restriction enzymes and cloning vectors.
- Differentiate between cDNA and genomic libraries.
- Describe the process of the polymerase chain reaction and its usefulness in genetics research.
- Explain DNA sequencing and describe the Human Genome Project
- Recognize the application of genomics to various fields.
- Explain what is meant by genetic engineering and describe the various biotechnology applications.

FOCAL POINTS

- Figure 15.2 [p.236] and Figure 15.4 [p.237] look at the use of restriction enzymes in producing recombinant DNA.
- Figure 16.6 [p.245] illustrates the cycles of polymerase chain reaction (PCR)
- Figure 16.8 [p.246] looks at DNA sequencing
- Figure 16.9 [p.247] look at DNA fingerprinting.
- Figures 16.12–16.15 [pp.251–253] look at genetic engineering and the production of transgenic organisms.

INTERACTIVE EXERCISES

15.1 PERSONAL DNA TESTING [p.235]

15.2. CLONING DNA [pp.236-237]

Boldfaced, Page-Referenced Terms

single nucleotide polymorphism _____

_____

restriction enzymes _____

_____

recombinant DNA _____

_____

DNA cloning _____

_____

cloning vectors _____

_____

reverse transcriptase_____

_____

cDNA _____

_____

Fill-in-the-Blanks [pp.236-237]

Years after the discovery of the structure of DNA, it was discovered that some bacteria resist infection with

(1) _____ due to the activity of (2) _____, enzymes that cut DNA at specific nucleotide sequences. These

enzymes are isolated from specific bacteria; for example, (3) _____ is isolated from *E. coli*. In order to produce

this (4) _____ DNA, (5) _____ enzymes make staggered cuts in DNA, leaving (6) _____, which

can base-pair with other DNA fragments produced by the same enzyme. This is the first step in (7) _____, of a

set of laboratory methods using living cells to make many copies of specific DNA fragments. Small circles of

bacterial DNA called (8) _____ are used to insert specific DNA into bacteria. These small circles replicate at

the same time as the bacterial chromosome and are easily transferred between bacteria, acting as (9) _____ for

transfer of the attached genes. The resulting cells are genetically identical (10) _____ that contain a copy of the

vector and the foreign DNA it carries. To study eukaryotic genes and their expression, researchers work with (11)

_____ because the (12) _____ have already been snipped out. However, restriction enzymes and (13)

_____ cut and paste only double-stranded DNA. To make a template for double-stranded DNA, mRNA is

placed in a test tube and used with (14) _____, a replication enzyme from certain types of viruses. The

resulting strand that base-pairs with the mRNA is called (15) _____. A DNA copy is then made from this

molecule using DNA polymerase. Restriction enzymes can then be used for cloning experiments just like an original

DNA strand might be used.

Matching [pp.236-237]

Match the steps in the formation of a recombinant DNA with the parts of the accompanying illustration.

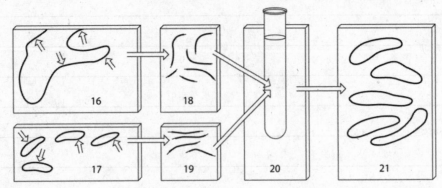

16. ____

17. ____

18. ____

19. ____

20. ____

21. ____

A. Joining of chromosomal and plasmid DNA using DNA ligase

B. Restriction enzyme cuts chromosomal DNA at specific recognition site

C. Cut plasmid DNA

D. Recombinant plasmids containing cloned library

E. Fragments of chromosomal DNA

F. Same restriction enzyme is used to cut plasmids.

Short Answer [p.236-237]

22. Why was the discovery of restriction enzymes such an important development in DNA technology studies?

_____

_____

_____

23. Why do scientists use cloning vectors to produce DNA products?

_____

_____

_____

15.3. ISOLATING GENES [pp.238-239]

Boldfaced, Page-Referenced Terms

genome_____

_____

DNA libraries_____

_____

probes_____

_____

nucleic acid hybridization _____

_____

polymerase chain reaction (PCR) _____

_____

primers_____

_____

Matching [pp.238-239]

Match the most appropriate letter with each numbered partner.

1. ____ DNA polymerase
2. ____ polymerase chain reaction
3. ____ nucleic acid hybridization
4. ____ *Thermus aquaticus*
5. ____ *Taq* polymerase
6. ____ primers

A. Bacterial species from which DNA polymerase used in PCR is extracted.

B. Assembles DNA wherever primers have hybridized with a template in PCR.

C. Oligomers that base-pair with DNA at a certain sequence.

D. Base paring between DNA (or DNA and RNA) from more than one source.

E. Cycled reaction using heat-tolerant DNA polymerase to copy DNA.

F. DNA polymerase used in PCR.

Sequence [p.239]

Order the following events of polymerase chain reaction (PCR) by placing the numbers 1-6 in the appropriate order. Then match the drawings below to the correct step by placing the letters of the figure in the parentheses for 8-12.

7. ____ The number of target DNA copies reaches the billions

8. __( ) DNA is reheated, separating the DNA into single strands; cooled primers hydrogen-bond to DNA

9. __( ) Mixture is heated originally causing DNA strands to separate; cooled primers hydrogen- bond to template DNA

10. __( ) *Taq* polymerase causes primers to form complimentary strands in the first complete round of PCR

11. __( ) *Taq* polymerase causes primers to form complimentary strands in the second complete round of PCR

12. __( ) DNA template is mixed with primers, free nucleotides, and heat-tolerant *Taq* DNA polymerase

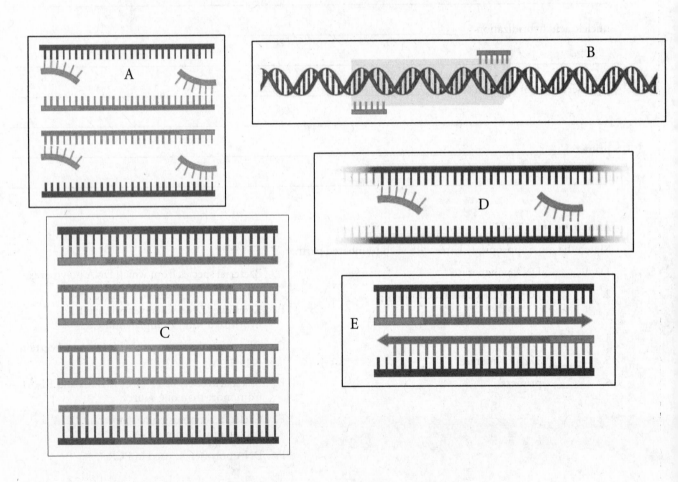

15.4. DNA SEQUENCING [pp.240-241]

Boldfaced, Page-Referenced Terms

DNA sequencing _____

_____

electrophoresis _____

_____

Short Answer [pp.240-241]

1. Why do the DNA fragments end up in different lengths in the process of gel electrophoresis?

_____

_____

_____

2. How is electrophoresis used to separate DNA fragments of different lengths to determine the DNA sequence?

_____

_____

_____

3. Describe the goals and outcomes of the Human Genome Project.

_____

_____

_____

15.5. GENOMICS [pp.242-243]

Boldfaced, Page-Referenced Terms

genomics _____

_____

DNA profiling _____

_____

Short tandem repeats _____

_____

Short Answer [pp.242-243]

1. Describe the process of DNA fingerprinting.

   _____

   _____

   _____

2. How is DNA fingerprinting used in crime cases?

   _____

   _____

   _____

3. In what other ways is genetic fingerprinting used?

   _____

   _____

   _____

Matching [pp.242-243]

Match the term with its description.

4. _____ Sequence of DNA nucleotides coding for a human being

5. _____ Gene used in mice knockout experiments looking at blood lipids

6. _____ Microscopic arrays of different DNA samples separated on small glass plates

7. _____ Used to determine the three-dimensional structure of proteins encoded by a genome

8. _____ Used to compare genomes of different species

9. _____ Consequence of mutations in the APOA5 gene

10. _____ Investigations into the genomes of humans and other species

A. APOA5

B. comparative genomics

C. DNA chips

D. genomics

E. high triglycerides

F. human genome

G. structural genomics

15.6. GENETIC ENGINEERING [p.244]

Boldfaced, Page-Referenced Terms

genetic engineering _____

_____

transgenic _____

_____

genetically modified organisms (GMOs) _____

_____

Short Answer [p.244]

1. List five ways that humans use genetic engineering to benefit humanity.

_____

_____

_____

_____

2. What benefit may be achieved from genetically modifying bacteria expressing a fluorescent protein from jellyfish?

_____

_____

_____

15.7. DESIGNER PLANTS [pp.244-245]

Sequence [pp.244-245]

Order the following events in producing a transgenic plant by placing the numbers 1-5 in the correct order.

1. _____ Plant cell divides to form an embryo that expresses the foreign gene

2. _____ A tumefaciens containing a Ti plasmid is produced to use in the transformation

3. _____ The transgenic plant expresses a foreign gene

4. _____ Transgenic plants are grown from genetically-engineered embryos

5. _____ A Ti plasmid-containing bacterium infects a plant cell

True/False [pp. 244-245]

Determine if each statement is true or false. For false statements, correct to make the statement true.

6. _____ "Frankenfood" is a scientific term used to promote genetic engineering.

7. _____ Genetically-modified cotton, canola, corn, alfalfa, soy and sorhum are commonly grown in the United States.

8. _____ Transgenic crops are used to incorporate drought tolerance.

9. _____ *Bt* plasmids are used as vectors to transfer foreign or modified genes into plants.

10. _____ Plants are protected from the devastating effects of insect larvae by insertion of the *Ti* plasmid into the plant cells.

15.8. BIOTECH BARNYARDS [pp.246-247]

15.9. SAFETY ISSUES [p.247]

Boldfaced, Page-Referenced Term

xenotransplantation _____

_____

Short Answer [pp.246-247]

In exercises 1–5, summarize the results/promise of the given experimentation dealing with genetic modifications of animals.

1. rat growth hormone: _____

_____

2. lysozyme: _____

_____

3. glucose metabolism: _____

_____

4. xenotransplantation: _____

_____

5. spider silk _____

_____

6. List the safety guidelines regulated by the USDA for genetic engineering research.

_____

_____

_____

15.10. GENETICALLY MODIFIED HUMANS [p.248]

Boldfaced, Page-Referenced Term

gene therapy _____

_____

Short Answer [p.248]

1. How might age-related diseases like cancer, Parkinson's disease, and diabetes be cured with gene therapy?

_____

_____

2. List five diseases currently in testing for gene therapy.

_____

_____

3. Summarize the treatment that allowed eighteen of twenty boys with SCID-X1 to be cured.

_____

_____

4. What happened to 25% of the boys with SCID-X1? Why is this a warning for future use of gene therapy with humans?

_____

_____

5. What are some of the ethical concerns for eugenic engineering? What are some of the arguments for eugenic engineering?

_____

_____

SELF-TEST

___ 1. Small, circular molecules of DNA in bacteria are called _____.
   a. plasmids
   b. desmids
   c. pili
   d. F particles

___ 2. Enzymes used to cut genes in recombinant DNA research are _____.
   a. ligases
   b. restriction enzymes
   c. transcriptases
   d. DNA polymerases

___ 3. The total DNA in a haploid set of chromosomes of a species is its _____.
   a. plasmid
   b. enzyme potential
   c. genome
   d. DNA library

___ 4. Any DNA molecule that is copied from mRNA is known as _____.
   a. cloned DNA
   b. cDNA
   c. DNA ligase
   d. hybrid DNA

___ 5. The study of the genomes of humans and other organisms is called _____.
   a. cloning
   b. genetic engineering
   c. gene therapy
   d. genomics

___ 6. A knockout cell is _____.
   a. a fertilized egg that can lead to an exceptionally attractive person
   b. a cell in which a particular gene sequence has been excised
   c. any cell that has been removed from an organism
   d. any cell that has been genetically engineered

___ 7. The most commonly used method of DNA amplification is _____.
   a. polymerase chain reaction
   b. gene expression
   c. genome mapping
   d. RFLPs

___ 8. Tandem repeats are valuable because _____.
   a. they reduce the risks of genetic engineering
   b. they provide an easy way to sequence the human genome
   c. they allow DNA fragmenting without enzymes
   d. they provide DNA fragment sizes unique to each person

___ 9. Which method can rapidly reveal the base sequence of cloned DNA or PCR-amplified DNA fragments?
   a. Polymerase chain reaction
   b. Nucleic acid hybridization
   c. Recombinant DNA technology
   d. Automated DNA sequencing

___ 10. A cDNA library is _____.
   a. a collection of DNA fragments derived from mRNA and free of introns
   b. cDNA plus the required restriction enzymes
   c. mRNA–cDNA
   d. composed of mature mRNA transcript

___ 11. Restriction enzymes produce unpaired nucleotides which are referred to as _____.
   a. sticky ends
   b. free bases
   c. single ends
   d. fragments

___ 12. In order to insert any gene into a plasmid, both the chromosome and plasmid must
   a. have the same DNA sequence
   b. be cut by the same restriction enzyme
   c. code for the same polypeptide
   d. have the same number of nucleotides

___ 13. Why is golden rice pale yellow in color?
   a. It is lacking nutrients.
   b. It is rich in beta-carotene.
   c. It is rich in chlorophyll.
   d. Scientists dye the rice.

___ 14. _____ is the process of cutting genes out, and then inserting and pasting them into other DNA sequences.
   a. cloning
   b. DNA fingerprinting
   c. gel electrophoresis
   d. recombinant DNA technology

15. In animal cloning, a _____ is needed to support the clone's development but is not genetically related to the clone.
    a. cell donor
    b. male
    c. gastrula
    d. surrogate mother

16. Which of the following has NOT been successfully genetically modified.
    a. rice
    b. chickens
    c. goats
    d. humans

17. Which of the following is NOT a beneficial product produced by transgenic species?
    a. insulin
    b. cheese
    c. human blood
    d. organs

18. _____ is a general term for a carrier used to transfer a foreign DNA fragment into a host cell.
    a. vector
    b. virus
    c. bacteria
    d. plasmid

CHAPTER OBJECTIVES/REVIEW QUESTIONS

1. Some bacteria produce enzymes that cut apart DNA molecules injected into the cell by viruses; such DNA fragments are cut in a staggered way that produces "sticky ends" capable of base-pairing with other DNA molecules cut by the same enzymes. Summarize how these enzymes are used to insert DNA from bacterial plasmids into host bacteria. [pp.236-237]
2. Base-pairing between chromosomal fragments and cut plasmids is made permanent by DNA . [pp.236-237]
3. Define cDNA. [pp.236-237]
4. _____ are small, circular, self-replicating molecules of DNA or RNA within a bacterial cell. [p.237]
5. A special viral enzyme, _____, presides over the process by which mRNA is transcribed into DNA. [p.237]
6. What is the difference between a genomic library and a cDNA library? [p.238]
7. Polymerase chain reaction is the most commonly used method of DNA _____. [pp.238-239]
8. Explain how gel electrophoresis is used in DNA fingerprinting. [pp.240-241]
9. List some practical genetic uses of DNA fingerprinting. [pp.242-243]
10. How could Golden Rice be beneficial to people in poor countries? [pp.244-245]
11. How is *Agrobacterium tumefaciens* used in gene transfer in plants? [pp.244-245]
12. How could gene therapy reduce the number of infant deaths and hospital admissions? [pp.248-249]
13. Describe how pig organs can be altered so as not to be rejected by a human recipient. [pp.246-247]

INTEGRATING AND APPLYING KEY CONCEPTS

1. How could scientists guarantee that *Escherichia coli*, the human intestinal bacterium, will not be transformed into a severely pathogenic form and released into the environment if researchers use the bacterium in recombinant DNA experiments?
2. Explain how knowing the genetic makeup of Earth's organisms can help us reconstruct the evolutionary history of life.
3. What problems might be involved in trying to clone extinct animals?
4. Explain how knowing the composition of genes can help scientists derive counterattacks against rapidly mutating organisms.

16

EVIDENCE OF EVOLUTION

INTRODUCTION

This chapter first presents a brief history of evolutionary thought from the time of the ancient Greeks up to Darwin. The chapter then brings Darwin's theory up to date by presenting supporting evidence from a variety of sources- fossils, biogeography, comparative morphology, and comparative biochemistry. As you progress through this chapter, you should note how each class of evidence supports Darwin's theory and helps explain the process of macroevolution.

STUDY STRATEGIES

- Read all of Chapter 16 with the goal of familiarizing yourself with the boldface terms.
- Realize the depth of geologic time and some of the ways we can understand the physical, chemical, and biological past.
- Explain the study of biogeography and comparative morphology.
- Recognize the incorrect hypotheses of Cuvier and Lamark.
- Compare and contrast the ideas of Darwin/Wallace and Lyell with those preceeding them.
- Explain the theory of evolution by natural selection.
- Describe the importance fossils have for paleontologists.
- Explain why the fossil record is incomplete.
- Describe the importance of radiometric dating and DNA sequencing for filling in "missing" links.
- Explain the theory of plate tectonics.
- Identify some of the major biological occurances in geologic time.

FOCAL POINTS

- The review of Table 17.1 [p.259] relates Darwin's initial theory to what is currently known from genetics and molecular biology.
- Figure 16.5 [p.256] illustrates Darwin's journeys.
- Figure 16.15 [pp.266-267] overlays major biological events with the geologic time scale and introduces the concepts of mass extinctions and adaptive radiations.

INTERACTIVE EXERCISES

16.1. REFLECTIONS OF A DISTANT PAST [p.253]

16.2. EARLY BELIEFS, CONFOUNDING DISCOVERIES [pp.254-255]

16.3. A FLURRY OF NEW THEORIES [pp.256-257]

16.4. DARWIN, WALLACE, AND NATURAL SELECTION [pp.258-259]

Boldfaced, Page-Referenced Terms

mass extinction _____

_____

naturalists _____

_____

biogeography _____

_____

comparative morphology _____

_____

fossils _____

_____

catastrophism _____

_____

evolution _____

_____

lineage _____

_____

theory of uniformity _____

_____

artificial selection _____

_____

fitness _____

_____

adaptation _____

_____

adaptive trait _____

_____

natural selection _____

_____

Matching [pp.254-259]

Choose the most appropriate answer for each term.

1. _____ comparative morphology
2. _____ biogeography
3. _____ fossils
4. _____ Chain of Being
5. _____ species
6. _____ Aristotle
7. _____ naturalists
8. _____ evolution
9. _____ lineage
10. _____ theory of uniformity
11. _____ fitness
12. _____ adaptation
13. _____ catastrophism
14. _____ artificial selection
15. _____ natural selection

A. The process of selective breeding by humans to enhance a desired trait in a given species

B. The person that came to view nature as a continuum of organization, from lifeless matter through complex forms of plants and animal life

C. Lyell's theory that stated the gradual, repetitive change was the answer to the shaping of the Earth

D. Each kind of being that represents a link in the Chain of Being

E. Relative genetic contribution to future generations

F. Darwin's theory that environmental pressures select for organisms that have adaptations more suitable to their current environment; these organisms will survive and reproduce more often

G. Extended from the lowest forms of life to humans and spiritual beings

H. Cuvier's theory that forces unlike the ones we see today were responsible for the shaping of the Earth

I. People who view life from a scientific perspective

J. A trait that enhances an individual's fitness

K. Study of similarities and differences in body plans

L. Another term for "line of descent"

M. Study of the world distribution of plants and animals

N. Theory of change in line of descent

O. Used now as evidence of life in ancient times

Choice [pp.262-263]

For each of the following, choose the individual from the list that is best associated with the statement.

   a. Georges Cuvier   b. Jean Lamarck   c. Charles Lyell

16. _____ Stretching directed "fluida" to the necks of giraffes, which lengthened permanently
17. _____ Supporter of the theory of uniformity
18. _____ Inheritance of acquired characteristics
19. _____ Catastrophism
20. _____ Over great spans of time, geological processes have sculpted the Earth

21. _____ Species improve over generations because of an inherent drive toward perfection

22. _____ First to suggest that the environment is a factor in lines of descent

23. _____ Many species that once existed are now extinct

Choice [pp.258-259]

For each of the following, choose a term from the list below that best completes the sentence. A term may be used more than once.

a. individual(s)   b. population(s)   c. alleles   d. natural selection

24. _____ As a _____ expands the resources become limited

25. _____ Natural _____ have an inherent reproductive capacity to increase in numbers through successive generations

26. _____ _____ will vary in their traits

27. _____ _____ is the outcome of differences in reproduction among individuals of a population that vary in shared traits

28. _____ When resources are limited, members of a _____ will compete

29. _____ An _____ associated with a positive trait tends to increase in frequency

30. _____ As resources become limited _____ will end up competing

Short Answer [pp.274-275]

31. Explain the long neck of a giraffe using principles of natural selection.

_____

_____

_____

16.5. FOSSILS: EVIDENCE OF ANCIENT LIFE [pp.260-261]

16.6. FILLING IN PIECES OF THE PUZZLE [pp.262-263]

Boldfaced, Page-Referenced Terms

half-life _____

_____

radiometric dating _____

_____

Matching [pp.260-263]

Choose the correct term for each of the following statements.

1. _____ fossilization
2. _____ fossil record
3. _____ stratification
4. _____ trace fossils
5. _____ lineage
6. _____ half-life
7. _____ radiometric dating

A. describes the formation of sedimentary rock

B. indirect evidence of past life, such as tracks or burrows

C. a single line of descent

D. a vertical series of fossils that serves as a record of past life

E. the transformation by which metal ions replace minerals in bones and hardened tissues

F. measuring the radioisotope & daughter element content

G. time it takes for half of a radioisotope's atoms to decay into a product

Fill-in-the-Blanks [pp.260-261]

The fossil record consists of approximately (8) _____ known specimens. The odds of finding fossil evidence is extremely (9) _____. The remains of an organism usually (10) _____ or are (11) _____ by scavengers. Fossils are more likely to form in the absence of (12) _____. Typically, (13) _____ organisms do not fossilize even though they were most likely present in ancient communities.

Short Answer [p.262]

14. Carbon-14 has a half-life of 5,370 years. Assume that a sample of organic material contains 4.0 grams of carbon-14. How many grams would be left after three half-lives? How many years would have elapsed?

_____

_____

_____

16.7. DRIFTING CONTINENTS, CHANGING SEAS [pp.264-265]

16.8. PUTTING TIME INTO PERSPECTIVE [pp.266-267]

Boldfaced, Page-Referenced Terms

Gondwana_____

_____

Pangea_____

_____

plate tectonics _____

_____

Matching [p.266-267]

Refer to Figure 16.15 in the text. For each of the following events, choose the most appropriate time period from the list provided.

1. _____ Origin of the vascular plants
2. _____ Origin of the mammals
3. _____ Origin of the flowering plants
4. _____ Origin of the bony fishes
5. _____ Origin of birds
6. _____ Origin of eukaryotic cells
7. _____ Modern humans evolve
8. _____ Origins of life

A. Cretaceous
B. Quaternary
C. Triassic
D. Ordovician
F. Proterozoic
G. Archean and earlier
H. Silurian
I. Jurassic

Labeling and Matching [p.264]

9. Match the letter in the figure on plate tectonics with the appropriate term from the following (some may be used more than once):

hot spot, trench, ridge, fault

A. _____
B. _____
C. _____
D. _____
E. _____

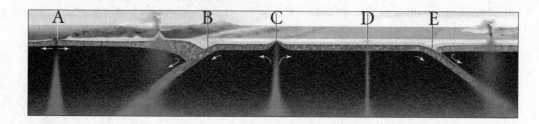

Short Answer [p.265-266]

10. Identify the previous 5 mass extinction events in the history of life on Earth and identify major changes that resulted.

_____

_____

_____

_____

_____

SELF-TEST

___ 1. The patterns, trends, and changes in lines of descent is known as _____.
   a. taxonomy
   b. classification
   c. biogeography
   d. evolution

___ 2. Which is not part of natural selection?
   a. Populations will increase in size over time.
   b. Resources will be limited.
   c. Individuals in a population will vary in their traits.
   d. The individual can use "fluida" to change their physical features.

___ 3. Who co-discovered the process of evolution by natural selection with Darwin?
   a. Lyell
   b. Cuvier
   c. Lamarck
   d. Wallace

___ 4. Study of the distribution of similar species around the world is called _____.
   a. comparative morphology
   b. biogeography
   c. stratification
   d. catastrophism

___ 5. The idea that gradual, repetitive changes shaped the surface of Earth is called _____.
   a. the theory of uniformity
   b. natural selection
   c. catastrophism
   d. the Chain of Being

___ 6. Which scientist developed catastrophism?
   a. Lyell
   b. Lamarck
   c. Darwin
   d. Cuvier

___ 7. The method of using half-lives of radioactive compounds to date fossils is called _____.
   a. nucleic acids hybridization
   b. radiometric dating
   c. catastrophism
   d. plate tectonics

___ 8. An increase in adaptation to an environment is called _____.
   a. morphological divergence
   b. mutation
   c. fitness
   d. speciation

9. How many mass extinctions have occurred on Earth?
   a. 5
   b. 4
   c. 2
   d. 1

10. Which of the following is used as evidence of plate tectonics?
    a. Spreading of the ocean floor at mid-ocean ridges
    b. Iron compasses
    c. Fossils of ancient organisms such as Lystrosaurus
    d. The supercontinents of Pangea and Gondwana
    e. All of the above

11. Which scientist believed that individual organisms gradually improve over time and pass their improvements onto offspring?
    a. Lyell
    b. Lamarck
    c. Darwin
    d. Cuvier

12. Which scientist was a proponent of the Theory of Uniformity?
    a. Lyell
    b. Lamarck
    c. Darwin
    d. Cuvier

13. The study of structures in different organisms to determine relatedness is _____.
    a. comparative morphology
    b. biogeography
    c. stratification
    d. catastrophism

14. Which of the following was one of the first recorded to try to make sense of the order of the natural world?
    a. Darwin
    b. Aristotle
    c. Lyell
    d. Lamark

15. A trait that helps an organism enhance its fitness is a(n) _____.
    a. mutation
    b. adaptation
    c. gene
    d. allele

16. Fossils form through _____.
    a. random, unknown processes
    b. organisms being covered in sediments or volcanic ash
    c. artificial selection
    d. mutations

17. The boundary layer in the fossil record between the Age of the Dinosaurs and the Age of the Mammals is known as the _____ layer.
    a. meteorite layer
    b. K-T
    c. D-M
    d. none of these

18. How many different species do we currently have fossils for?
    a. unknown
    b. 2,500
    c. 2.5 million
    d. 250,000

19. The ancient supercontinent that existed approximately 270 million years ago is named _____.
    a. Pangaea
    b. Gondwanaland
    c. Archaea
    d. Africa

20. According to fossil, chemical, and physical evidence, the Earth is approximately _____ years old.
    a. 3.8 million
    b. 3.8 billion
    c. 4.5 billion
    d. 6,000

CHAPTER OBJECTIVES / REVIEW QUESTIONS

1. Explain how biogeography contributes to the study of evolution. [pp.254-255]
2. Give the basic premises of Cuvier's and Lamarck's theories to explain the change in organisms over time. [pp.256-257]
3. Explain the theory of uniformity and why it played an important role in the study of evolution. [p.257]
4. Describe the evidence Darwin used to develop the theory of natural selection. [pp.256-257]
5. Explain how fitness relates to natural selection. [pp.258-259]
6. Explain how stratification and fossilization are related. [pp.260-261]
7. Define *lineage*. [p.261]
8. Describe the importance of radiometric dating to the study of fossils. [p.262]
9. Calculate the amount of radioactive material that would remain after a given number of half-lives. [p.262]
10. List the evidence that supports the plate tectonics theory. [pp.264-265]
11. Be able to place major biological events on a geologic time scale. [pp.266-267]

INTEGRATING AND APPLYING KEY CONCEPTS

1. Species have previously been defined as groups of interbreeding organisms that are reproductively isolated from other groups. What problem does this definition hold for scientists who study macroevolutionary processes?
2. As a researcher, you have discovered two groups of plants on two isolated islands in the Pacific. These plants have very similar characteristics, but also minor differences. You are not sure whether your observations can be explained by divergence or convergence. How would you determine if these two plants are members of the same lineage?
3. Draw an evolutionary timeline and indicate when the appearance of the major groups of animals occurred (fish, reptiles, amphibians, mammals, birds, etc.). Also indicate when the Earth experienced the mass extinctions.

17

PROCESSES OF EVOLUTION

INTRODUCTION

Having examined the "big" processes that influenced evolution in the previous chapter, it is time now to focus on how populations change over time. This is called microevolution. This chapter first presents some important terminology regarding variation, and then establishes how to measure change in a population using the Hardy-Weinberg equilibrium equation. Following the introduction to microevolution, the chapter proceeds into a discussion of the many forms of selection, from sexual to disruptive. Perhaps one of the more important aspects of the chapter is the discussion of genetic drift, interbreeding, and gene flow and how these factors work on the allele frequencies in a population. This is a detailed chapter, but one that provides a good working knowledge of how scientists study evolutionary processes.

STUDY STRATEGIES

- Read all of Chapter 17 with the goal of familiarizing yourself with the boldface terms.
- Define population and describe the role that mutations lead to changes in populations over time.
- Identify the various effects mutations may have on an individual and/or population.
- Recognize and be able to solve a Hardy-Weinberg equation given an allelic frequency.
- Explain what is necessary for genetic equilibrium to occur.
- Compare and contrast the various forms and characteristics of natural selection.
- Describe what types of selection pressures foster and maintain genetic and species diversity.
- Explain genetic drift and gene flow.
- Identify the various types of reproductive isolation and when each may occur.
- Compare and contrast the different forms of speciation.
- Define macroevolution and describe the patterns of macroevolution.

FOCAL POINTS

- For the Hardy-Weinberg equilibrium [pp.274-275], note especially the five conditions of a population at equilibrium, as this provides the foundation for studying evolution in populations.
- Figure 17.4 [p.275] shows three modes of natural selection. The text devotes several sections towards understanding the material presented in this figure.
- Figure 17.16 [p.276] illustrates various modes of reproductive isolation.
- Section 17.12 [p.290] discusses processes of macroevolution.

INTERACTIVE EXERCISES

17.1 RISE OF THE SUPER RATS [pp.271]

17.2. INDIVIDUALS DON'T EVOLVE, POPULATIONS DO [pp.272-273]

17.3. A CLOSER LOOK AT GENETIC EQUILIBRIUM [pp.274-275]

Boldfaced, Page-Referenced Terms

population _____

_____

lethal mutation _____

_____

neutral mutation _____

_____

gene pool _____

_____

allele frequency _____

_____

microevolution _____

_____

genetic equilibrium _____

_____

Matching [pp.272-275]

1. _____ microevolution
2. _____ gene pool
3. _____ population
4. _____ dimorphism
5. _____ alleles
6. _____ polymorphism
7. _____ lethal mutations
8. _____ morphological traits
9. _____ neutral mutation
10. _____ allele frequencies
11. _____ genetic equilibrium
12. _____ physiological traits
13. _____ behavioral traits

A. Genes with a slightly different molecular form
B. A trait with three or more forms in a population
C. Traits that define the form of the body
D. A change in the DNA that does not affect survival or reproduction
E. Traits that determine how an individual will respond to stimuli
F. Small scale change in a population's allele frequencies
G. A group of individuals of the same species in a specified area
H. A change in a trait that disrupts a phenotype and causes death
I. A population's genetic makeup
J. Traits that help the body function in its environment such as metabolism
K. A trait with two forms in a population
L. The relative abundances of an allele among all individuals of a population
M. A population that is not evolving with regard to a certain allele

14. List the five conditions that must be met if the gene pool is to remain stable and the population is not evolving.

_____

_____

_____

15. What is meant by, "individuals don't evolve but populations do"?

_____

_____

16. For the following situation, assume that the conditions listed in question 14 do exist; therefore, there should be no change in gene frequency generation after generation. Consider a population of hamsters in which the dominant gene B produces a black coat color and the recessive gene b produces a gray coat color (two alleles will produce the coat color). The dominant gene has a frequency of 80% or (.80). The recessive gene will then have a frequency of 20% or (.20). From this, the assumption is made that 80% of all sperm and eggs have gene B and 20% of the sperm and eggs have gene b.

    a. Calculate the probabilities of all possible matings in the Punnett square. (see left-hand diagram)

_____

_____

    b. Summarize the genotype and phenotype frequencies on the F_1 generation. (see center diagram)

_____

_____

    c. Further assume that the individuals of the F_1 generation produce another generation and the assumptions of the Hardy-Weinberg rule still hold. What are the frequencies of the sperm produced? (see right-hand diagram)

_____

_____

Processes of Evolution   **203**

	Sperm	
	0.80 B	0.20 b
Eggs 0.80 B	BB	Bb
0.20 b	Bb	bb

Genotypes	Phenotypes
____ BB	
____ Bb	____ % black
____ bb	____ % gray

Parents	B sperm	b sperm
____ BB	____	____
____ Bb	____	____
____ bb	____	____
Totals= ____	____	____

The egg frequencies may be similarly calculated. Note that the gamete frequencies of the F_2 generation are the same as the gamete frequencies of the previous generation. Phenotype percentage also remains the same. Thus, the gene frequencies did not change between the F_1 and F_2 generation. Again, given the assumptions of the Hardy-Weinberg equilibrium, gene frequencies do not change generation after generation.

17. In a population, 81% of the organisms are homozygous dominant, and 1% are homozygous recessive. Find the following:

   a. The percentage of heterozygotes

   _____

   b. The frequency of the dominant allele

   _____

   c. The frequency of the recessive allele

   _____

18. In a population of 200 individuals, determine the following for a particular locus if $p = 0.80$.

   a. The number of homozygous dominant individuals

   _____

   b. The number of heterozygous individuals

   _____

   c. The number of heterozygous individuals

   _____

19. If the percentage of gene D is 70% in a gene pool, find the percentage of gene d.

   _____

20. If the frequency of gene *R* in a population is 0.60, what percentage of the individuals will be heterozygous (*Rr*)?

_____

_____

21. If the frequency of gene *t* in a population is 0.3, what percentage of the individuals will be homozygous recessive (*tt*)?

_____

_____

17.4. PATTERNS OF SELECTION REVISITED [p.275]

17.5. DIRECTIONAL SELECTION [pp.276-277]

17.6. STABILIZING & DISRUPTIVE SELECTION [pp. 278-279]

Boldfaced, Page-Referenced Terms

directional selection _____

_____

stabilizing selection _____

_____

disruptive selection _____

_____

Labeling [p.275]

1. For each of the following three curves, first identify the curve as an example of stabilizing selection, directional selection, or disruptive selection; then give the general characteristics of each form of selection.

a. _____

_____

_____

b. _____

_____

_____

c. _____

_____

_____

Choice [pp.276-279]

For each of the following, choose from one of the forms of selection in the lettered list.

    a. stabilizing selection   b. directional selection c. disruptive selection

2. _____ Intermediate forms of a trait in a population are favored
3. _____ The range of variation in a phenotype tends to shift in a consistent direction
4. _____ The most frequent wing color of peppered moths shifted from a light form to a dark form as tree trunks became soot-darkened because coal was being used for fuel during the English industrial revolution
5. _____ Intermediate forms of a phenotype are selected against, while forms at the end of the range of variation are favored
6. _____ Antibiotic resistance favors resistant bacterial populations
7. _____ An example is the body weight of sociable weavers
8. _____ An example is the selection of bill size in finches of West Africa
9. _____ Coat color of pocket mice favors their environmental surroundings

Short Answer [p.277]

10. Explain the problems that are arising from our misuse of antibiotics.

_____

_____

_____

11. Explain the case of the peppered moth as it relates to directional selection.

_____

_____

_____

17.7. FOSTERING DIVERSITY [pp.280-281]

17.8. GENETIC DRIFT AND GENE FLOW [pp.282-283]

17.9. REPRODUCTIVE ISOLATION [pp.284-285]

Boldfaced, Page-Referenced Terms

sexual selection _____

_____

balanced polymorphism _____

_____

genetic drift _____

_____

fixation _____

_____

bottleneck _____

_____

founder effect _____

_____

inbreeding _____

_____

gene flow _____

_____

Choice [pp.280-283]

For each of the following, choose the most appropriate category form the list below.

a. balanced polymorphism   b. genetic drift   c. sexual selection   d. gene flow   e. inbreeding

1. _____ The random change of allele frequencies over time
2. _____ Immigration and emigration
3. _____ Adaptive traits increase reproductive success
4. _____ The physical flow of alleles between populations
5. _____ This is leading to the spread of genes from transgenic organisms to wild species
6. _____ May result in the fixation of alleles over time
7. _____ The relationship in hemoglobin structure between sickle-cell anemia and malaria
8. _____ An example is the courtship rituals of many species
9. _____ Examples are bottleneck and founder effect
10. _____ The increased frequency of Ellis-van Creveld syndrome is caused by this in Amish populations
11. _____ Is primarily driven by female choice
12. _____ Increases the frequency of homozygous individuals and lowers genetic diversity

Fill-in-the-Blanks [pp.282-283]

Random change in allele (13) _____ leads to the (14) _____ condition and a(n) (15) _____ of genetic diversity over time. This is genetic drift's outcome in all (16) _____; it simply happens faster in (17) _____ ones. Once alleles from the parent population have become (18) _____, their (19) _____ will not change again unless mutation or (20) _____ flow introduces new alleles.

Choice [pp.284-285]

For the following statements, choose the appropriate type of isolating mechanism

    a. prezygotic isolating mechanism   b. postzygotic isolating mechanism

21. _____ temporal isolation
22. _____ hybrid sterility
23. _____ mechanical isolation
24. _____ behavioral isolation
25. _____ hybrid inviability
26. _____ ecological isolation

Short Answer [p.280-281]
27. How does nonrandom mating foster diversity?

_____

_____

_____

28. How does balanced polymorphism maintain diversity?

_____

_____

_____

29. Describe how the founder effect and a population bottleneck can reduce diversity.

_____

_____

_____

_____

30. Why is the *Hb*^{S} allele maintained in human populations if the homozygous recessive condition is lethal?

_____

_____

_____

Labeling and Matching [p.284]

For each of the possible steps in reproductive isolation, identify the term with the letter of the appropriate definition in the figure to the right.

31. _____ behavioral isolation

32. _____ gamete incompatibility

33. _____ temporal isolation

34. _____ ecological isolation

35. _____ hybrid sterility

36. _____ mechanical isolation

37. _____ hybrid inviability

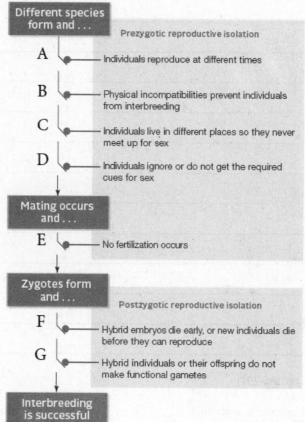

Different species form and ...

Prezygotic reproductive isolation

A — Individuals reproduce at different times

B — Physical incompatibilities prevent individuals from interbreeding

C — Individuals live in different places so they never meet up for sex

D — Individuals ignore or do not get the required cues for sex

Mating occurs and ...

E — No fertilization occurs

Zygotes form and ...

Postzygotic reproductive isolation

F — Hybrid embryos die early, or new individuals die before they can reproduce

G — Hybrid individuals or their offspring do not make functional gametes

Interbreeding is successful

17.10. ALLOPATRIC SPECIATION [pp.286-287]

1711. OTHER SPECIATION MODELS [pp.288-289]

17.12. MACROEVOLUTION [pp.290-291]

Boldfaced, Page-Referenced Terms

allopatric speciation _____

_____

sympatric speciation _____

_____

parapatric speciation_____

_____

exaptation _____

_____

stasis _____

_____

extinct _____

_____

adaptive radiation _____

_____

key innovation _____

_____

coevolution _____

_____

Matching [pp.286-291]

1. _____ allopatric speciation
2. _____ sympatric speciation
3. _____ extinction
4. _____ parapatric speciation
5. _____ coevolution
6. _____ stasis
7. _____ exaptation
8. _____ key innovation
9. _____ microevolution
10. _____ macroevolution

A. species does not change over time
B. the process of species being permanently lost to the planet
C. physical barrier separates a population
D. new structure evolves from an existing one
E. change in allele frequencies within a species or population
F. two species evolve in response to each other
G. a population extends across a variety of habitats
H. large-scale patterns of change leading to a multitude of new species
I. new species form within the home range of the original species
J. structural modification that increases an organism's chance for survival

Choice [pp.286-289]

For the following statements, choose the appropriate type of speciation.

    a. allopatric speciation   b. sympatric speciation   c. parapatric speciation

11. _____ Genetic differences in trees and shrubs due to the Great Wall of China

12. _____ Giant velvet walking worm & blind velvet walking worm

13. _____ Cichlids of Lake Victoria

14. _____ Hawaiian honeycreepers

15. _____ Greenish warblers in central Asia

16. _____ Polyploidy resulting in *T. aestivum*

17. _____ Snapping shrimp near the Isthmus of Panama

Dichotomous Choice [pp.290-291]

Choose the correct one of the two possible answers given between parentheses in each statement.

18. _____ (Stasis/genetic drift) leads to lineages persisting for millions of years with little-to-no change.

19. _____ (Preadaptation/microevolution) refers to the adaptation of an existing structure for a completely different purpose.

20. _____ A lineage rapidly diversifies into several new species through (adaptive radiation/allopatric speciation).

21. _____ A new trait that allows an organism to exploit a habitat more efficiently or in a novel way is a (founder effect/key innovation).

22. _____ More than 99 percent of all species that have ever existed are now (retired/extint) and are no longer found on the planet.

Short Answer [p.286-291]

23. Explain the differences between sympatric, allopatric, and parapatric speciation.

_____

_____

_____

24. What is a mass extinction event and how many of these have occurred in the past?

_____

_____

25. Describe coevolution and give an example.

_____

_____

_____

SELF-TEST

___ 1. Selection for the intermediate form of a trait is called _____.
   a. stabilizing selection
   b. directional selection
   c. disruptive selection
   d. wayward selection

___ 2. Differences in the molecular structure of a gene are called _____.
   a. a gene pool
   b. a bottleneck
   c. gene flow
   d. alleles

___ 3. The sum of all genes in the entire population is the _____
   a. gene pool
   b. genetic variation
   c. gene flow
   d. allele frequency

___ 4. The relative abundance of each type of allele in a population is the _____.
   a. gene pool
   b. genetic variation
   c. gene flow
   d. allele frequency

___ 5. According to the Hardy-Weinberg rule, the allele frequencies of a population will not change over successive generations if which of the following is true?
   a. The population is infinitely large and all individuals survive and produce equal number of offspring.
   b. There is random mating.
   c. There is no mutation.
   d. The population is isolated.
   e. All of the above.

___ 6. A trait that exists in only two forms in a population is said to be _____.
   a. lethal
   b. dimorphic
   c. polymorphic
   d. fixed

___ 7. A _____ is a group of individuals that can interbreed.
   a. gene pool
   b. balanced polymorphism
   c. population
   d. species

___ 8. An insect population that becomes increasingly resistant to a class of insecticides is an example of _____ selection.
   a. directional
   b. sexual
   c. disruptive
   d. stabilizing

___ 9. If one sex of a species favors a trait in the opposite sex of the species, this is called _____.
   a. sexual selection
   b. directional selection
   c. a bottleneck
   d. genetic drift

___ 10. Selection for traits at both ends of a range of variations is called _____ selection.
   a. polymorphic
   b. directional
   c. sexual
   d. disruptive
   e. balancing

___ 11. Which isolating mechanism is considered a postzygotic mechanism?
   a. hybrid sterility
   b. ecological
   c. behavioral
   d. mechanical

___ 12. The ultimate source of all new genetic variation in a species is _____.
   a. cloning
   b. stabilizing selection
   c. fertilization
   d. mutation

___ 13. A mutation that leads to the death of an organism is a _____ mutation.
   a. neutral
   b. silent
   c. lethal
   d. beneficial

___ 14. This type of evolution occurs when the gene frequency of a population changes.
   a. mesoevolution
   b. microevolution
   c. macroevolution
   d. nanoevolution

15. This type of evolution occurs when there are large fluctuations in the number of species on the planet.
    a. mesoevolution
    b. microevolution
    c. macroevolution
    d. nanoevolution

16. The definition of a population includes which of the following?
    a. same species
    b. same location
    c. same timeframe
    d. interbreeding
    e. all of the above

17. Which of the following fosters genetic diversity?
    a. nonrandom mating
    b. balanced polymorphism
    c. genetic bottleneck
    d. both A and B

18. Which of the following can drastically reduce genetic diversity?
    a. nonrandom mating
    b. balanced polymorphism
    c. genetic bottleneck
    d. both A and B

19. Gene flow happens in which of the following cases?
    a. A tiger roams over a mountain range to find new territory and mates
    b. A bird flies from a nesting site on one island to another in order to find a mate
    c. A wind-pollinated plant sends its pollen tens of miles downwind
    d. all of the above

20. The _____ speciation pattern occurs when one group is separated from another by a physical barrier.
    a. allopatric
    b. sympatric
    c. parapatric
    d. none of the above

CHAPTER OBJECTIVES / REVIEW QUESTIONS

1. What is a population? [p.272]
2. Distinguish among morphological, physiological, and behavioral traits. [p.272]
3. Distinguish between dimorphism and polymorphism. [p.272]
4. Explain the difference between a lethal and a neutral mutation. [pp.272-273]
5. Explain the difference between gene pool and alleles. [p.273]
6. Define microevolution. [p.273]
7. List the five conditions that must be met for the Hardy-Weinberg rule to apply. [p.274]
8. Calculate allele and other genotype frequencies when provided with the homozygous recessive genotype frequency. [pp.274-275]
9. Define and provide an example of directional selection, stabilizing selection, and disruptive selection. [pp.275-279]
10. Define and give an example of sexual selection. [pp.275-279]
11. Define balanced polymorphism and explain why the relationship between malaria and sickle-cell anemia is used as an example. [p.281]
12. Explain how genetic drift can lead to the fixation of alleles in a population. [pp.282-283]
13. Distinguish the founder effect from a bottleneck. [pp.282-283]
14. Explain the consequences of inbreeding. [p.283]
15. Explain the influence that gene flow has on the gene pool of a population. [p.283]
16. List the different types of isolating mechanisms. [pp.284-285]
17. Explain the difference between allopatric, sympatric, and parapatric speciation. [pp.286-289]
18. Explain the difference between microevolution and macroevolution. [p.290]

INTEGRATING AND APPLYING KEY CONCEPTS

1. What type of selection do diseases such as HIV and SARS exhibit on the human population? What effects do vaccines have on the gene pool?
2. Endangered animals are frequently confined to zoos for protection. How does this contribute to a bottleneck effect? How could a gene pool be increased and what would be the benefit of doing so?
3. What are the similarities between bacterial resistance and insect resistance to pesticides? What should society do in order to avoid the production of "super" bacteria and insects?
4. Give an example of a neutral trait in humans. For this trait predict what conditions may have existed in the past that would have made this trait advantageous.
5. Indicate the positives and negatives of creating pure bred breeds of dogs, horses, cattle, etc.
6. Explain how mutations and gene flow can counter the effects of genetic drift.

18

ORGANIZING INFORMATION ABOUT SPECIES

INTRODUCTION

This chapter explains how recent evidence has caused many species to be reclassified according to shared ancestry instead of the traditional ranking systems. The various lines of evidence include comparing body form, patterns of development, and biochemistry to help form evolutionary trees.

STUDY STRATEGIES

- Read all of Chapter 18 with the goal of familiarizing yourself with the boldface terms.
- Describe the science of phylogeny.
- Explain the process of parsimony in cladistics.
- Describe how morphology, development, amino acid sequences, and DNA sequences can be used to determine evolutionary relationships.
- Discuss the usefulness of phylogeny in research and conservation.

FOCAL POINTS

- Figure 18.2 [p.296] is an example of parsimony analysis.
- Figure 18.6 [p.300] shows the genetic similarity, based upon amino acid sequences, between various species.

INTERACTIVE EXERCISES

18.1. BYE BYE BIRDIE [p.295]

18.2. PHYLOGENY [pp.296-297]

18.3. COMPARING FORM AND FUNCTION [pp.298-299]

Boldfaced, Page-Referenced Terms

phylogeny _____

_____

character _____

_____

derived trait _____

_____

clade _____

_____

monophyletic group _____

_____

cladistics _____

_____

cladogram _____

_____

evolutionary tree _____

_____

sister groups _____

_____

homologous structures _____

_____

morphological divergence _____

_____

morphological convergence _____

_____

analogous structures _____

_____

Matching [pp.296.299]

1. _____ phylogeny
2. _____ cladistics
3. _____ cladogram
4. _____ clade
5. _____ homologous structures
6. _____ analogous structures
7. _____ morphological divergence
8. _____ morphological convergence
9. _____ derived trait
10. _____ characters
11. _____ monophyletic group
12. _____ parsimony
13. _____ sister groups

A. Evolution of similar body parts in different lineages

B. Evolutionary history of a species or a group of them

C. Similar body parts that reflect shared ancestry

D. Quantifiable features of an organism

E. Group of species that share a set of derived traits

F. Body parts are similar in appearance but did not evolve from a common ancestor

G. The theory that the simplest explanation describing the evolutionary relationship between clades is the correct one

H. Study of the evolutionary relationships between groups; hypotheses made regarding clades

I. Change in body form from that of a common ancestor

J. Diagram that shows a network of evolutionary relationships

K. Evolutionary group that includes an ancestor and all of its descendants

L. Character present in a group under consideration, but not in the group's ancestors

M. Two different lineages emerging out of one node on a cladogram

Fill-in-the-Blanks [pp.296-297]

In order to understand evolutionary relationships better biologists collect data about species. They are looking for (14) _____ in specific characters. Collecting and analyzing this type of data is referred to as (15) _____. The basic principle is that the (16) _____ number of steps is the most likely pathway to have occurred. Finding the simplest pathway is called (17) _____. A (18) _____ or grid of information is established to compare specific characters. The proposed pattern of evolution creates an evolutionary diagram called a (19) _____.

Organizing Information About Species   **217**

For each of the vertebrate derived traits below, match it to the letter corresponding to its place on the cladogram to the right. Some of traits may have more than one letter.

20. _____ wings

21. _____ limbs with 5 digits

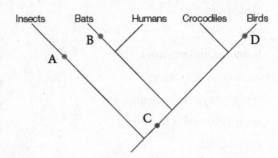

18.4. COMPARING BIOCHEMISTRY [pp.300-301]

18.5. COMPARING PATTERNS OF DEVELOPMENT [pp.302-303]

18.6. APPLICATIONS OF PHYLOGENY RESEARCH [pp.304-305]

Boldfaced, Page-Referenced Terms

molecular clock _____

_____

Fill-in-the-Blanks [pp.302-303]

The development of an (1) _____ into a full plant or animal is layer by layer through (2) _____. An altered body plant can occur whenever there is a failure in (2) _____, usually with drastic results. These genes tend to be highly (3) _____, which means that (4) _____ are not likely to affect them. This gene has been retained, even among (5) _____ that diverged long ago. For example, the majority of vertebrate embryos will develop in similar ways. All will have a stage in which they have four (6) _____ and a (7) _____. The buds form when the (8) _____ gene is expressed. This gene encodes for a signal that tells the cells to (9) _____ and give rise to an (10) _____. The (11) _____ gene helps sculpt the details of the body's form. When (12) _____ is expressed the (13) _____ gene is suppressed.

Answer the following questions based upon the DNA sequences.

Species	DNA Sequence
A	TTATCGTACCGTAT
B	TTATTGTAGCGTAT
C	TTATTGTACCGTAT
D	TTTTTGTACCGAAT

14. Which species are most closely related? _____

15. Which species are the least related? _____

16. Assuming that mutations occur individually and at constant intervals, indicate the evolutionary timeline for the species. _____

Choice [pp.302-303]

Indicate the correct letter next to each choice.

    a. homeotic gene   b. *Dlx* gene   c. *Hox* gene

17. _____ Forms legs, arms and wings of various species

18. _____ Suppressor gene that blocks the formation of appendages

19. _____ A master gene that guides the formation of body parts

20. _____ Apetala 1 gene

21. _____ Encodes for the formation of appendages

22. _____ Reason why pythons do not possess legs

23. _____ Most likely the gene that evolved later

Short Answer [pp.304-305]

24. Explain two ways in which conservation biology can benefit from phylogeny research.

_____

_____

_____

25. How might research into influenza profit from phylogeny research?

_____

_____

_____

SELF-TEST

1. Who established a simple naming system for species in the 18th century?
   a. Darwin
   b. Lyell
   c. Linnaeus
   d. Wallace

2. Evolutionary relationships are referred to as _____.
   a. phylogeny
   b. cladistics
   c. taxons
   d. sister groups

3. A _____ is a diagram that shows evolutionary relationships.
   a. cladogram
   b. clade
   c. monophyletic group
   d. taxon

4. Structures that are similar due to common ancestry are referred to as _____ structures.
   a. analogous
   b. homologous
   c. vestigial
   d. morphological

5. The wing surfaces of birds, bats, and insects are _____ structures.
   a. homologous
   b. vestigial
   c. morphological
   d. analogous

6. Which gene signals embryonic cells to give rise to an appendage?
   a. Hox
   b. Apetala 1
   c. Dlx
   d. Control

7. A _____ is a measurable, heritable trait.
   a. character
   b. clade
   c. sister group
   d. taxon

8. Which type of DNA is least likely to be used in a nucleotide comparison?
   a. Mitochondrial
   b. Nucleic
   c. Chloroplast
   d. Ribosomal

9. _____ is the process of finding the simplest explanation for evolutionary relatedness.
   a. heredity
   b. phylogeny
   c. parsimony
   d. cladistics

10. Change in the body form from an ancestor is called _____.
    a. morphological divergence
    b. morphological convergence
    c. physiological divergence
    d. physiological convergence

11. Scientists can count the number of neutral mutations in DNA and use them as a type of molecular _____.
    a. forensics
    b. timebomb
    c. clock
    d. determination

12. These type of genes are the master genes for development
    a. homeotic
    b. introns
    c. extrons
    d. mastrons

13. Which gene regulates limb development in animals?
    a. Hox
    b. Apetala 1
    c. Dlx
    d. Control

14. Which of the following areas of study are examples of benefits of phylogeny?
    a. Hawaii 'an honeycreepers
    b. African antelope populations
    c. Influenza research
    d. All of the above have benefited

15. All of these EXCEPT _____ are reasons for the losses in honeycreeper populations in Hawai'i.
    a. mosquitoes
    b. dingos
    c. rats
    d. habitat loss

16. A _____ is a group of monophyletic organisms.
    a. clade
    b. taxon
    c. phylogeny
    d. character

CHAPTER OBJECTIVES / REVIEW QUESTIONS

1. Explain the advantages of using cladistics to develop an evolutionary diagram. [pp.296-297]
2. Distinguish between morphological divergence and morphological convergence. Explain how each process results in homologous or analogous structures. [pp.298-299]
3. Compare the functions of the *Hox* gene to the *Dlx* gene. [pp.302-303]
4. Analyze the DNA sequences in Figures 18.6 & 18.7 for similarities and differences between the species. [pp.300-301]
5. Explain what information the tree of life can provide for us. [pp.304-305]

INTEGRATING AND APPLYING KEY CONCEPTS

1. Design a matrix with 10 characters that you feel are important to determine the relatedness of 4 different mammal species. Fill in the matrix and evaluate the relatedness of the species. Based upon your results, construct an evolutionary tree for the species.
2. Consider the various groups that are currently classified under Kingdom Protista. Develop new Protista into several new kingdoms and explain the features necessary for an organism to belong in the new kingdoms.

19

LIFE'S ORIGIN AND EARLY EVOLUTION

INTRODUCTION

As scientists study our planet they continue to find new forms of life in a wide array of habitats. One of the most challenging questions for them to answer is "How did life first arise on Earth"? Some scientists believe that life initially arose from nonliving matter while others believe the earliest forms of life were carried to Earth by a meteor. In this chapter you will take a quick tour through the history of life while examining the key evolutionary events that have occurred throughout the Earth's history. What you should recognize is how each evolutionary event was in response to an environmental change and how that event changed life on the planet.

STUDY STRATEGIES

- Read all of Chapter 19 with the goal of familiarizing yourself with the boldface terms.
- Identify the various environments where scientists look for evidence of early life.
- Describe the Big Bang Theory, development of our solar system, and conditions on the early Earth.
- Explain some of the scientific theories about the origins of life.
- Discuss the possible origins of cells from organic molecules.
- Explain the timeline of the origins of prokaryotic life through eukaryotic organisms.

FOCAL POINT

- Figure 19.6 [p.312] illustrates the evolution of cells.
- Figure 19.13 [pp.318-319] provides a summary of the major concepts of the chapter. You should understand the key events portrayed on the diagram, as well as the general time line of evolution.

INTERACTIVE EXERCISES

19.1. LOOKING FOR LIFE [p.309]

19.2. THE EARLY EARTH [p.310]

19.3. FORMATION OF ORGANIC MONOMERS [p.311]

19.4 FROM POLYMERS TO PROTOCELLS [pp.312-313]

Boldfaced, Page-Referenced Terms

astrobiology _____

ozone layer _____

_____

big bang theory _____

_____

hydrothermal vent _____

_____

iron-sulfur world hypothesis _____

_____

RNA world hypothesis _____

_____

ribozymes _____

_____

protocell _____

_____

Choice [pp.309-313]

For each of the following statements choose the category about the study of the early evolution of life that fits the best.

    a. conditions on the early Earth   b. origin of building blocks of life

    c. origin of protein and metabolism   d. origin of genetic material

    e. origin of the plasma membranes   f. life in extreme environments

1. _____ Proto-cells were membrane enclosed sacs that captured energy, concentrated materials, and replicated themselves

2. _____ Organisms such as *Thermus aquaticus* and *Cyanidium caldarium* are examples showing that life can exist under these conditions

3. _____ There was little or no free Oxygen in the atmosphere

4. _____ Stanley Miller's experiment attempted to replicate this

5. _____ One hypothesis predicts that simple organic compounds may have formed in outer space

6. _____ Early life may have existed as an RNA world

7. _____ The first proteins may have originated on clay rich tidal flats

8. _____ The first biomolecules may have originated near thermal vents

9. _____ Water vapor, carbon dioxide, and nitrogen gas were present with hydrogen gas

10. _____ Simple metabolic pathways evolved at hydrothermal vents

11. _____ A bilayer is the basis of all cell membranes

12. _____ Life can adapt to nearly any environment that has carbon and energy

Short Answer [pp.309-313]

13. Discuss the origins of prokaryotes and the major evolutionary steps during the first 2 billion years of the Earth's history.

_____

_____

_____

14. Explain the origins of metabolism, the genome, and the plasma membrane.

_____

_____

_____

_____

15. What are the three leading theories for the origin of life on Earth?

_____

_____

_____

19.5. LIFE'S EARLY EVOLUTION [pp.314-315]

19.6. HOW DID EUKARYOTIC TRAITS EVOLVE? [pp. 316-317]

19.7. TIME LINE FOR LIFE'S ORIGIN AND EVOLUTION [pp.318-319]

Boldfaced, Page-Referenced Terms

stromatolites _____

_____

biomarkers _____

_____

endosymbiont hypothesis _____

_____

Choice [pp.314-315]

For each of the following statements, choose from one of the two categories.

    a. rise of eukaryotes   b. age of prokaryotes

1. _____ An atmosphere containing oxygen starts to form
2. _____ May have originated near deep-sea thermal vents
3. _____ The red alga *Bangiomorpha pubescens* is an early example
4. _____ The first fossils of these are evident in the Proterozoic
5. _____ Stromatolites are fossilized remains
6. _____ Earliest evidence is in 2.7 billion year old rocks full of lipid biomarkers.
7. _____ Evolution of sexual reproduction and multicellularity.
8. _____ Sponge-like animals arose around 870 million years ago.

Fill-in-the-Blanks [p.315]

Life first arose approximately (9) _____ years ago. As the first (10) _____ and other photosynthetic bacteria flourished, (11) _____ began to accumulate in the atmosphere. An atmosphere enriched with free (12) _____ had three irreversible effects. First, it stopped the further (13) _____ origin of living cells. Second, (14) _____ respiration evolved and in time became the dominant energy-(15) _____ pathway. Third, as oxygen enriched the atmosphere, an (16) _____ formed. This layer blocks much of the sun's (17) _____ from reaching the Earth's surface. This layer enabled life to move onto (18) _____ .

Labeling [pp.318-319]

For each of the following, match each label from the diagram with the correct event in the history of life.

19. _____

20. _____

21. _____

22. _____

23. _____

24. _____

25. _____

26. _____

27. _____

A. Start of aerobic respiration
B. Cyclic pathway of photosynthesis
C. Noncyclic pathway of photosynthesis
D. Origin of prokaryotes
E. Endomembrane systems and nucleus
F. Ancestors of eukaryotes
G. Endosymbiotic origin of the chloroplasts
H. Endosymbiotic origin of the mitochondria
I. Origin of the first protists

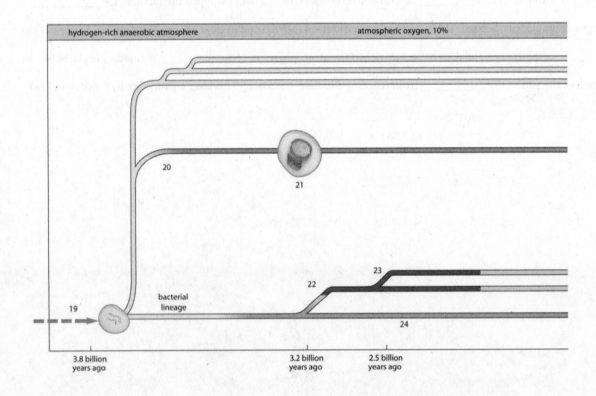

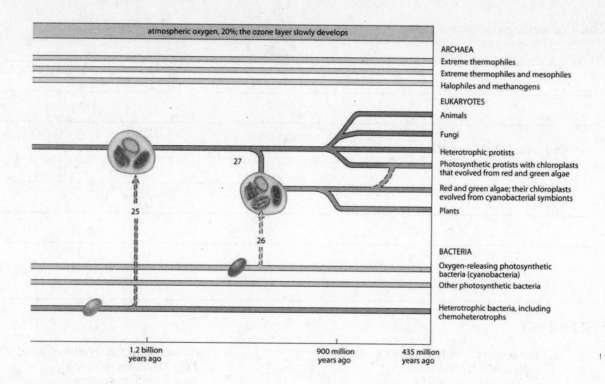

atmospheric oxygen, 20%; the ozone layer slowly develops

ARCHAEA
Extreme thermophiles
Extreme thermophiles and mesophiles
Halophiles and methanogens

EUKARYOTES
Animals

Fungi

Heterotrophic protists
Photosynthetic protists with chloroplasts that evolved from red and green algae

Red and green algae; their chloroplasts evolved from cyanobacterial symbionts

Plants

BACTERIA
Oxygen-releasing photosynthetic bacteria (cyanobacteria)
Other photosynthetic bacteria

Heterotrophic bacteria, including chemoheterotrophs

1.2 billion years ago

900 million years ago

435 million years ago

Choice [pp.316-317]

For the following statements, choose the most appropriate category from the list below.

a. Evolution of mitochondria and chloroplasts

b. Origin of the nucleus and ER

c. Evidence of endosymbiosis

28. _____ Amoeba discoides is a living example of this process

29. _____ Occurred after the noncyclic pathway of photosynthesis changed the atmosphere

30. _____ Photosynthetic cells may have been engulfed by aerobic bacteria

31. _____ May have evolved to protect metabolic machinery from uninvited guests

32. _____ May have originated as infoldings of the plasma membrane

33. _____ This evolved to help protect the cell's hereditary material from foreign DNA

Short Answer [pp.314-319]

34. Describe the timeline of life and the major points in the development of various types of organisms.

_____

_____

_____

Life's Origin and Early Evolution **227**

35. Explain the endosymbiosis theory.

_____

_____

_____

36. What is the likely origin of internal membranes?

_____

_____

_____

_____

SELF-TEST

1. The concept of endosymbiosis is used to explain which of the following?
   a. The origin of Archaebacteria
   b. The evolution of organelles such as mitochondria and chloroplasts
   c. The major trend in the evolution of the animal kingdom
   d. The formation of the first replicating proteins
   e. None of the above

2. When did the first Prokaryotic cells arise?
   a. 2.8 – 2.0 billion years ago
   b. 4.3 – 3.2 billion years ago
   c. 2.0 – 1.6 billion years ago
   d. 1.4 – 1.1 billion years ago

3. Which of the following gases was not present in the atmosphere of the early Earth?
   a. Hydrogen
   b. Carbon dioxide
   c. Oxygen
   d. Methane
   e. All of the above were present

4. The hereditary material of early cells was most likely _____.
   a. DNA
   b. Proteins
   c. Lipids
   d. RNA
   e. Carbohydrates

5. When did the fist Eukaryotic cell arise?
   a. 2.8 – 2.0 billion years ago
   b. 4.3 – 3.2 billion years ago
   c. 2.0 – 1.6 billion years ago
   d. 1.4 – 1.1 billion years ago

6. Stromatolites are_____.
   a. the first eukaryotes
   b. remnants of mitochondria in living cells
   c. an example of the first plasma membrane
   d. fossilized remains of early prokaryotes
   e. none of the above

7. Thermophiles and halophiles are examples of _____.
   a. early prokaryotes
   b. life that exists in extreme environments
   c. endosymbiotic events
   d. organisms that use RNA as hereditary material

8. What substance is essential for life to exist?
   a. Water
   b. Sunlight
   c. Sugar
   d. Oxygen

9. Evidence suggests the Earth is _____ years old.
   a. 4.5 million
   b. 4.5 billion
   c. 6,000
   d. 3.8 billion

10. This theory summarizes the origins of the universe.
   a. cell theory
   b. evolutionary theory
   c. big bang theory
   d. plate tectonics theory

11. Development of the ozone layer allowed for
   a. evolution of life in the ocean
   b. development of animals
   c. reduction in the number of species
   d. evolution of life on land

12. This theory summarizes the origins of mitochondria and chloroplasts.
   a. cell theory
   b. evolutionary theory
   c. big bang theory
   d. endosymbiont theory

13. The three domains of life include _____.
   a. Bacteria
   b. Archaea
   c. Eukaryota
   d. all of the above

14. Photosynthesis evolved approximately _____ ago.
   a. 2.8 – 2.0 billion years ago
   b. 4.3 – 3.2 billion years ago
   c. 3.2 – 2.7 billion years ago
   d. 1.4 – 1.1 billion years ago

15. Mitochondria and chloroplasts evolved approximately _____ ago.
   a. 2.8 – 2.0 billion years ago
   b. 4.3 – 3.2 billion years ago
   c. 3.2 – 2.7 billion years ago
   d. 1.5 – 1.2 billion years ago

16. The first cells must have been _____ as there was little-to-no oxygen present in the early Earth atmosphere.
   a. aerobic
   b. anaerobic
   c. photosynthetic
   d. commensalistic

17. Early life probably converted from RNA to DNA as the genetic material because
   a. DNA allows for larger, more stable material.
   b. RNA is more fragile than DNA.
   c. DNA mutates less often than RNA.
   d. all of the above.

18. Which of the following researchers are not correctly matched with the target of their research?
   a. Wächtershäuser - iron–sulfur world
   b. Crick and Orgel - RNA
   c. Miller and Urey – early atmosphere
   d. Deamer – hydrothermal vents

CHAPTER OBJECTIVES / REVIEW QUESTIONS

1. Explain the concept of the big bang model. [p.310]
2. List the probable chemical constituents of Earth's first atmosphere. [p.310]
3. Give two locations where the abiotic synthesis of organic molecules may have first occurred. [p.311]
4. Explain the probable origins of the agents of metabolism, genome, and plasma membranes. [p.312-313]
5. Explain why researchers believe that early organisms may have used RNA as their hereditary material. [p.312]
6. Explain the significance of stromatolites. [pp.314-315]
7. Explain the environmental changes that occurred due to the accumulation of oxygen in the atmosphere. [p.315]
8. Describe how the endosymbiosis theory may help explain the origin of eukaryotic cells. [p.316]
9. Understand the modern evidence supporting the theory of endosymbiosis. [pp.316-317]
10. Understand the basic timeline of the evolution of life and key events along the time line. [pp.318-319]
11. Explain how studying conditions on other planets can provide clues to how life may have arisen on Earth. [p.320]

INTEGRATING AND APPLYING KEY CONCEPTS

1. As Earth's atmosphere and oceans become increasingly loaded with carbon dioxide and various industrial waste products, how do you think life on Earth will evolve?
2. The evolution of life on this planet has been shaped by the dynamic nature of the Earth's surface. Scientists are currently studying the landscape of other planets in our solar system. What conditions do you think they should be looking for if they want to find a planet that has or had life on it?
3. Endosymbiosis was a major event in the evolution of life on the planet. What do you think life would look like today if endosymbiosis had never occurred?

20

VIRUSES, BACTERIA, AND ARCHAEA

INTRODUCTION

This chapter discusses viruses, bacteria, and the Archaea. The features of each group are discussed as well as the evolution of disease causing agents.

STUDY STRATEGIES

- Read all of Chapter 20 with the goal of familiarizing yourself with the boldface terms.
- Recognize the characteristics of viruses and viroids.
- Describe the lytic and lysogenic pathways of virus replication.
- Identify common viral diseases, their symptoms, and means of prevention.
- Describe the basic prokaryotic structure, function, and diversity.
- Recognize the benefits of many bacteria and identify those that are pathogenic.
- Discuss the diversity of the extremeophile Archaea domain.

FOCAL POINTS

- The introduction [p.323] discusses the impact that HIV/AIDS is having upon the human population.
- Figure 20.2-20.3 [pp.324-335] show the characteristics of viruses
- Table 20.1 [p.326] and Figure 20.5 & 20.6 [pp.326-327] illustrate how viruses replicate.
- Table 20.2 [p.328] lists major viral pathogens and the diseases they cause.
- Table 20.3 [p.332] lists the traits of Bacteria and Archaea.
- Figure 20.13, Figure 20.14 and Figure 20.15 [pp.332-335] illustrate DNA and cell replication in bacteria, and the DNA movement between bacteria that occurs in conjugation.
- Section 20.8 [pp.336-337] provides an overview of the Archaea Domain.

INTERACTIVE EXERCISES

20.1. EVOLUTION OF A DISEASE [p.323]

20.2. VIRUSES AND VIROIDS [pp.324-325]

20.3. VIRAL REPLICATION [pp.326-327]

20.4. VIRUSES AS HUMAN PATHOGENS [pp.328-329]

Boldfaced, Page-Referenced Terms

emerging disease _____

pathogen _____

_____

virus _____

_____

bacteriophages _____

_____

viroids _____

_____

lytic pathway _____

_____

lysogenic pathway _____

_____

reverse transcriptase _____

_____

vector _____

_____

viral reassortment _____

_____

epidemic _____

_____

pandemic _____

_____

Matching [pp324-329]

1. _____ Adenovirus

2. _____ Bacteriophage

3. _____ HIV

4. _____ pathogen

5. _____ Viroid

6. _____ Herpes virus

A. Enveloped DNA virus

B. Disease-causing agent

C. Naked virus that infects animals

D. Infects bacteria or Archaeans

E. Likely evolved from simian immunodeficiency virus

F. Circle of RNA without a protein coat

Short Answer [pp.324-327]

7. List the 3 hypotheses about viral origins. _____

_____

_____

8. List the steps of viral replication. _____

_____

_____

9. Explain the difference between the lytic and lysogenic cycle of viral reproduction. _____

_____

_____

10. Describe the principal features of a virus. _____

_____

_____

Matching [pp.323, 328-329]

Match the following terms with the health concerns.

11. _____ AIDS	A.	Disease that spreads worldwide
12. _____ tuberculosis	B.	Disease that abruptly spreads through a large section of the population in a limited region
13. _____ malaria	C.	Caused by a virus, has lead to 2.7 million deaths per year
14. _____ epidemic	D.	This brief pandemic began in China in 2002–2003 and infected 8000 people
15. _____ pandemic	E.	Caused by a bacteria, has lead to 1.6 million deaths per year
16. _____ emerging disease	F.	Pathogen that has not coevolved with humans
17. _____ influenza	G.	Made up of many different subtypes including H1N1 and H5N1
18. _____ SARS	H.	Caused by a Protists, has lead to 1.3 million deaths per year

Short Answer [pp.328-329]

19. What is the best way to reduce the frequency of disease?

_____

_____

_____

20. What are some symptoms of viral diseases?

_____

_____

_____

Labeling and Matching [pp.326-327]

21. Match the following steps in a retrovirus process with the steps in the figure below.

A. _____ Transcription produces viral RNA.

B. _____ New virus buds from the host cell, with an envelope of host plasma membrane.

C. _____ Viral proteins and viral RNA self-assemble at the host plasma membrane.

D. _____ Other viral RNA forms the new viral genome.

E. _____ Viral protein binds to proteins at the surface of a white blood cell.

F. _____ Some viral RNA is translated to produce viral proteins.

G. _____ Viral RNA and enzymes enter the cell.

H. _____ Viral reverse transcriptase uses viral RNA to make doublestranded viral DNA.

I. _____ Viral DNA enters the nucleus and becomes integrated into the host genome.

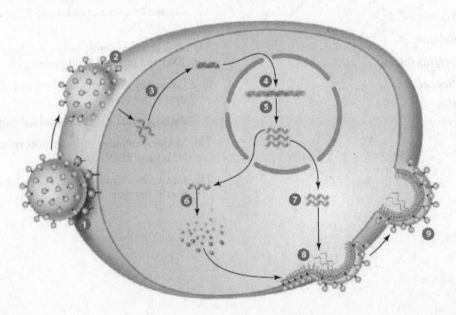

20.5. "PROKARYOTES" – ENDURING, ABUNDANT, & DIVERSE [pp.330-331]

20.6. STRUCTURE AND FUNCTION OF PROKARYOTES [pp.332-333]

Boldfaced, Page-Referenced Terms

prokaryotes _____

_____

Bacteria _____

_____

Archaea _____

_____

metagenomics _____

_____

photoautotrophs _____

_____

chemoautotrophs _____

_____

chemoheterotrophs _____

_____

Matching [p.331]

Match the correct mode of nutrition to the energy and carbon source.

1. _____ Photoautotrophic   A. Light & organic compounds

2. _____ Chemoautotrophic   B. Organic compounds & organic compounds

3. _____ Photoheterotrophic   C. Inorganic substances & carbon dioxide

4. _____ Chemoheterotrophic   D. Light & carbon dioxide

Short Answer [pp.332-333]

5. List the four main features found in Prokaryotic cells. _____

_____

_____

_____

6. Explain the difference between transduction and transformation. _____

_____

_____

_____

7. How do bacteria maintain genetic diversity in their populations? _____

_____

_____

_____

Labeling [p.332]

For each of the following, match each label from the diagram with the correct prokaryotic cell structure.

8. _____   A. Outer capsule

9. _____   B. Cytoplasm

10 _____   C. Nucleiod region

11. _____   D. Cilia

12. _____   E. Plasma membrane

13. _____   F. Cell wall

14. _____   G. Flagellum

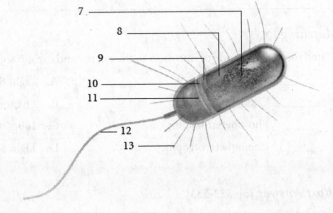

Prokaryotes have an amazing reproductive capacity. One cell can divide into two, these two can then divide into 4, the four can divide into eight, and so on. In some prokaryotes the process may only take (14) _____ minutes. Each descendant cell will contain one (15) _____. The most common method of cell reproduction is called (16) _____. Some prokaryotes can inherit DNA directly from another type; this process is called (17) _____. When a plasmid is exchanged between two prokaryotic cells it is called (18) _____. Having such a variety of reproductive methods has enabled the survival of prokaryotes for billions of years.

20.7. BACTERIAL DIVERSITY [pp.334-335]

20.8. ARCHAEA [pp.336-337]

Boldfaced, Page-Referenced Terms

cyanobacteria _____

_____

nitrogen fixation _____

_____

proteobacteria _____

_____

gram-positive bacteria _____

_____

endospores _____

_____

spirochetes _____

_____

chlamydias _____

_____

extreme thermophile _____

_____

extreme halophile _____

_____

methanogens_____

_____

Choice [pp.336-337]

For the following statements, choose the most appropriate category from the list below.

    a. Archaean   b. Methanogens   c. Extreme halophiles   d. Extreme thermophiles

1. _____ Live in the guts of termites, cattle and other animals
2. _____ Lives in extremely salty environments
3. _____ Strictly anaerobic, free oxygen will kill them
4. _____ Most recent domain that is recognized by the scientific community
5. _____ Contains a purple pigment called bacteriorhodopsim
6. _____ Release carbon-containing gas that is impacting the global carbon cycle
7. _____ Live beside hydrothermal vents
8. _____ Can use aerobic reactions & photosynthesis to produce ATP
9. _____ Among the smallest known cells
9. _____ Group of cells that are the most closely related to eukaryotes

Short Answer [pp.334-337]

10. What are three benefits humanity or science has obtained from bacteria?_____

_____

_____

_____

11. What are some human diseases caused by bacteria?_____

_____

_____

_____

12. In what types of extreme conditions have scientists found the Archaea? _____

_____

_____

_____

SELF-TEST

___ 1. Which of the following is a noncellular particle that consists of protein & nucleic acids?
   a. virus
   b. bacteria
   c. archaean
   d. Protista

___ 2. Which ones are single-celled organisms that do not contain a nucleus?
   a. Eukaryotes
   b. Prokaryotes
   c. Viruses
   d. Protistas

___ 3. In which pathway is the host cell destroyed in order to release the new copies of the virus?
   a. lysogenic
   b. dormant
   c. lytic
   d. reverse transcriptase

___ 4. Which virus will specifically attack bacteria?
   a. adenovirus
   b. HIV
   c. tobacco mosaic
   d. bacteriophage
   e. all of the above

___ 5. Which Prokaryotic cell will use organic compounds for energy?
   a. photoautrophic
   b. chemoautotrophic
   c. chemoheterotrophic
   d. none of the above
   e. all of the above

___ 6. Organisms that breakdown wastes or remains are called _____..
   a. saprobes
   b. photosynthesizers
   c. heterotrophs
   d. autotrophs
   e. none of the above

___ 7. What is the correct term for a bacterial cell that has a round shape?
   a. bacillus
   b. spirillium
   c. helical
   d. coccus
   e. none of the above

___ 8. Which process involves the transfer of a plasmid between two Prokaryotic cells?
   a. conjugation
   b. transduction
   c. transformation
   d. translocation

___ 9. Which is an aquatic cyanobacteria that carries out nitrogen fixation?
   a. *Helicobacter pylori*
   b. *Thermus aquaticus*
   c. *Anabaena*
   d. *Chondromyces crocatus*

___ 10. What term describes an agent that causes disease?
   a. infection
   b. pathogen
   c. virus
   d. bacteria

___ 11. What term is used to describe when a pathogen breaches the body's surface?
   a. infection
   b. disease
   c. pandemic
   d. epidemic

___ 12. Adenoviruses are examples of this type of viral structure.
   a. helical
   b. complex
   c. polyhedral
   d. envelope

___ 13. In which pathway does the host cell make copies of the viral DNA in its own cell process?
   a. lysogenic
   b. dormant
   c. lytic
   d. reverse transcriptase

___ 14. Which of the following viral diseases causes the most deaths worldwide per year?
   a. measels
   b. acute respiratory infections
   c. AIDS
   d. viral diarrhea

___ 15. This is a study currently working to catalog all of the living organisms in/on humans.
   a. Human Genome Project
   b. Human Microbiome Project
   c. Biosphere Project
   d. none of these

Viruses, Bacteria, and Archaea   **239**

16. Which of the following is not a similarity between bacteria and archaea?
    a. lack of a nucleus
    b. semirigid, porous cell wall
    c. peptidoglycan in cell wall
    d. single chromosome

17. Which process involves acquiring of DNA from the environment from a bacteria or archaea?
    a. conjugation
    b. transduction
    c. transformation
    d. translocation

18. The _____ group of bacteria are the most diverse.
    a. proteobacteria
    b. cyanobacteria
    c. nitrogen fixers
    d. heat loving

19. Which bacteria causes Lyme disease?
    a. *Bordetella pertussis*
    b. *Mycobacterium tuberculosis*
    c. *Borrelia burgdorferi*
    d. *Streptococcus aureus*

20. Archaea can be categorized into all of the following groups except _____.
    a. extreme thermophile
    b. extreme halophile
    c. methanogen
    d. Gram-positive

CHAPTER OBJECTIVES / REVIEW QUESTIONS

1. Draw examples of the various types of viruses and explain how they are adapted to infecting a specific host. [pp.324-325]
2. Distinguish between the lytic and lysogenic cycle of viral reproduction. Give examples of diseases that follow each of the cycles. [pp.326-327]
3. Distinguish between chemoautotrophs and photoautotrophs. [pp.330-331]
4. Describe the principal body forms of bacteria (inside and outside). [pp.332-333]
5. Explain how, with no nucleus and very few membrane bound organelles, bacteria are capable of reproduction as well as running metabolism. [pp.332-334]
6. State the ways in which Archaea differ from bacteria. [pp.336-337]

INTEGRATING AND APPLYING KEY CONCEPTS

1. Various types of Archaea produce methane as a waste product of their metabolism. Methane is a combustible gas that could be harvested and used as a potential source of energy. Design a methane production system that could be used as a form of alternate energy. Be sure to identify the limiting factors and come up with realistic solutions to overcome them.
2. What is the most realistic explanation for the origin of various viral pathogens? What concerns should the human species have in regards to the emerging diseases?

21

"PROTISTS"—THE SIMPLEST EUKARYOTES

INTRODUCTION

This chapter introduces you to the most varied group of eukaryotes, the Protista. Their metabolism, structure, and chromosomes are often more closely related to members of other kingdoms than members of their own.

STUDY STRATEGIES

- Read all of Chapter 21 with the goal of familiarizing yourself with the boldface terms.
- Recognize the diversity in structure, function, and life cycles of the Kingdom Protista.
- Describe the various subgroups of the protozoans, including what types of diseases they can cause in humans.
- Discuss the foraminferans and the radiolarians and their role in science and in planetary systems.
- Describe the three groups of ciliates.
- Explain the role that dinoflagellates play in ecosystems, as well as their structure.
- Compare and contrast the life cycles of two amicomplexans: malaria and toxoplasmosis.
- Discuss the structure and life cycles of the stramenopiles.
- Compare and contrast brown, red, and green algae.
- Describe the amoebans and the choanoflagellates.

FOCAL POINTS

- Figure 21.2 [p.342] shows a family tree of the Protista and gives examples of organisms in each.
- Figure 21.3 [p.343] shows the generalized Protist life cycles.
- Table 21.1 [p.343] illustrates the characteristics of protist groups.
- Figures 21.4-21.7 [p.354] illustrate the structure of flagellated Protists.
- Figure 21.14 [p.349] depicts the life cycle of the plasmodia that cause malaria.
- Figure 21.21 [p.353] diagrams the life cycle of the green algae *Chlamydomonas*, which is similar to the life cycle of green plants.
- Figure 21.23 [p.354] illustrates at the slime mold life cycle.

INTERACTIVE EXERCISES

21.1 MALARIA – FROM TUTANKHAMUN TO TODAY [p.351]

21.2. THE MANY PROTIST LINEAGES [pp.342-343]

Boldfaced, Page-Referenced Terms

protists _____

alternation of generations _____

_____

Choice [pp.342-343]

For each of the following terms choose P if it is a Prokaryotic (bacterial) characteristic, E if it is a Eukaryotic characteristic, and B if it is a characteristic of both groups.

1. _____ Multicellular
2. _____ Circular DNA (recall Ch. 20)
3. _____ Can be photoautotrophs (recall Ch. 20)
4. _____ Can develop into a cyst
5. _____ Contain a nucleoid region (recall Ch. 20)
6. _____ Can reproduce sexually and asexually

Fill-in-the-Blanks [pp.342-343]

The vast majority of Protists are (7) _____ celled organisms. Most of them are (8) _____ that live in (9) _____ or (10) _____. Although tiny, they have had a huge impact upon the Earth. (11) _____ Protists contain chloroplasts that enable them to run (12) _____. Some protists live in other organisms and are considered (13) _____. The (14) _____ will determine which mode of energy production that will be used. Protists will reproduce (15) _____ when environmental conditions are favorable and (16) _____ when environmental conditions become less favorable. (17) _____ cells will dominate the life cycle of most Protists while only the (18) _____ is diploid.

21.3. FLAGELLATED PROTOZOANS [pp.344-345]

21.4. FORAMINIFERANS & RADIOLARIANS [p.346]

21.5. THE CILIATES [p.347]

21.6. DINOFLAGELLATES [p.348]

21.7. CELL-DWELLING APICOMPLEXANS [p.349]

21.8. THE STRAMENOPILES [pp.350-351]

Boldfaced, Page-Referenced Terms

flagellated protozoans _____

_____

pellicle _____

_____

hydrogenosomes _____

_____

trypanosomes _____

_____

euglenoids _____

_____

contractile vacuole _____

_____

plankton _____

_____

foraminifera _____

_____

radiolarians _____

_____

alveolates _____

_____

ciliates _____

_____

dinoflagellates _____

_____

algal bloom _____

_____

bioluminescence _____

_____

apicomplexans _____

_____

stramenopiles _____

_____

water molds _____

_____

dinoflagellates _____

_____

brown algae _____

_____

Fill-in-the-Blanks [pp.344-349]

(1) _____ & (2) _____ have multiple flagella and are capable of living in oxygen poor water and will typically lack (3) _____. (4) _____ are long, tapered cells with an undulating membrane. The tsetse fly will act as a (5) _____, helping spread African sleeping sickness from one person to the next. Photosynthetic euglenoids can detect light levels with an (6) _____ that is located near the base of the flagellum. Forams have a (7) _____ shell that makes up their body plan. Many forams form a significant component of the marine (8) _____. Ciliates, dinoflagellates, & apicomplexans are all members of a group known as the (9) _____. (10) _____ are freshwater ciliates that are covered with cilia. Some of the photosynthetic (11) _____ will have an endosymbiotic relationship with corals. Various (12) _____ will infect (parasitize) a variety of animals. The female (13) _____ mosquito acts as a vector and transmits the (14) _____ to a human host.

Matching [p.344]

Match the organism or type of organism to the letter of the second column and place the letter in the answer blank. Then, match each organism or type of organism to as many roman numerals from the second column as are applicable and place them in the parentheses.

15. _____ ( )Euglenoids

16. _____ ( )Foraminiferans

17. _____ ( )*Paramecium*

18. _____ ( )Plasmodium

19. _____ ( )*Giardia lamblia*

20. _____ ( )*Trichomonas vaginalis*

21. _____ ( )*Trypanosoma brucei*

22. _____ ( )Dinoflaellates

A. red tide

B. African sleeping sickness

C. malaria

D. transmitted by contaminated water

E. apicomplexan

F. primary component of many ocean sediments

G. transmitted by biting insects

H. sexually transmitted

I. primarily freshwater

J. uses trichocysts to give it shape

I. photosynthetic

II. flagellated

III. ciliated

IV. hetertrophic

V. parasitic

VI. endosymbiont

Labeling and Matching [p.345]

Match each indicated structure of a *Euglena gracilis* with the corresponding letter in the illustration.

23. _____ flagellum

24. _____ pellicle

25. _____ Golgi body

26. _____ eyespot

27. _____ contractile vacuole

28. _____ mitochondrion

29. _____ endoplasmic reticulum (ER)

30. _____ chloroplast

31. _____ nucleus

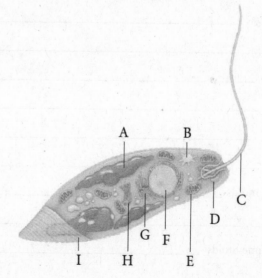

Labeling and Matching [p.347]

Match each indicated structure of a *Paramecium* with the corresponding letter in the illustration.

32. _____ cilia

33. _____ pellicle

34. _____ micronucleus

35. _____ alveolus

36. _____ full food vacuole

37. _____ macronucleus

38. _____ gullet

39. _____ emptying food vacuole

40. _____ trichocysts

41. _____ empty food vacuole

21.9. RED ALGAE [p.352]

21.10. GREEN ALGAE [pp.352-353]

21.11. AMOEBOZOANS AND CHOANOFLAGELLATES [pp.354-355]

Boldfaced, Page-Referenced Terms

red algae _____

_____

chlorophyte algae _____

_____

charophyte algae _____

_____

amoebozoans _____

_____

amoebas _____

_____

cellular slime molds _____

_____

plasmodial slime molds _____

_____

choanoflagellates _____

_____

Labeling and Matching [p.353]

Match each indicated stage of the following life cycle of a *Chladymonas* with the corresponding letter in the illustration. More than one letter may match a stage.

1. _____ thick-walled zygospore develops
2. _____ new zoospores form by mitosis
3. _____ zoospores are released
4. _____ zygote
5. _____ gametes develop, meet up

Match each different reproductive strategy with its appropriate definition:

6. _____ a strategy to survive unfavorable environmental conditions
7. _____ a strategy to maintain successful genetic makeup during consistent environmental conditions

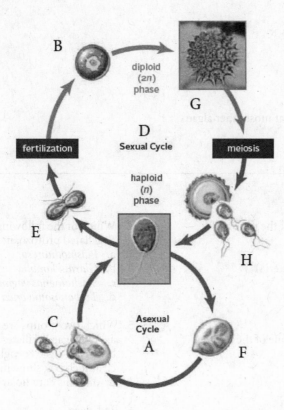

Choice [pp.352-355]

For the following statements, choose the most appropriate category from the list below.

    a. green algae   b. red algae   c. amoeboid cells

8. _____ Chlamydomonas

9. _____ Carrageenan extraction

10. _____ Use pseudopodia for movement

11. _____ Single-cell organisms can cooperate to form a multicelled mass

12. _____ Chorella

13. _____ *Dictyostelium discoideum*

14. _____ Agar is extracted from the cell walls

15. _____ Porphyra

16. _____ Cladophora

17. _____ Contains phycobilins

18. _____ *Entamoeba histolytica*

19. _____ Volvox

20. _____ Ancestor of land plants

21. _____ Survive at deeper depths that most other algae

SELF-TEST

__ 1. Which is not regarded as one of the major protist lineages?
   a. Green algae
   b. Blue-green algae (cyanobacteria)
   c. Red algae
   d. Ciliates, dinoflagellates, & apicomplexans

__ 2. Which of the following is not one of the stramenopiles?
   a. Diatom
   b. Euglenoid
   c. Oomycote
   d. Brown algae

__ 3. Population "blooms" of _____ cause "red tides" and extensive fish kills?
   a. *Euglena*
   b. specific dinoflagellates
   c. diatoms
   d. *Plasmodium*

__ 4. Which of the following is not a parasitic flagellated protozoan?
   a. *Plasmodium sp.*
   b. *Giardia lamblia*
   c. *Trichomonas vaginalis*
   d. *Trypanosoma cruzi*

__ 5. Which two groups are most closely related?
   a. red algae & ciliates
   b. brown algae & euglenoids
   c. amoebas & slime molds
   d. diatoms & radiolarians

__ 6. Red algae _____.
   a. are primarily marine organisms
   b. contain chlorophylls a & b
   c. contain xanthophylls as their main accessory pigment
   d. all of the above

7. Because of their pigmentation, cellulose walls, and starch storage similarities, the _____ algae are thought to be ancestral to more complex plants.
   a. Red
   b. Brown
   c. Blue-green
   d. Green

8. Which group contains calcium carbonate shells?
   a. Foraminiferans
   b. Diatoms
   c. Dinoflagellates
   d. Ciliates

9. Which is not a chlorophyte?
   a. Volvox
   b. Chlamydomonas
   c. *Ulva*
   d. Spirogyra

10. Which of the following is not one of the alveolates?
   a. apicomplexans
   b. dinoflagellates
   c. ciliates
   d. red algae

11. Of the following, which is multicellular?
   a. brown algae
   b. amoeba
   c. diatoms
   d. radiolarians

12. The _____ use hydrogenosomes instead of mitochondria.
   a. oomyocytes
   b. anaerobic flagellates
   c. green algae
   d. choanoflagellates

13. Corals form a mutualistic relationship with _____.
   a. dinoflagellates
   b. foraminiferans
   c. red algae
   d. amoeba

14. Malaria is a disease caused by a _____.
   a. brown algae
   b. ampicomplexan
   c. radiolarian
   d. diatom

15. Which of the following was responsible for the Irish potato famine of the mid-1800s?
   a. *Phytophthora sp.*
   b. *Saprolegnia sp.*
   c. *Karenia brevis*
   d. *Balantidium coli*

16. All of these but _____ is a commercial use of red algae.
   a. agar
   b. carrageenan
   c. algins
   d. nori

17. This type of organism uses pseudopods for locomotion.
   a. dinoflagellates
   b. foraminiferans
   c. red algae
   d. amoeba

18. This type of organism has a very similar structure to a sponge.
   a. oomyocytes
   b. anaerobic flagellates
   c. green algae
   d. choanoflagellates

CHAPTER OBJECTIVES / REVIEW QUESTIONS

1. Two flagellated protozoans that cause human disease and misery are _____ & _____. [pp.344-345]
2. State the principal characteristics of the radiolarians and foraminiferans. Indicate how they generally move from one place to another and how they obtain food. [pp.346-347]
3. List the features common to most ciliated protozoans. [p.347]
4. Explain what causes red tides and how they can impact human health. [pp.348-349]
5. Characterize the apicomplexan group, identify the group's most prominent representative, and describe the life cycle of that organism. [p.349]
6. State the outstanding characteristics of the red, brown, and green algae. [pp.351-353]
7. List the common features of amoebans and choanoflagellates [pp.354-355]

INTEGRATING AND APPLYING KEY CONCEPTS

1. Explain why classifying the Protists is such a difficult process. Which phyla are currently recognized by the scientific community? Indicate the possible evolutionary connection between these phyla and phyla from other Kingdoms.

2. What problems could arise if various algae were introduced into communities that they are not native to?

22

THE LAND PLANTS

INTRODUCTION

This chapter introduces you to the evolutionary relationships that exist between members of the plant kingdom. In order to understand the major divisions of plants we must first examine the evolutionary trends that exist in plants, presented in Section 22.2. An understanding of the general life cycle of a plant [Figure 22.2] is also important. As you progress through the chapter, you should appreciate the diversity and relationship that plants have to the other kingdoms.

STUDY STRATEGIES

- Read all of Chapter 22 with the goal of familiarizing yourself with the boldface terms.
- Recognize the diversity in structure, function, and life cycles of the Kingdom Plantae.
- Identify the major structure and physiological adaptations of plants.
- Describe the characteristics of the bryophytes and their life cycle: liverworts, hornworts, and mosses.
- Discuss the characteristics of the seedless vascular plants and their life cycle: club mosses, whisk ferns, horsetails, and ferns.
- Explain the history of the vascular plants and rise of the seed plants.
- Describe how coal was formed millions of years ago.
- Describe the characteristics of the gymnosperms and their life cycle: conifers, cycads, gnetophytes, and ginko.
- Discuss the characteristics of the angiosperms, their life cycle, and some economically important flowering plants.
- Identify the major structures of a flower.

FOCAL POINTS

- Figure 22.2 [p.360] shows the general life cycle of plants. A general understanding of this diagram will help with the more detailed diagrams later in the chapter.
- Figure 22.25 [p.373] outlines the life cycle of a flowering plant. Of particular importance is the introduction of double fertilization and the formation of endosperm.

INTERACTIVE EXERCISES

22.1. SPEAKING FOR THE TREES [p.359]

22.2. PLANT ANCESTRY & DIVERSITY [pp.360-361]

22.3. EVOLUTIONARY TRENDS AMONG PLANTS [pp.362-363]

Boldfaced, Page-Referenced Terms

plants _____

_____

embryophytes _____

_____

sporophyte _____

_____

gametophyte _____

_____

bryophytes _____

_____

vascular plants _____

_____

seed plants _____

_____

cuticle _____

_____

stomata _____

_____

vascular tissues _____

_____

xylem _____

_____

phloem _____

_____

lignin _____

_____

pollen grain _____

_____

seed _____

_____

Choice [p.368]

Choose the most appropriate description for each of the numbered time frames in the following diagrams.

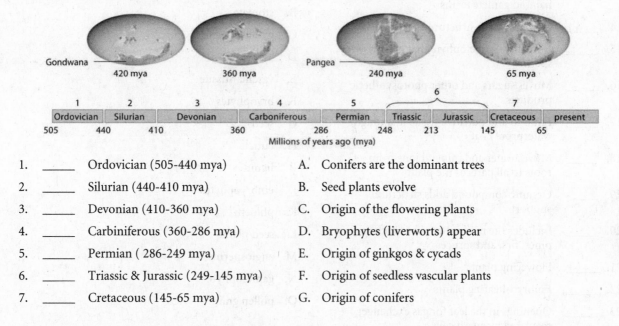

Gondwana — 420 mya   360 mya   Pangea — 240 mya   65 mya

1	2	3	4	5	6	7	
Ordovician	Silurian	Devonian	Carboniferous	Permian	Triassic	Jurassic	Cretaceous	present

505   440   410   360   286   248   213   145   65

Millions of years ago (mya)

1. _____ Ordovician (505-440 mya)   A. Conifers are the dominant trees

2. _____ Silurian (440-410 mya)   B. Seed plants evolve

3. _____ Devonian (410-360 mya)   C. Origin of the flowering plants

4. _____ Carbiniferous (360-286 mya)   D. Bryophytes (liverworts) appear

5. _____ Permian (286-249 mya)   E. Origin of ginkgos & cycads

6. _____ Triassic & Jurassic (249-145 mya) F. Origin of seedless vascular plants

7. _____ Cretaceous (145-65 mya)   G. Origin of conifers

Labeling, Matching, and Sequencing [p.373]

Label each of the following plant groups with the letter of their characteristics from the figure. Then, order the groups from 1-5 in order of evolutionary development.

8. _____ seedless vascular plants   11. _____ ancient algae ancestor

9. _____ gymnosperms   12. _____ bryophytes

10. _____ angiosperms

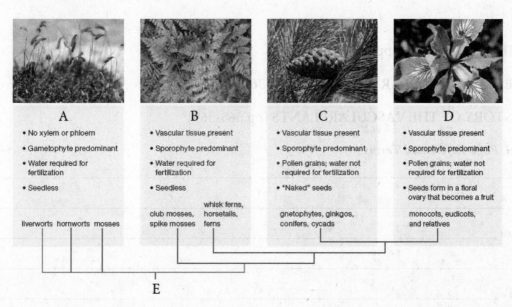

A
- No xylem or phloem
- Gametophyte predominant
- Water required for fertilization
- Seedless

liverworts hornworts mosses

B
- Vascular tissue present
- Sporophyte predominant
- Water required for fertilization
- Seedless

club mosses, spike mosses   whisk ferns, horsetails, ferns

C
- Vascular tissue present
- Sporophyte predominant
- Pollen grains; water not required for fertilization
- "Naked" seeds

gnetophytes, ginkgos, conifers, cycads

D
- Vascular tissue present
- Sporophyte predominant
- Pollen grains; water not required for fertilization
- Seeds form in a floral ovary that becomes a fruit

monocots, eudicots, and relatives

E

Matching [pp.362-363]

Match each of the following statements to the appropriate term.

13. _____ Haploid multicelled body in which the
haploid gamete forms

14 _____ Spore forming structure

15. _____ Immature gametophyte that gives rise to
sperm

16. _____ Moves sugars and other photosynthetic
products

17. _____ Embryo and nutritive tissue in a tough,
waterproof coat

18. _____ Moves water and mineral ions from
roots to all parts of the plant

19. _____ Organic compound adds structural
support

20. _____ Includes the modern conifers such as
pines, firs, and spruces

21. _____ Flowering plants

22. _____ Embryo-bearing plants

23. _____ Openings in the leaf for gas exchange;
regulated by guard cells

24. _____ Protective covering of a leaf

25. _____ Transports nutrients and water;
includes xylem and phloem

26. _____ Vascular plants that hold onto spores
and deliver them via seeds

27. _____ Mosses, hornworts, and liverworts

A. sporophyte

B. stomata

C. gymnosperms

D. xylem

E. vascular tissue

F. bryophytes

G. seed

H. cuticle

I. lignin

J. embryophytes

K. phloem

L. seed plants

M. angiosperms

N. gametophyte

O. pollen grain

22.4. THE BRYOPHYTES [pp.364-365]

22.5. SEEDLESS VASCULAR PLANTS [pp.366-367]

22.6. HISTORY OF THE VASCULARPLANTS [pp.368-369]

Boldfaced, Page-Referenced Terms

rhizoids _____

_____

sporangium _____

_____

gametangia _____

_____

peat bogs _____

_____

rhizome _____

_____

strobilus _____

_____

sori _____

_____

epiphytes _____

_____

pollen sacs _____

_____

microspores _____

_____

ovules _____

_____

megaspores _____

_____

pollination _____

_____

Fill-in-the-Blanks [pp.364-365]

Modern bryophytes contain 24,000 species of (1) _____, (2) _____, and hornworts. None of these (3) _____ plants is taller that twenty centimeters due to the lack of (4) _____. They have leaflike, stemlike, and usually rootlike parts called (5) _____. These are elongated cells or threadlike structures that absorb (6) _____ and nutrients, and anchor the (7) _____ to substrates. All three groups will produce motile (8) _____ that are flagellated and swim through water to the egg. The (9) _____ are attached and generally dependant on the (10) _____. (11) _____ will take place when the sperm and egg unite. This generally leads to the formation of a (12) _____.

Choice [p.364]

Indicate if the step / structure occurs during the (H) haploid or (D) diploid stage of the life cycle.

13. _____ Sporophyte

14. _____ Gametophyte

15. _____ Rhizoids

16. _____ Spores germinate

17. _____ Zygote develops into a sporophyte

18. _____ Rain will disperse the spores

Choice [pp.366-367]

For questions 19-29, choose from the following:

     a. whisk ferns   b. lycophytes   c. horsetails   d. ferns   e. all of the above

19. _____ The members of the genus Equisetum are the surviving species

20. _____ The club mosses belong to this group

21. _____ Seedless vascular plants

22. _____ The genus Psilotum is an example

23. _____ Clusters of sporangia called sori are found on the lower surface of the fronds

24. _____ Stems were used by pioneers of the American West to scrub cooking pots

25. _____ The sporophyte contains xylem and phloem

26. _____ The sporophyte is the predominant phase of the life cycle

27. _____ The genus Lycopodium is an example

28. _____ The young leaves are coiled into the shape of a fiddlehead

29. _____ "Amphibians" of the plant kingdom, life cycle requires water

Labeling [p.367]

Each of the following numbers corresponds to the indicated structure in the diagram. For each item, name the structure and within the parentheses indicate whether it is diploid (2*n*) or haploid (*n*).

30. _____ ( )   33. _____ ( )

31. _____ ( )   34. _____ ( )

32. _____ ( )   35. _____ ( )

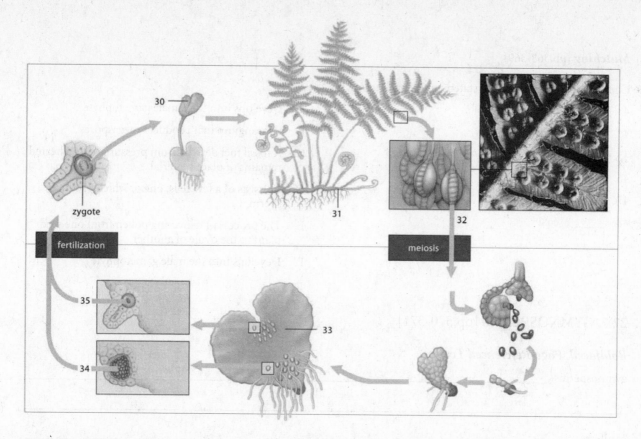

Sequencing [p.364-365]

Using the figure below as a guide, order the following stages of the bryozoan life cycle from 1-7.

36. _____ Spores germinate and develop into male or female gametophytes with gametangia that produce eggs or sperm by mitosis.

37. _____ The zygote grows and develops into a new sporophyte while remaining attached to and nourished by the female gametophyte.

38. _____ Haploid spores form by meiosis in the capsule, are released, and drift with the winds.

39. _____ Sperm swim to eggs.

40. _____ The leafy green part of a moss is the haploid gametophyte.

41. _____ Fertilization produces a zygote.

42. _____ The diploid sporophyte has a stalk and a capsule (sporangium). It is not photosynthetic.

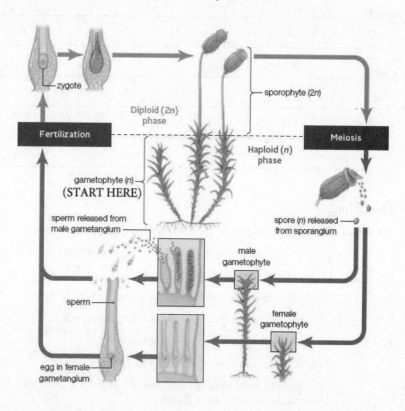

Matching [pp.368-369]

Choose the most appropriate statement for each term.

43. _____ microspore

44. _____ pollen

45. _____ megaspore

46. _____ ovule

47. _____ pollination

48. _____ coal

A. Develops into the female gametophyte

B. Sporangium that produces megaspores

C. A fossil fuel derived from pressurized and heated remains of plant material.

D. Consists of a few cells, one of which produces sperm

E. The process of delivering pollen from one seed plant to the ovule of another

F. Develops into the male gametophyte

22.7. GYMNOSPERMS [pp.370-371]

Boldfaced, Page-Referenced Terms

gymnosperms _____

_____

conifers _____

_____

cycads _____

_____

ginko _____

_____

gnetophytes _____

_____

Choice [pp.370-371]

For the following questions choose the most appropriate category from the following:

    a. cycads   b. ginkgos   c. gnetophytes   d. conifers

1. _____ A deciduous gymnosperm

2. _____ Includes redwoods and bristlecone pines

3. _____ Look like palms or ferns but are not that closely related

4. _____ Only a single species, Ginkgo biloba, survives

5. _____ Includes the genera Welwitschia and Ephedra

6. _____ The favored male trees are now planted; they have attractive, fan-shaped leaves and are resistant to insects, disease, and air pollutants

7. _____ The tallest of the trees belong to this group

8. _____ Most species are "evergreen" with needlelike or scale-like leaves

9. _____ Includes "sago palms"

10. _____ Shaggy-looking group in which some are over 1,000 years old

Labeling [p.371]

Label each of the following steps in the life cycle of a gymnosperm with the appropriate letter.

11. _____ A pollen grain alights on a scale of an ovulate cone. It germinates and a pollen tube grows toward the ovule. Sperm form as the tube grows.

12. _____ Meiosis of cells in ovules yields megaspores (n).

13. _____ Fertilization produces a zygote that develops into a seed.

14. _____ Microspores develop into pollen grains (male gametophytes) that are released and travel on the wind.

15. _____ Scales of pollen cones hold pollen sacs.

16. _____ Scales of ovulate cones contain ovules.

17. _____ The seed germinates and grows into a new sporophyte.

18. _____ Meiosis of cells in pollen sacs yields microspores (n).

19. _____ Megaspores develop into eggbearing female gametophytes inside the ovule.

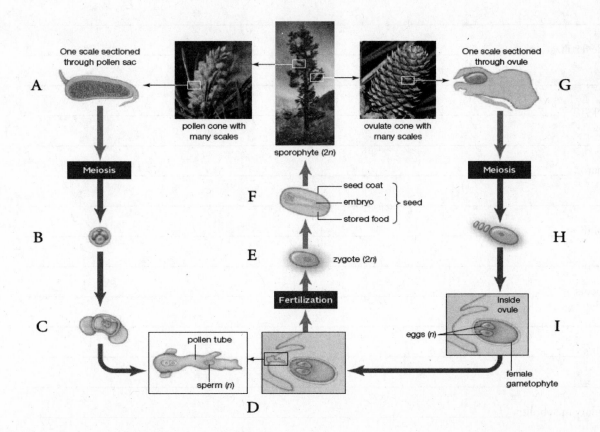

22.8. ANGIOSPERMS—THE FLOWERING PLANTS [pp.372-373]

22.9. ANGIOSPERM DIVERSITY AND IMPORTANCE [pp.374-375]

Boldfaced, Page-Referenced Terms

angiosperms _____

_____

flower _____

_____

stamens _____

_____

carpel _____

_____

ovary _____

_____

fruit _____

_____

double fertilization _____

_____

endosperm _____

_____

pollinator _____

_____

coevolution _____

_____

monocots _____

_____

eudicots _____

_____

secondary metabolite _____

_____

Matching [pp.372-375]

Choose the most appropriate definition for each term.

1. _____ coevolution
2. _____ flowers
3. _____ pollinators
4. _____ examples of magnoliid plants
5. _____ examples of eudicot plants
6. _____ examples of monocots

A. Orchids, palms, lilies, and grasses, including sugarcane, corn, rice, and wheat

B. Specialized reproductive shoots

C. Agents that deliver pollen of one species to female parts of the same species

D. Most herbaceous plants, such as cabbages and dandelions, most flowering shrubs and trees, such as oaks, maples, and fruit trees

E. Avocados and magnolias

F. The evolution of two or more species due to ecological interactions

Labeling and Matching [p.372]

Match the following parts of a flower with the appropriate letter in the figure.

7. _____ filament
8. _____ petal
9. _____ anther
10. _____ carpal
11. _____ receptacle
12. _____ stigma
13. _____ stamen
14. _____ ovule
15. _____ sepal
16. _____ ovary
17. _____ style

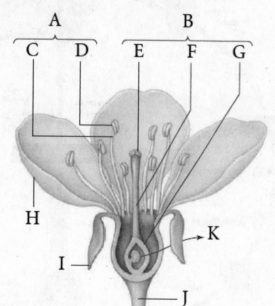

Labeling [p.373]

Label each of the following structures indicated in the diagram of a flowering plant life cycle.

18. _____

19. _____

20. _____

21. _____

22. _____

23. _____

24. _____

25. _____

26. _____

27. _____

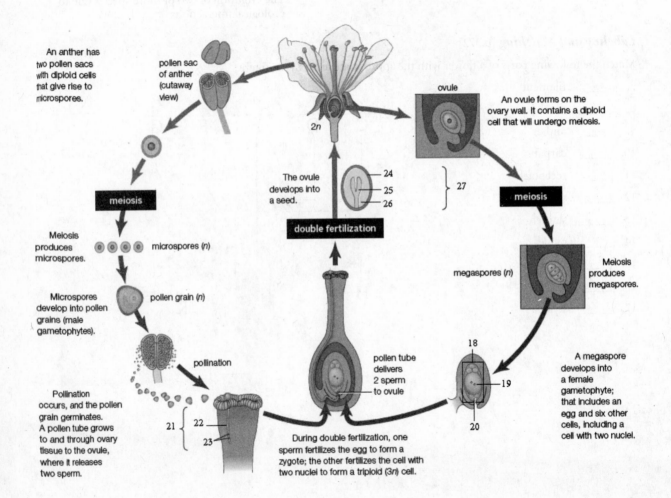

An anther has two pollen sacs with diploid cells that give rise to microspores.

pollen sac of anther (cutaway view)

2n

ovule

An ovule forms on the ovary wall. It contains a diploid cell that will undergo meiosis.

The ovule develops into a seed.

24
25
26

27

meiosis

meiosis

double fertilization

Meiosis produces microspores.

microspores (n)

megaspores (n)

Meiosis produces megaspores.

Microspores develop into pollen grains (male gametophytes).

pollen grain (n)

pollination

pollen tube delivers 2 sperm to ovule

18

19

20

A megaspore develops into a female gametophyte; that includes an egg and six other cells, including a cell with two nuclei.

Pollination occurs, and the pollen grain germinates. A pollen tube grows to and through ovary tissue to the ovule, where it releases two sperm.

21
22
23

During double fertilization, one sperm fertilizes the egg to form a zygote; the other fertilizes the cell with two nuclei to form a triploid (3n) cell.

Fill-in-the-Blanks [pp.374-375]

Quinoa is a leafy (28) _____ that originated in the Andes. It produces highly nutritious (29)

_____. Quinoa seeds contain (30) _____ protein compared to 12% found in wheat and 8% found in

rice. It also contains all of the necessary (31) _____ that humans require. Quinoa is also highly resistant to

(32) _____, (33) _____, and (34) _____ which makes it an easy plant to cultivate.

SELF-TEST

__ 1. The plants around you today are descendants of ancient species of _____.
   a. brown algae
   b. green algae
   c. bryophytes
   d. red algae

__ 2. The _____ move(s) water and mineral ions from the roots to all plant parts.
   a. spores
   b. seeds
   c. flowers
   d. phloem
   e. xylem

__ 3. Which of the following contain vascular tissue?
   a. Horsetails
   b. Mosses
   c. Liverworts
   d. Hornworts

__ 4. The _____ distributes sugars that are made in the photosynthetic tissue.
   a. spores
   b. seeds
   c. flowers
   d. phloem
   e. xylem

__ 5. Bryophytes _____.
   a. have vascular systems that enable them to live on land
   b. include lycophytes, horsetails, and ferns
   c. have true roots but not stems
   d. include mosses, liverworts, and hornworts

__ 6. Which is a seedless vascular plant?
   a. Mosses
   b. Gymnosperms
   c. Angiosperms
   d. Hornworts
   e. Ferns

__ 7. _____ are underground stems found in ferns.
   a. Rhizomes
   b. Rhizoids
   c. Strobilus
   d. Sori

__ 8. Which of the following is an example of a gymnosperm?
   a. Cycad
   b. Conifer
   c. Ginkgo
   d. All of the above

__ 9. Coal deposits are formed primarily from the remains of _____.
   a. angiosperms
   b. bryophytes
   c. green algae
   d. gymnosperms
   e. seedless vascular plants such as club mosses

__ 10. Flowers are the defining characteristics of the _____.
   a. gymnosperms
   b. bryophytes
   c. ferns
   d. angiosperms
   e. all of the above

__ 11. Which of the following is not an angiosperm?
   a. Magnoliid
   b. Cycad
   c. Eudicot
   d. Monocot

12. The _____ is a structural adaptation of plants to gain water and nutrients from the soil.
    a. leaves
    b. roots
    c. stems
    d. flowers

13. This aspect of land plants puts them in a clade together as compared to aquatic photosynthesizers.
    a. embryophytes
    b. chlorophylls
    c. xanthophylls
    d. zygote

14. These plants are representative of those that are vascular and release seeds.
    a. hornworts
    b. club mosses
    c. ferns
    d. conifers

15. The angiosperms evolved during the _____ period.
    a. Carboniferous
    b. Silurian
    c. Cretaceous
    d. Jurassic

16. The most diverse form of byrophytes are the
    a. hornworts
    b. liverworts
    c. mosses
    d. quinoa

17. One of the most important environmental issues affecting forests around the world is
    a. acid rain
    b. deforestation
    c. wildfires
    d. fungal infections

18. The most diverse form of the seedless vascular plants are the
    a. club mosses
    b. ferns
    c. horsetails
    d. whisk ferns

19. Fossils of vascular plants date back to _____ years old
    a. 450 million
    b. 45 million
    c. 1.5 billion
    d. 3.8 billion

20. The _____ of a flower produces pollen (sperm).
    a. ovule
    b. filament
    c. anther
    d. stigma

CHAPTER OBJECTIVES / REVIEW QUESTIONS

1. Define *sporophyte* and *gametophyte*. [p.360]
2. Distinguish between xylem and phloem. [p.362]
3. List the common features shared by all of the bryophytes. [pp.364-365]
4. State the functions of the rhizoids. [p.364]
5. Understand the general life cycle of a bryophyte. [pp.364-365]
6. Explain the function of a rhizome. [p.366]
7. List the four groups of seedless vascular plants. [pp.366-367]
8. Understand the major aspects of a fern life cycle. [p.367]
9. Explain the meaning of the general term *epiphyte*. [p.367]
10. Describe the process by which coal was formed. [p.368]
11. Understand the evolutionary time line of plants. [pp.368-369]
12. Distinguish between microspores and megaspores. [pp.368-369]
13. Explain the differences between pollination and fertilization. [pp.368-369]
14. Describe the key features of conifers, cycads, ginkgos, and gnetophytes. [pp.370-371]
15. Describe the general life cycle of a typical gymnosperm. [p.371]
16. Explain how coevolution relates to pollination. [p.372]
17. Understand the key aspects of an angiosperm life cycle. [p.373]
18. Describe the sequence of events in a typical angiosperm life cycle. [p.373]
19. Name and cite examples of the three major classes of flowering plants. [pp.374-375]
20. Explain the benefits Quinoa can provide to human society. [p.375]

INTEGRATING AND APPLYING KEY CONCEPTS

1. Why is coal considered a nonrenewable resource? Describe the environmental conditions necessary to renew the Earth's supply of coal.

2. Explain the ecological consequences if one species of a coevolved pair were to go extinct? What type of ripple effect can this produce throughout an entire ecosystem?

23
FUNGI

INTRODUCTION

Fungi represent one of the four kingdoms of eukaryotic organisms, yet they are frequently overlooked in the study of biology. In this chapter, you will be introduced to the anatomy, physiology, and classification of the fungi. Typically, the greatest challenge for students is understanding the basic life cycle of fungus with its haploid and diploid states. As you progress through the chapter you should notice the role fungi play in various ecosystems as well as the relationship humans and various fungi have.

STUDY STRATEGIES

- Read all of Chapter 23 with the goal of familiarizing yourself with the boldface terms.
- Recognize the diversity of the Kingdom Fungi.
- Describe the major fungal traits and life cycles.
- Discuss the zygote fungi, their life cycle, and their major characteristics.
- Recognize the roles of microsporidia and glomeromycetes in nature.
- Identify the major characteristics of the ascomycetes, their life cycle, and human uses.
- Describe the major characteristics of the basidiomycetes, their life cycle, and diversity.
- Discuss some of the various partnerships that plants form with other organisms and the types of fungal pathogens that exist in the environment.

FOCAL POINTS

- Figure 23.3 [p. 380] illustrates the life cycle of a typical fungus.
- Figure 23.5 [p.382] shows the life cycle of a zygote fungi (zygomycetes).
- Figure 23.13 [p.386] depicts the general life cycle of a club fungus. Note the similarities to and differences from that of the other two classes of fungi.
- Figure 23.15 [p. 388] illustrates the structure of a lichen and its internal organization

---

INTERACTIVE EXERCISES

23.1. HIGH-FLYING FUNGI [p.379]

23.2. FUNGAL TRAITS AND CLASSIFICATION [pp.380-381]

Boldfaced, Page-Referenced Terms

fungi _____

_____

saprobes _____

_____

mycelium _____

_____

hypha _____

_____

dikaryotic _____

_____

mycosis _____

_____

Matching [pp.380-381]

1. _____ lichen
2. _____ chytrids
3. _____ glomeromycetes
4. _____ fungi
5. _____ mycelium
6. _____ hypha
7. _____ spores

A. Reproductive cells or multicelled structures of fungi

B. Meshes of branched filaments with a high surface area for food absorption

C. Live inside plant roots without harming the plant

D. The filaments in a mycelium; each consists of cells of interconnecting cytoplasm reinforced walls

E. Spore producing heterotroph with chitin cell walls

F. Fungal association with photosynthetic cells

G. Produce flagellated spores

Fill-in-the-Blanks [pp.380-381]

Fungi are spore-producing (8) _____ that contain (9) _____ within their cell wall. Some are single celled like (10) _____. Those that are multicelled grow as a mesh of filaments called the (11) _____. Each filament is called a (12) _____. Fungi will grow over organic matter, secrete (13) _____, and absorb the nutrients. This method of obtaining nutrients classifies them as a (14) _____. Due to this role, fungi play a major part in keeping nutrients (15) _____ in any ecosystem.

Short Answer [pp.380-381]

16. Describe the major characteristics of fungi. _____

_____

_____

17. List the 5 major groups of fungi. _____

_____

23.3. THE FLAGELLATED FUNGI [p.381]

23.4. ZYGOSPORE FUNGI AND RELATED GROUPS [pp.382-383]

Boldfaced, Page-Referenced Terms

chytrids _____

_____

mycosis _____

_____

zygote fungi_____

_____

microsporidia _____

_____

glomeromycetes _____

_____

mycorrhiza _____

_____

Labeling [p.382]

The numbered items in the following illustration (*Rhizopus* life cycle) represent missing information. Indicate the name of each numbered structure or process on the corresponding line. For each structure, indicate in the parentheses whether it is haploid (n) or diploid (2n).

1. _____ ( )

2. _____ ( )

3. _____ ( )

4. _____ ( )

5. _____ ( )

6. _____ ( )

7. _____ ( )

8. _____ ( )

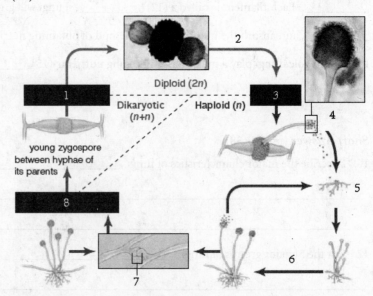

Choice [pp.381-383]

For the following statements chose the correct group;

    a. chytrid   b. zygote   c. microsporidian   d. glomeromycetes

9. _____ Ancient fungal lineage

10. _____ *Nosema ceranae*

11. _____ Pilobus

12. _____ Intracellular parasites

13. _____ Flagellated spores

14. _____ Form zygospore during sexual reproduction

15. _____ Can survive adverse conditions for years.

16. _____ Parasite of amphibians (*Batrachochytrium dendobatidis*)

17. _____ *Rhizopus stolonifer*

18. _____ All involved in root-fungus partnership

23.5. SAC FUNGI- ASCOMYCETES [pp.384-385]

23.6. CLUB FUNGI- BASIDIOMYCETES [pp.386-387]

23.7. FUNGAL PARTNERS AND PATHOGENS [pp.388-389]

Boldfaced, Page-Referenced Terms

sac fungi _____

_____

budding _____

_____

club fungi _____

_____

lichen _____

_____

mutualism _____

_____

Matching [pp.384-389]

Match the correct term with the definition or example.

1. _____ sac fungi   A. Mitotically-produced spores of sac fungus

2. _____ club fungi   B. Fungi that lives in association with tree roots

3. _____ lichen   C. Ascomycota

4. _____ conidia

5. _____ mycorrhiza

6. _____ button mushrooms

7. _____ jelly fungi

8. _____ *Armillaria ostoyae*

D. Mutualistic relationship between fungus and green algae

E. Basidiomycetes

F. The basidiocarps of club fungi

G. A type of club fungus that parasitizes other fungus as they feed on wood

H. The honey mushroom that breaks down wood in the forest.

Labeling [p.386]

The numbered items in the following illustration of the club fungi life cycle represent missing information. Indicate the name of each numbered structure or process on the corresponding line. For each structure, indicate in the parentheses whether it is haploid (n) or diploid (2n).

9. _____

10. _____

11. _____ ( )
_____

12. _____

13. _____ ( )
_____

14. _____ ( )
_____

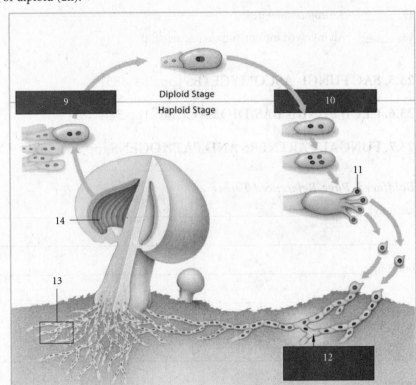

Matching [pp.384-387]

Choose the most appropriate statement for each term.

15. _____ *Saccharomyces cerevisiae*

16. _____ truffles

17. _____ *Candida albicans*

18. _____ *Penicillium*

19. _____ *Aspergillus*

20. _____ *Trichoderma*

A. "Flavor" Camembert and Roquefort cheeses; produce antibiotics

B. Helps prevent rejection of transplanted organs

C. The "death cap" mushroom; often mistaken for a puffball.

D. Causes infections in humans

E. Type of mushroom eaten for its hallucinogenic effects.

21. _____ *Amanita phalloides*
22. _____ psilocybin

F. Highly prized edible fungi
G. Makes citric acid for preservatives and soft drinks; ferments soybeans for soy sauce
H. Baking yeast

Matching [p.388-389]

Choose the most appropriate statement for each of the following terms.

23. _____ ergotism
24. _____ powdery mildew
25. _____ ringworm
26. _____ apple scab; peach leaf curl
27. _____ athlete's foot
28. _____ *Claviceps purpurea*

A. Keratin-dissolving skin infection caused by several different species of fungus
B. A sac fungus that feeds on leaf tissue
C. Examples of crop fungal infections
D. A disease that causes hysteria, convulsions, and hallucinations
E. Fungal infections of the skin characterized by a ring-shaped red spot; highly contagious
F. The species responsible for the disease ergotism

Short Answer [pp.388-389]

29. Describe three types of fungal partnerships in nature. _____

_____

_____

31. List five types of fungal pathogens on plants. _____

_____

_____

31. Describe three types of fungal pathogens on humans. _____

_____

_____

SELF-TEST

___ 1. Most true fungi send out cellular filaments called ____.
   a. mycelia
   b. hyphae
   c. mycorrhizae
   d. asci

___ 2. The cell wall of fungi is composed of _____.
   a. cellulose
   b. starch
   c. chitin
   d. glycogen

3. Fungi are most closely related to which of the following groups?
   a. Animals
   b. Plants
   c. Protista
   d. Bacteria

4. Fungi utilize _____ to breakdown organic material outside of their bodies.
   a. acid
   b. digestive enzymes
   c. lipids
   d. sugars

For questions 5 – 20, choose from the following;
   a. basidiomycetes
   b. chytrids
   c. ascomycetes
   d. zygomycetes
   e. glomeromycetes
   f. microsporidia

5. The only group able to break down lignin in wood.

6. Possible connection to honeybee collapse disorder.

7. The group that includes the commercial button mushrooms.

8. The group that includes *Penicillium*, which has a variety of species that produce penicillin and substances that flavor Camembert and Roquefort cheeses.

9. The group whose spore-producing structure form inside a saclike structure called an ascus.

10. The group that forms a thin, clear covering around the zygote called the zygospore.

11. Plays an important role as a decomposer of forest plants.

12. Parasite of amphibians, may be contributing to the decline in amphibians worldwide.

13. Lives in the intestines of sheep and cattle, breaking down cellulose for their hosts.

14. Used to ferment grains into beer and wine.

15. The group that includes edible morels and truffles as well as bakers' and brewer's yeast.

16. All species in this group function as mycorrhiza.

17. One species acts as a facilitator of lungworm.

18. A predatory fungus of roundworms in the soil; uses a special hyphae as a noose.

19. The group that includes *Rhizopus stolonifer*, the notorious black bread mold.

20. Insects are a major dispersers of a species in this group due to their rotting corpse smell.

CHAPTER OBJECTIVES / REVIEW QUESTIONS

1. Explain the process by which fungi absorb their nutrients. [p.380]
2. Distinguish among the meanings of the following terms: *hypha, spore,* and *mycelium.* [p.380]
3. List the common names for the major groups of fungi. [p.381]
4. Explain the significance of the chytrids, microsporidians, and glomeromycetes. [pp.381-383]
5. Understand the general life cycle of a zygomycete. [pp.382-383]
6. Understand the economic importance of the sac fungi. [pp.384-385]
7. Understand the general life cycle of a club fungus. [pp.386-387]
8. Define mutualism. [p.388]
9. Describe the mycorrhizae relationship. [p.388]
10. Recognize selected disease-causing fungi by name. [pp.388-389]

INTEGRATING AND APPLYING KEY CONCEPTS

1. In the past fungi have been classified with the plants, but are now considered to be more animal-like. List several characteristics of the fungi that would lead scientists to consider them more closely related to the animals than the plants.
2. Explain what could happen to a pond ecosystem if there was an explosion in the chytrid population.
3. Since fungal spores can be dispersed by wind, what are some potential ways to prevent widespread fungal invasions from Africa?

24

ANIMAL EVOLUTION—THE INVERTEBRATES

INTRODUCTION

This chapter is a quick survey of the various invertebrate phyla. Animal traits are reviewed as are the origins of the Kingdom Animalia.

STUDY STRATEGIES

- Read all of Chapter 24 with the goal of familiarizing yourself with the boldface terms.
- Recognize the diversity of the Kingdom Animalia.
- Describe the origins of animal life and the explosion of diversity during the Cambrian.
- Describe the major characteristics and life cycle of the sponges.
- Discuss the cnidarians and their fundamental characteristics.
- Recognize the simple organ systems of the flatworms and their roles in the ecosystem.
- Explain the major characteristics of the annelids and their roles in the ecosystem.
- Describe the mollusks and their diversity.
- Discuss the features of the rotifers.
- Identify the major organs of the roundworms.
- Recognize the key adaptations of the arthropods.
- Describe the major characteristics of the chelicerates, myriapods, and crustations.
- Identify the origins of the insects, their importance, and their characteristic features
- Explain the difference between the protostomes and deuterostomes.
- Discuss the echinoderm characteristics and body plan.

FOCAL POINTS

- Figure 24.2 [p.394] shows two proposed evolutionary trees for animal classification.
- Figures 24.3 [p.395] describe the body symmetry and the different types of body cavities of various animals.
- Each section after that looks at the incredible diversity of invertebrate animals.

INTERACTIVE EXERCISES

24.1. OLD GENES, NEW DRUGS [p.393]

24.2. ANIMAL TRAITS & BODY PLANS [pp.394-395]

24.3. ANIMAL ORIGINS & ADAPTIVE RADIATION [p.396]

Boldfaced, Page-Referenced Terms

invertebrate_____

_____

animals _____

_____

radial symmetry _____

_____

bilateral symmetry _____

_____

cephalization _____

_____

gut_____

_____

coelom _____

_____

pseudocoelom_____

_____

protostomes _____

_____

deuterostomes_____

_____

Fill-in-the-Blanks [p.393]

Cone snails produce a powerful substance, known as (1) _____, that helps them subdue their prey, in

a manner similar to snakes. These peptides shut down the perception of (2) _____ in their victims without

damaging the (3) _____ of the snail. Biologists have studied this peptide compound of the cone snails and

have produced a synthetic painkiller based on it, known generically as (4) _____. The gene that codes for this

protein sequence is also found in (5) _____ and humans but has different functions in each. It is estimated the

gene has been around for at least (6) _____years, when these three groups shared a common ancestor.

Matching [pp.394-396]

Choose the most appropriate description for each term.

7. _____ animals

8. _____ gut

9. _____ ectoderm, endoderm, mesoderm

10. _____ anterior end

11. _____ radial symmetry

12. _____ bilateral symmetry

13. _____ pseudocoel

14. _____ coelom

15. _____ posterior end

16. _____ segmentation

17. _____ cephalization

18. _____ protostomes

19. _____ deuterostomes

20. _____ choanoflagellates

A. An evolutionary process whereby sensory structures and nerve cells became concentrated in a head

B. Digestive sac or tube that opens at the body surface

C. Animal body cavity lined with tissue derived from mesoderm

D. Head end

E. Animals having body parts arranged regularly around a central axis

F. Primary tissue layers that give rise to all adult animal tissues and organs

G. Cavity that is not fully lined by mesoderm

H. Animal lineage in which the 1st opening becomes the mouth

I. Animals having appendages that are paired

J. Series of animal body units that may or may not be similar to one another

K. Tail end

L. Animal lineage in which the 1st opening becomes the anus

M. Multicellular organisms, diploid body cells, hetertorphic, most are motile at some point in their lives

N. Modern protests most closely related to animals

Labeling and Matching [p.394]

Match the indicated linage of the evolutionary tree with its appropriate evolutionary development.

21. _____

22. _____

23. _____

24. _____

25. _____

26. _____

27. _____

28. _____

A. tissues

B. molting

C. bilateral symmetry

D. trocophore larva

E. protostome development

F. multicellularity

G. deuterostome development

H. radial symmetry

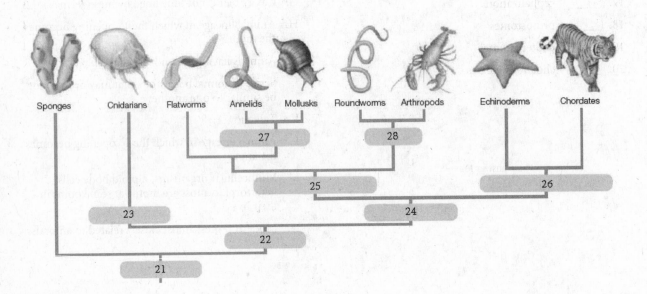

24.4. THE SPONGES [p.397]

24.5. CNIDARIANS—PREDATORS WITH STINGING CELLS [pp.398-399]

Boldfaced, Page-Referenced Terms

sponges _____

_____

sessile animal _____

_____

hermaphrodite _____

_____

larva _____

_____

cnidarians _____

_____

gastrovascular cavity _____

_____

medusa _____

_____

polyp _____

_____

nerve net _____

_____

hydrostatic skeleton _____

_____

Matching [pp.397-399]

Choose the most appropriate answer for each term.

1. _____ medusa
2. _____ larva
3. _____ amoeboid cells
4. _____ sponge skeletal elements
5. _____ *Trichoplax*
6. _____ scyphozoans
7. _____ collar cells

A. Flagellated cells that absorb and move water through a sponge as well as engulf food

B. Distribute food to cells throughout the sponge's body

C. Closest animal relative to choanoflagellates

D. Jellyfish shaped like a bell

E. Group known as true jellyfish

F. Protein fibers and glasslike spicules of silica

G. Free living sexually immature stage

Fill-in-the-Blanks [pp.398-397]

All (8) _____ posses radial symmetry; they include the jellyfishes, sea anemones, corals, and animals

such as *Hydra*. Most of these animals live in a marine environment. (9) _____ are one of the distinguishing

features of the Cnidarians. They are capsules capable of discharging threads that will entangle or pierce prey.

Cnidarians have two common body plans, the (10) _____ looks like a bell or an umbrella while the (11)

_____ has a tube-like body with a tentacle fringed mouth at one end. The saclike cnidarian gut processes food

within the (12) _____. Cells of the (13) _____ will secrete digestive enzymes. The nervous system is

a(n) (14) _____, a simple nervous system to control movement and changes in shape. The (15) _____ is

a layer of gelatinous secreted material that lies between the epidermis and gastrodermis. Some cnidarians will have a

symbiotic relationship with photosynthetic (16) _____. If the coral loses its protist symbiont, this will result in

an event called (17) _____.

Labeling [p.399]

Identify each indicated part of the following illustration.

18. _____

19. _____

20. _____

21. _____

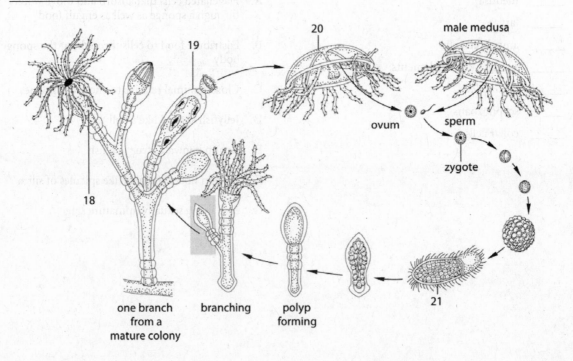

Labeling and Matching [p.397]

Match each indicated part of the following illustration of a sponge with its appropriate structure.

22. _____

23. _____

24. _____

25. _____

26. _____

27. _____

28. _____

29. _____

30. _____

A. semifluid matrix

B. collar of microvilli

C. pore

D. nuclue

E. flagellum

F. glassy structural elements

G. collar cell

H. amoeboid cell

I. flattened surface cells

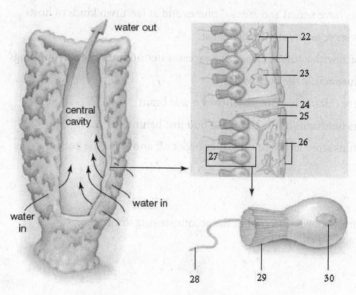

24.6. FLATWORMS—SIMPLE ORGAN SYSTEMS [pp.400-401]

Boldfaced, Page-Referenced Terms

flatworms _____

_____

planarians _____

_____

pharynx_____

_____

ganglion _____

_____

nerve cords _____

_____

Choice [pp.400-401]

For questions 1 – 12, choose from the following. Letters can be used more than once and blanks can have more than one letter.

a. turbellarians   b. flukes   c. tapeworm

1. _____ Parasitic

2. _____ Posses a scolex

3. _____ Ancestral forms probably had a gut but later lost it during their evolution in animal intestines

4. _____ Use a pharynx to capture food

5. _____ Only a few types (including planarians) live in freshwater habitats

6. _____ Their life cycles have sexual and asexual phases and at least two kinds of hosts

7. _____ Aquatic snails serve as an intermediate host

8. _____ Flame cells, each with a tuft of cilia, drive excess water out into the surroundings

9. _____ Nutrients diffuse across the body wall

10. _____ Cluster of nerve cell bodies, ganglia, form a crude brain

11. _____ Proglottids are new units of the body that bud just behind the head

12. _____ Older proglottids store fertilized eggs; they break off and leave the body in feces

Labeling [p.400]

Identify the parts of the animal shown dissected in the accompanying drawings.

13. _____

14. _____

15. _____

16. _____

17. _____

18. _____

Short Answer [p.400]

Answer exercises 19-22 with reference to the drawings accompanying the exercise set above.

19. What is the common name of the animal dissected? _____

20. Is the animal parasitic? _____

21. Is the animal hermaphroditic? _____

22. Name the coelom type exhibited by this animal._____

23. What type of symmetry does this animal have? _____

Labeling [p.401]

The numbered items in the following illustration represent missing information; fill in the corresponding answer blanks to complete the narrative of the life cycle of the beef tapeworm.

24. _____   29. _____

25. _____   30. _____

26. _____   31. _____

27. _____   32. _____

28. _____   33. _____

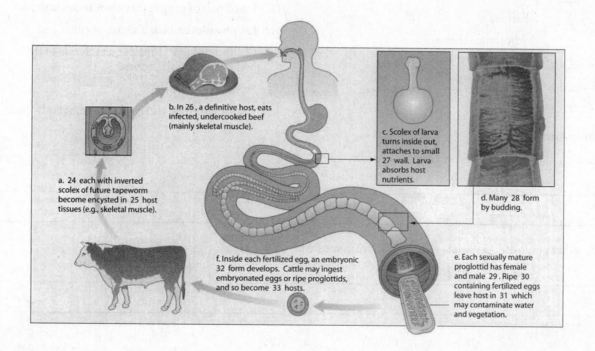

a. 24 each with inverted scolex of future tapeworm become encysted in 25 host tissues (e.g., skeletal muscle).

b. In 26 , a definitive host, eats infected, undercooked beef (mainly skeletal muscle).

c. Scolex of larva turns inside out, attaches to small 27 wall. Larva absorbs host nutrients.

d. Many 28 form by budding.

e. Each sexually mature proglottid has female and male 29. Ripe 30 containing fertilized eggs leave host in 31 which may contaminate water and vegetation.

f. Inside each fertilized egg, an embryonic 32 form develops. Cattle may ingest embryonated eggs or ripe proglottids, and so become 33 hosts.

24.7. ANNELIDS—SEGMENTED WORMS [pp.402-403]

Boldfaced, Page-Referenced Terms

annelids_____

_____

trochophore larva _____

_____

nephridium_____

_____

Matching [pp.402-403]

Choose the appropriate answer for each term.

1. _____ cuticle
2. _____ annelids
3. _____ earthworms
4. _____ marine polychaetes
5. _____ brain
6. _____ nephridia
7. _____ nerve cords
8. _____ chaetae
9. _____ hydrostatic skeleton
10. _____ leeches
11. _____ clitellum

A. Fluid filled coelomic chambers
B. Oligochaete
C. Extensions of nerve cell bodies leading away from the brain
D. Contain parapodia
E. Secretory region that produces mucus
F. Outer layer of secreted protein
G. Fused pair of ganglia that coordinates activities
H. Lack bristles and has a sucker at either end
I. Chitin-reinforced bristles on each side of the body on nearly all segments
J. Segmented worms
K. Regulate volume and composition of body fluid

Short Answer [p.403]

12. Describe how an earthworm burrows into the soil.

_____

_____

13. What purpose does *Hirudo medicinalis* have in medicine?

_____

_____

Labeling [p.403]

Identify each part of the following illustrations of an earthworm's anatomy.

14. _____

15. _____

16. _____

17. _____

18. _____

19. _____

20. _____

21. _____

22. _____

23. _____

24. _____

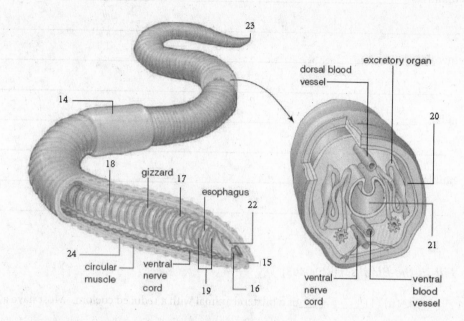

Short Answer [p.406]

Answer items 25-30 with reference to the drawings labeled in the preceding section.

25. Name this animal._____

26. Name this animal's phylum. _____

27. Name two distinguishing characteristics of this group.

_____

28. Is this animal segmented? (yes / no)

29. What type of symmetry does the adult have? _____

30. Does this animal have a true coelom? (yes / no)

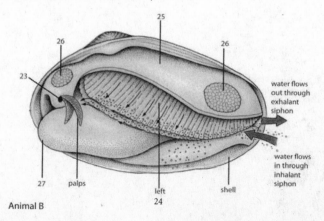

Animal B

24.8. MOLLUSKS—ANIMALS WITH A MANTLE [pp.404-405]

Boldfaced, Page-Referenced Terms

mollusks _____

_____

mantle_____

_____

gills _____

_____

radula _____

_____

Fill-in-the-Blanks [pp.404-405]

A(n) (1) _____ is a bilateral animal with a reduced coelom. Most have a(n) (2) _____ of

calcium carbonate and protein, which were secreted from cells of a tissue that drapes over the body mass. This

tissue, the (3) _____, is unique to mollusks. Special respiratory organs, the (4) _____, contain cilia that

help cause water to flow through the cavity. Most mollusks have a fleshy (5) _____. Many have a(n) (6)

_____, a tongue-like organ hardened with chitin. The mollusks with a well-developed head often posses (7)

_____ and (8) _____. To protect themselves from predators, (9) _____ sometimes incorporate

(10) _____ into their tissues, while cephalopods use (11) _____ to quickly escape predators.

Matching [pp.404-405]

Identify the animals in the following drawings by matching each with the appropriate description.

12. _____ Animal A   I. Bivalve

13. _____ Animal B   II. Cephalopod

14. _____ Animal C   III. Gastropod

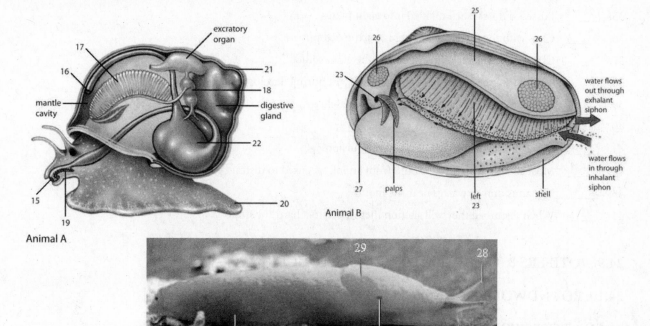

Animal A

Animal B

Animal C

Labeling [pp.404-405]

Identify each numbered part in the preceding drawings by writing its name in the appropriate blank.

15. _____   21. _____   27. _____

16. _____   22. _____   28. _____

17. _____   23. _____   29. _____

18. _____   24. _____   30. _____

19. _____   25. _____   31. _____

20. _____   26. _____   32. _____

Choice [pp.404-405]

For questions 35-46, choose from the following classes of the mollusks.

    a. chitons   b. gastropods   c. bivalves   d. cephalopods

33. _____ Class with the swiftest invertebrates, the squids

34. _____ The "belly foots"

35. _____ Possess a dorsal shell divided into eight plates

36. _____ Class with the smartest invertebrates, the octopuses

37. _____ Class with the largest invertebrates, the giant squids

38. _____ Torsion causes most internal organs to twist during development

39. _____ Includes clams, scallops, oysters, and mussels

40. _____ Most diverse of the mollusks

41. _____ Move rapidly with a system of jet propulsion

42. _____ Most can discharge a dark fluid from an ink sac, used to distract predators

43. _____ The anus dumps wastes near the mouth

44. _____ When disturbed they will suction themselves to a hard substrate

24.9. ROTIFERS & TARDIGRADES—TINY & TOUGH [p.406]

24.10. ROUNDWORMS—UNSEGMENTED WORMS THAT MOLT [p.407]

24.11. ARTHROPODS—ANIMALS WITH JOINTED LEGS [p.408]

Boldfaced, Page-Referenced Terms

rotifers _____

_____

tardigrades _____

_____

roundworms _____

_____

arthropods _____

_____

exoskeleton _____

_____

antennae _____

_____

metamorphosis _____

_____

Matching [pp.406-408]

Match the group with the appropriate term.

1. _____ rotifer
2. _____ tardigrades
3. _____ roundworms
4. _____ *Trichinella spiralis*
5. _____ *Ascaris lumbricoides*
6. _____ Pinworms
7. _____ *Wuchereria bancrofti*

A. Causes lymphatic filariasis

B. Posses cilia on their head

C. Forms cysts within the muscle of pork

D. Common infection in children

E. Commonly called the water bears

F. Bilateral, unsegmented worm with a cuticle

G. A large infection can clog the digestive tract of the host

Short Answer [p.407]

Answer exercises 8-10 for the following drawing of a dissected animal.

8. What is the common name of the dissected animal? _____

9. Is the animal hermaphroditic? _____

10. Name the coelom type exhibited by this animal. _____

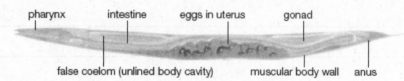

false coelom (unlined body cavity)   muscular body wall   anus

Choice [p.408]

For questions 11-21, choose from the following six adaptations that contributed to the success of arthropods:

    a. hardened exoskeleton   b. highly modified segments   c. jointed appendages

    d. specialized developmental stages   e. specialized sensory structures

11. _____ Tissues get remodeled as the juveniles mature into adults

12. _____ A cuticle of chitin, proteins, and waxes

13. _____ Each stage of the life cycle is specialized for a different task

14. _____ Might have evolved as a defense against predation

15. _____ The immature forms are considered eating machines

16. _____ In the ancestor of insects, different segments became fused into a head, thorax, and an abdomen

17. _____ This is shed in order to provide room for new growth

18. _____ The appendages became modified for specialized tasks

19. _____ The exoskeleton of the legs is modified for walking

20. _____ These structures can detect touch & waterborne chemicals

21. _____ The eyes are compound with many lenses

24.12. CHELICERATES AND THEIR RELATIVES [p.409]

24.13. MYRIAPODS [p.410]

24.14. CRUSTACEANS [pp.410-411]

Boldfaced, Page-Referenced Terms

chelicerates _____

_____

arachnids_____

_____

myriapod _____

_____

crustaceans _____

_____

Dichotomous Choice [pp.409-411]

Circle one of two possible answers given between parentheses in each statement.

1. Nearly all arthropods possess (strong claws / an exoskeleton).

2. (Spiders / Crabs) exchange gases through book lungs.

3. Chelicerae are found in (ticks / barnacles).

4. The largest arthropods are (lobsters and crabs / barnacles and pillbugs).

5. (Barnacles / Copepods) are part of the marine zooplankton.

6. Of all the arthropods, only (copepods / barnacles) have a calcified "shell."

7. Adult (barnacles / copepods) cement themselves to wharf pilings, rocks, and similar surfaces.

8. As is true of other arthropods, crustaceans undergo a series of (rapid feedings / molts) and so shed the exoskeleton during their lifecycle.

9. (Millipedes / Centipedes) have fused segments and each one contains two pairs of legs.

10. Adult (millipedes / centipedes) have a flattened body with a pair of walking legs per segment.

11. (Millipedes / Centipedes) are fast moving, aggressive predators, outfitted with fangs and venom glands.

12. Spiders posses (Malphigian tubules / gills) that filter waste from the body tissues.

13. (Horseshoe crabs / Lobsters) posses a spine-like telson for steering.

Labeling [p.409-410]

Identify each numbered body part in this illustration, then answer question 18.

14. _____

15. _____

16. _____

17. _____

18. Name the subgroup of arthropods to which the animal belongs. _____

Identify each numbered part of the animal pictured at the right.

19. _____

20. _____

21. _____

22. _____

23. _____

24. _____

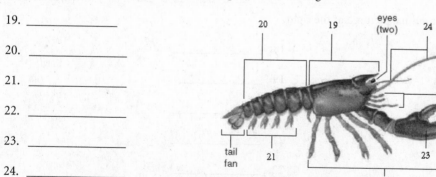

24.15. INSECTS – DIVERSE AND ABUNDANT [pp.412-413]

Boldfaced, Page-Referenced Terms

insects _____

_____

malphigian tubules _____

_____

Matching [pp.412-413]

Choose the most appropriate answer for each term.

1. ___ complete metamorphosis

2. ___ the most successful species of insect

3. ___ incomplete metamorphosis

4. ___ malphigian tubules

5. ___ insect life-cycle stages

6. ___ shared insect adaptations

A. Head, thorax, and abdomen; paired sensory antennae and mouthparts; three pairs of legs and two pairs of wings

B. Involves gradual, partial change from the first immature form until the last molt

C. Winged insects, also the only winged invertebrates

D. Larva-nymph-pupa-adult

E. Structures that help eliminate toxic metabolic wastes

F. Larva grows into a pupa and then reorganizes into the adult form

Matching [pp.412-413]

7. _____ scavenger with a flattened body

8. _____ wingless blood suckers

9. _____ winged and sucks plant juices

10. _____ has prokaryotic and protistan symbionts

11. _____ deadliest animal

A. Mosquito

B. Aphid

C. Lice

D. Termites

E. Earwig

Identify each numbered part of the animal pictured at the right.

12. _____

13. _____

14. _____

15. _____

16. _____

17. _____

18. _____

19. _____

20. _____

21. _____

22. _____

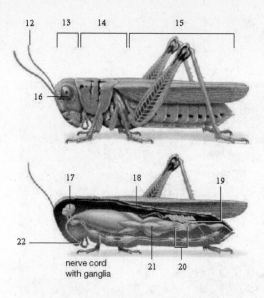

12 13 14 15
16

17 18 19
22
nerve cord
with ganglia
21 20

24.16. THE SPINY-SKINNED ECHINODERMS [pp.414-415]

Boldfaced, Page-Referenced Terms

echinoderms _____

_____

water-vascular system _____

_____

Fill-in-the-Blanks [pp.414-415]

The second lineage of coelomate animals is referred to as the (1) _____. The body wall of all echinoderms will contain protective spines, spicules, or plates made rigid with (2) _____. Oddly, adult echinoderms have (3) _____ symmetry with some bilateral features, but many produce larvae with (4) _____ symmetry. Adult echinoderms have no (5) _____, but a decentralized (6) _____ system allows them to respond to information about food, predators, and so forth. Sea stars feed on mollusks by sliding their (7) _____ into the bivalves shell. They then secrete (8) _____ that will kill the mollusk.

The (9) _____ feet of sea stars are used for walking, burrowing, clinging to rocks, or gripping a meal of clam or snail. These "feet" are part of a(n) (10) _____ system that is unique to echinoderms. Tube feet change shape constantly as (11) _____ action redistributes fluid through the water-vascular system. It is the (12) _____ that will redistribute the fluid among the tube feet. Coarse, indigestible remnants are regurgitated back through the mouth. Their small (13) _____ is of no help in getting rid of empty clam or snail shells. Another echinoderm, the (14) _____, has long pointed spines. Another, the (15) _____, sometimes evades predation by shooting its (16) _____ out of its anus.

Labeling [p.414]

Identify each indicated part of the two following illustrations.

17. _____

18. _____

19. _____

20. _____

21. _____

22. _____

23. _____

Short Answer [p.414]

Answer questions 25 and 26 for the animal just pictured.

24. Name the animal shown. _____

25. Is it a protostome or deuterostome? _____

_____

SELF-TEST

___ 1. Which of the following is true for sponges?
They have _____.
a. distinct cell types
b. symmetry
c. muscles
d. complete digestive systems

___ 2. Bilateral symmetry is characteristic of
_____.
a. cnidarians
b. sponges
c. jellyfish
d. flatworms

___ 3. Flukes and tapeworms are parasitic
_____.
a. leeches
b. flatworms
c. jellyfish
d. roundworms

___ 4. In which group does the mouth appear
before the anus during development?
a. Protostomes
b. Deuterostomes
c. Neither
d. Both groups

___ 5. Which two groups are the closest related?
a. Chordates & echinoderms
b. Chordates & cnidarians
c. Arthropods & echinoderms
d. Mollusks & roundworms

___ 6. Which group of mollusks are the most
intelligent?
a. Gastropods
b. Bivalves
c. Cephalopods
d. Chitons

___ 7. Torsion is a process characteristic of
_____.
a. chitons
b. bivalves
c. gastropods
d. cephalopods
e. echinoderms

___ 8. A complete digestive tract with a mouth
and an anus is not seen in _____.
a. annelids
b. flatworms
c. mollusks
d. roundworms

9. The _____ have bilateral symmetry, cylindrical bodies tapered on both ends, a tough protective cuticle, and a false coelom, and they present the simplest example of a complete digestive system.
   a. roundworms
   b. cnidarians
   c. flatworms
   d. echinoderms

10. What are the hair-like structures found on polychaetes called?
    a. Setae
    b. Parapodia
    c. Cilia
    d. Fimbrae

11. Insects are likely most closely related to the
    a. echinoderms.
    b. crustaceans.
    c. flatworms.
    d. myriapods.

12. One of the most ancient species, the horseshoe crab, belongs to the
    a. echinoderms.
    b. crustaceans.
    c. chelicerates.
    d. myriapods.

13. *Ascaris lumbricoides* currently infects more than 1 billion people and is in the group the
    a. roundworms.
    b. crustaceans.
    c. chelicerates.
    d. myriapods.

14. The tardigrades are missing the following
    a. coelom
    b. respiratory system
    c. digestive system
    d. excretory system

15. The cephalopods include all of the following except _____.
    a. squid
    b. octopus
    c. nautilus
    d. clams

16. These flatworms are parasitic
    a. tapeworm
    b. fluke
    c. planarian
    d. both a and b

17. The _____ use tentacles to capture and sting their prey
    a. mollusks
    b. cnidarians
    c. roundworms
    d. echinoderms

18. Which of the following invertebrate groups is the most diverse?
    a. cnidarians
    b. mollusks
    c. arthropods
    d. echinoderms

19. The roundworms have all of the following except
    a. nervous system
    b. respiratory system
    c. digestive system
    d. excretory system

20. The largest group of invertebrate deuterostomes are the
    a. cnidarians
    b. mollusks
    c. arthropods
    d. echinoderms

Matching

Match each of the following phyla with the corresponding characteristics (letters a-i) and representatives (A-M). A phylum may match with more than one letter from the group of representatives.

21. _____ , _____ Annelids

22. _____ , _____ Arthropods

23. _____ , _____ Cnidarians

24. _____ , _____ Echinoderms

25. _____ , _____ Mollusks

26. _____ , _____ Nematodes

27. _____ , _____ Flatworms

28. _____ , _____ Sponges

29. _____ , _____ Rotifers

a. Choanovytes (collar cells) + spicules

b. Jointed legs + an exoskeleton

c. Pseudocoelomate + wheel organ + soft body

d. Soft body + mantle; may or may not have radula or shell

e. Bilateral symmetry + blind-sac gut

f. Radial symmetry + blind sac-gut; stinging cells

g. Body compartmentalized into repetitive segments; coelom containing nephridia (primitive kidneys)

h. Tube feet + calcium carbonate structures in skin

i. Complete gut + bilateral symmetry = cuticle; includes many parasitic species, some of which are harmful to humans

A. Small animals with a crown of cilia and two "exuding toes"

B. Corals, sea anemones, and *Hydra*

C. Tapeworms and planaria

D. Insects

E. Jellyfish and the Portuguese man-of-war

F. Sand dollars and starfishes

G. Earthworms and leeches

H. Lobsters, shrimp, and crayfish

I. Organisms with spicules and collar cells

J. Scorpions and millipedes

K. Octopuses and oysters

L. Flukes

M. Hookworms, pinworms

30. Explain why Placozoans are classified as animals.

_____

_____

_____

_____

CHAPTER OBJECTIVES / REVIEW QUESTIONS

1. List the six general characteristics that define an "animal." [pp.394-395]
2. Distinguish between radial symmetry and bilateral symmetry, describe the various animal gut types that are found in both groups. [pp.394-395]
3. Describe, by their characteristics, *protostome* and *deuterostome* lineages, and cite examples of animal groups belonging to each lineage. [pp.396, 414]
4. List the characteristics that distinguish sponges from other animal groups. [p.397]
5. State what nematocysts are used for and explain how they function. [pp.398-399]
6. Two cnidarian body types are the _____ and the _____. [pp.398-399]
7. Describe the structure typical of a cnidarian, using terms such as *epithelium, nerve cells, nerve net, mesoglea,* and *hydrostatic skeleton.* [pp.398-399]
8. Define *organ-system* level of construction and relate this to flatworms. [pp.400-401]
9. List the three main types of flatworms and briefly describe each; name groups that are parasitic. [pp. 400-401]
10. Describe the body plan of roundworms, comparing its various systems with those of the flatworm body plan. [pp.402-403, 407]
11. List and generally describe the major groups of mollusks and their members. [pp.404-405]
12. Explain why cephalopods came to have such a well-developed sensory and motor system and are able to learn. [p.405]
13. List four different lineages of arthropods; briefly describe each. [pp.408-413]
14. Name some common types of crustaceans. [pp.410-411]
15. Name the features shared by all insects. [pp.412-413]
16. List five examples of animals known as echinoderms. [pp.414-415]
17. Describe how locomotion and predation occur in sea stars. [pp.414-415]

INTEGRATING AND APPLYING KEY CONCEPTS

1. Most highly evolved invertebrates have bilateral symmetry, a complete gut, a true coelom, and segmented bodies. Why do you suppose that having a true coelom and a segmented body is considered more highly evolved than the condition of lacking a coelom or possessing a false coelom and having an unsegmented body? Cite evidence from the chapter that would support or reject this idea.
2. What features make arthropods so successful?
3. What is the advantage for an organism to go through metamorphosis?

25

ANIMAL EVOLUTION—THE CHORDATES

INTRODUCTION

This chapter looks at trends in the development of vertebrates. The major groups of vertebrates are examined for their unique adaptations.

STUDY STRATEGIES

- Read all of Chapter 25 with the goal of familiarizing yourself with the boldface terms.
- Recognize the diversity in structure, function, and life cycles of the chordates.
- Describe the featured characteristics of the jawless and jawed fishes.
- Explain the theoretical evolution of the modern jaw.
- Discuss the different types of jawed fishes.
- Identify the major characteristics and life history patterns of the amphibians.
- Recognize the importance of the development of the amniote egg to chordate evolutionary history.
- Identify the major characteristics and life history patterns of the nonbird reptils.
- Describe the featured characteristics of the birds.
- Discuss the theoretical development of the mammals and the different mammalian groups.

FOCAL POINTS

- Figure 25.4 [p.421] looks at the family tree of chordates with some of the innovations at branch points.
- Figure 25.8 [p.424] illustrates the possible relationship between jawed vertebrates.
- Figure 25.16 [p.428] shows the family tree of amniotes.
- Figure 25.25 [p.432] shows dispersal of the various types of mammals over land masses over time.

INTERACTIVE EXERCISES

25.1. TRANSITIONS WRITTEN IN STONE [p.419]

25.2. CHORDATE TRAITS AND EVOLUTIONARY TRENDS [pp.420-421]

25.3. JAWLESS FISHES [p.422]

25.4. THE EVOLUTION OF JAWED FISH [p.423]

25.5. MODERN JAWED FISH [pp.424-425]

Boldfaced, Page-Referenced Terms

chordates _____

_____

notochord _____

_____

vertebrates _____

_____

lancelets _____

_____

tunicates _____

_____

endoskeleton _____

_____

tetrapods _____

_____

amniotes _____

_____

fishes _____

_____

ectotherms _____

_____

cartilage _____

_____

scales _____

_____

cartilaginous fishes _____

_____

swim bladder _____

_____

ray-finned _____

_____

lobe-finned _____

_____

Fill-in-the-Blanks [pp.419-425]

One obstacle in Darwin's time to the acceptance of evolution was the lack of fossils of (1) "_____" (2) _____ is one such fossil that bridges the gap between reptiles and birds. (3) _____ dating has dated fossils of this organism to 150 million years ago. It and many similar fossils provide support for (4) _____.

Four major features distinguish the embryos of chordates from those of all other animals: a tubular dorsal (5) _____, a pharynx with (6) _____ in its wall, a(n) (7) _____, and a tail that extends past the anus during some point of the organism's lifetime. The invertebrate chordates are represented today by tunicates and (8) _____, which obtain their food by (9) _____. Lancelets draw in plankton-laden water through the mouth, pass it over mucus coated cells that will trap the particulate matter and direct it towards the (10) _____. The water is then forced out through the gill slits in the pharynx. (11) _____ are among the most primitive of all living chordates. This group derives its name from the carbohydrate rich (12) _____ that covers the adult body. Most will remain attached to rocks or hard substrates in marine habitats after their larvae undergo (13) _____.

In ancestors of the vertebrate line, the notochord was replaced by a bony (14) _____, which was the foundation for fast moving (15) _____. The evolution of (16) _____ started an evolutionary arms race between predators and prey. Fish evolved (17) _____ to help them swim faster as well as paired (18) _____ evolved in early aquatic vertebrates to help enhance the exchange of gases.

As the ancestors of land vertebrates began spending less time immersed in water and more time on land exposed to air, their dependency on gill declined and (19) _____ evolved; more elaborate and efficient (20) _____ systems evolved along with more complex and efficient lungs.

Labeling [p.420]

21. Name the organism in the accompanying diagram. _____

22. Is this an invertebrate of vertebrate chordate? _____

Name the numbered structures in the diagram.

23. _____

24. _____

25. _____

26. _____

27. _____

28. _____

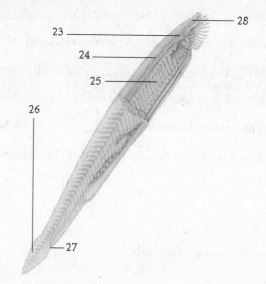

Labeling [p.421]

Label the following characteristics for each lineage on the Chordate family tree.

29. _____

30. _____

31. _____

32. _____

33. _____

34. _____

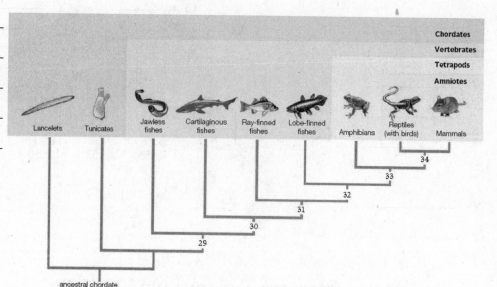

Fill-in-the-Blanks [p.422-425]

Existing jawless fish include the scavenging (35) _____ and the often parasitic (36) _____.

Cartilaginous fishes include about 850 species of rays, skates, and (37) _____. They have a conspicuous fins

and five to seven (38) _____. (39) _____ fish are the most diverse vertebrates on the planet. (40)

_____ have gills and lung-like sacs that are modified outpouchings of the gut wall. Bony fish (osteichthyes)

have a gas filled flotation device called a(n) (41) _____ which helps with buoyancy in water. (42) _____

are the only living representatives of the lobe-finned fish. These lobe-finned fish are most likely the closest living

relatives of modern (43) _____.

Labeling [p.425]

Label the following structures of a bony fish.

44. _____

45. _____

46. _____

47. _____

48. _____

49. _____

50. _____

51. _____

52. _____

53. _____

54. _____

25.6. AMPHIBIANS—FIRST TETRAPODS ON LAND [pp.426-427]

25.7. THE AMNIOTES [p.428]

25.8. NON-BIRD REPTILES [p.429]

Boldfaced, Page-Referenced Terms

amphibian _____

_____

amniote egg _____

_____

reptile _____

_____

dinosaur _____

_____

Fill-in-the-Blanks [pp.426-429]

Natural selection acting on lobe-finned fishes during the Devonian period favored adaptations that enabled fish to survive on (1) _____. These adaptations include a (2) _____, (3) _____, and changes to the (4) _____ and (5) _____. Moving onto land allowed amphibians to escape predators as well as presented them with a new food source, (6) _____.

There are three groups of existing amphibians: (7) _____, frogs and toads, and caecilians, All of which are (8) _____ as adults. Amphibians require free-standing (9) _____ or at least moist habitats to (10) _____. Amphibian skin must also be kept moist since it acts as an additional (11) _____. Many amphibian species are declining worldwide due to (12) _____ activity.

(13) _____ were the first truly land animals. They produced water-conserving (14) _____ with four protective membranes. They also had dry, tough, or scaly skin, which helped prevent (15) _____ from the body. One group of amniotes, the (16) _____, gave rise to the mammals, the (17) _____ gave rise to the crocodilians and eventually modern (18) _____. At the end of the (19) _____, a huge asteroid slammed into Earth and obliterated many amniotes, including the (20) _____ that had been the dominate animals for almost 125 million years.

Matching [p.429]

Match each of the "reptiles" with its characteristics.

21. _____ crocodilians

22. _____ lizards

23. _____ snakes

24. _____ turtles

A. Posses a bony, scale-covered shell

B. The most diverse and numerous "reptiles"

C. Evolved from a short-legged, long bodied lizard

D. Closest living relatives of birds; posses a 4 chambered heart

Animal Evolution–The Chordates   **301**

Labeling [p.429]

Label the following structures of a crocodilian.

25. _____

26. _____

27. _____

28. _____

29. _____

30. _____

31. _____

32. _____

33. _____

34. _____

35. _____

36. _____

37. _____

38. _____

25.9. BIRDS—THE FEATHERED ONES [pp.430-431]

25.10. THE RISE OF MAMMALS [pp.432-433]

25.11. MODERN MAMMALIAN DIVERSITY [pp.434-435]

Boldfaced, Page-Referenced Terms

birds _____

_____

endotherms _____

_____

mammals _____

_____

monotremes _____

_____

marsupials _____

placental mammals _____

_____

placenta _____

_____

Labeling [p.431]

Label the structures pictured at the right within the chicken egg.

1. _____

2. _____

3. _____

4. _____

5. _____

6. _____

7. _____

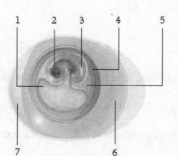

Labeling [pp. 430]

Label the structures pictured at the right.

8. _____

9. _____

10. _____

11. _____

12. _____

13. _____

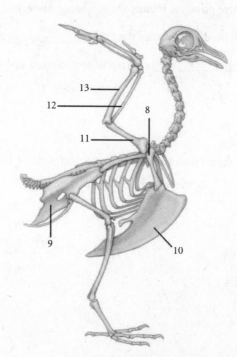

Fill-in-the-Blanks [pp.430-433]

Birds descended from (14) _____ that ran around on two legs during the (15) _____ era. Like their closest relatives, birds have (16) _____ on their legs. All birds have (17) _____ that insulate and help with flight. Generally, birds have a greatly enlarged (18) _____ to which flight muscles are attached. Bird bones contain (19) _____, which decrease weight. They also have lungs with attached airs sacs that increase gas exchange. Flight demands high (20) _____ rates, which requires an abundant supply of (21) _____ being pumped to all parts of the body by means of a large, durable (22) _____ chambered heart. Almost 9,000 species of birds show amazing variation in body structure. The smallest known species of bird weighs only (23) _____ grams. The largest existing bird is the (24) _____, which weighs about 150 kilograms and is a flightless sprinter. One complex behavior seen in birds is (25) _____, a seasonal movement from one region to another. The champion of this behavior is the artic tern, which travels from the Arctic to the (26) _____ and back each year.

Most mammals have (27) _____ as a means of insulation, and (28) _____ with which females will nurse their young. Unlike "reptiles," mammals have 4 different types of (29) _____ in their jaws. The ancestors of today's mammals diverged from small, hairless reptiles called (30) _____ at the same time that the dinosaurs were becoming the dominant species on Earth. Mammals coexisted with the diverse groups of (31) _____ through the Cretaceous. When the dinosaurs became extinct, diverse adaptive zones awaited exploitation by the three principal mammalian lineages: those that lay eggs, also called the (32) _____; those that use a pouch for reproduction called the (33) _____; and those that use a (34) _____ for reproduction. (35) _____ has occurred as evolutionary distinct, geographically isolated lineages evolved similar physical features.

Choice [pp.432-435]

Label each of the numbered mammals with one of the following lettered classifications.

a. monotreme   b. marsupial   c. placental

36. _____ bat
37. _____ human
38. _____ kangaroo
39. _____ koala

40. _____ platypus
41. _____ spiny anteater
42. _____ Tasmanian devil
43. _____ whale

SELF-TEST

___ 1. In true fishes, the gills serve primarily
_____ function.
   a. a gas exchange
   b. a feeding
   c. a water elimination
   d. both a feeding and a gas exchange

___ 2. Jawed fish have paired _____ that help
provide lift while swimming.
   a. fins
   b. scales
   c. gills
   d. kidneys

___ 3. Which of the following is an important
adaptation to life on land?
   a. amniote egg
   b. lungs
   c. 3 chambered heart
   d. all of the above

___ 4. What is believed to be the most likely
reason for the extinction of the dinosaurs?
   a. Increase in the Earth's temperature
   b. K-T asteroid impact
   c. Disease
   d. Competition with mammals
   e. Killed off by Neanderthals

___ 5. Which group is the closest living relative of
birds?
   a. Turtles
   b. Snakes
   c. Dinosaurs
   d. Crocodiles

___ 6. Which of these terms indicates the other
lineage of deuterostomes besides
echinoderms?
   a. proteostomes
   b. chordates
   c. amniotes
   d. marsupials

___ 7. Most chordates are
   a. lancelets
   b. tunicates
   c. vertebrates
   d. birds

___ 8. Fish are characterized by the following
characteristics
   a. aquatic
   b. nontetrapod
   c. gills throughout lifetime
   d. all of these

___ 9. Jawed fishes were the first to have _____
and paired fins.
   a. gills
   b. scales
   c. eyes
   d. notochord

___ 10. Modern jawed fishes can be categorized
into the _____ fishes and the bony
fishes.
   a. cartilagenous
   b. marsupial
   c. reptilian
   d. amphibian

___ 11. Eventually, the lobe-finned fishes gave rise
to the _____.
   a. reptiles
   b. dinosaurs
   c. ray-finned fishes
   d. tetrapods

___ 12. Most reptiles and amphibians have a ____-
chambered heart.
   a. 1
   b. 2
   c. 3
   d. 4

___ 13. While amphibians are mostly terrestrial,
they must _____ in aquatic environments.
   a. reproduce
   b. die
   c. breathe
   d. none of these

___ 14. Birds are _____, meaning they can
regulate their own body temperature.
   a. ectotherms
   b. poikilotherms
   c. endotherms
   d. heterotherms

___ 15. Mammals are characterized by organisms
that _____.
   a. lose gills in infancy
   b. have mammary glands
   c. give birth to live young
   d. lay eggs

Animal Evolution–The Chordates   **305**

Matching [pp.420-435]
Match the following groups and classes with the corresponding characteristics (a-i) and representatives (A-I).

16. ___ , ___ Amphibians

17. ___ , ___ Birds

18. ___ , ___ Bony fishes

19. ___ , ___ Cartilaginous fishes

20. ___ , ___ Jawless fishes

21. ___ , ___ Mammals

22. ___ , ___ "Reptiles"

23. ___ , ___ Urochordates

a. Hair + mammary glands

b. Feathers + hollow bones

c. Jawless + cartilaginous skeleton (in existing species)

d. Two pairs of limbs (usually) + glandular skin + "jelly" covered eggs

e. Amniote eggs + scaly skin + bony skeleton

f. Invertebrate + sessile adult (cannot swim)

g. Jaws + cartilaginous skeleton + vertebrae

h. In adult, notochord stretches from head to tail; mostly burrowed-in, adult can swim

i. Bony skeleton + skin covered with scales, adapted to aquatic environment

A. Lancelet

B. Loons, cardinals, and eagles

C. Tunicates and sea squirts

D. Sharks and manta rays

E. Lampreys

F. Sea horses and groupers

G. Lizards and turtles

H. Caecilians and salamanders

I. Platypuses and opossums

CHAPTER OBJECTIVES / REVIEW QUESTIONS

1. List four characteristics found only in chordates. [p.420]
2. State what sort of changes occurred in the primitive chordate body plan that could have promoted the emergence of vertebrates. [pp.420-421]
3. Describe the differences between primitive and advanced fishes in terms of skeleton, jaws, special senses, and brain. [pp.422-425]
4. Describe the changes that enabled aquatic fishes to give rise to amphibians. [pp.424-428]
5. What key features are present in birds that show their reptilian heritage? [pp.430-431]

INTEGRATING AND APPLYING KEY CONCEPTS

1. Birds and mammals both have four-chambered hearts, high metabolic rates, and can efficiently regulate their body temperatures. Evidence indicates that both of these groups evolved from a reptilian ancestor. Data suggests that reptiles have a heart that is between three and four-chambered, a lower metabolic rate, and are not that efficient at internally regulating their body temperature. If the bird / mammalian traits had evolved in reptiles how might their evolution been different?

26

HUMAN EVOLUTION

INTRODUCTION

This chapter looks at trends in the development of primates and the most recent evidence of human origins.

STUDY STRATEGIES

- Read all of Chapter 26 with the goal of familiarizing yourself with the boldface terms.
- Recognize the diversity and evolution of the primates, including humans.
- Discuss the differences between the anthropoids and the hominoids.
- Compare and contrast the tailless primates.
- Describe what traits separate humans from the other primates.
- Identify the ancestors of *Homo sapiens*.
- Discuss what is meant by "culture" and what evidence exists to tell us about the culture of early human-like species.
- Compare and contrast the multi-regional and replacement models of *Homo sapien* evolution from *Homo erectus*.

FOCAL POINTS

- Sections 26.2-26.4 [pp.440-445] describe the evolution of humans from early primates and our closest relatives, the apes.
- Figure 26.2 [p.440] illustrates an evolutionary tree for modern primates, including humans.
- Figure 26.7 [p.444] looks at the appearance and extinction of various hominid species.

INTERACTIVE EXERCISES

26.1. A BIT OF A NEANDERTHAL [p.439]

26.2. PRIMATES: OUR ORDER [pp. 440-441]

26.3. THE APES [pp.442-443]

26.4. RISE OF THE HOMININS [pp.444-445]

26.5. EARLY HUMANS [pp.446-447]

26.6. RECENT HUMAN LINEAGES [pp.448-449]

Boldfaced, Page-Referenced Terms

primate _____

_____

arthropoid _____

_____

hominoid _____

_____

bipedal _____

_____

hominins _____

_____

australopiths _____

_____

humans _____

_____

culture _____

_____

multiregional model _____

_____

replacement model _____

_____

Fill-in-the-Blanks [pp.440-449]

During the Cenozoic era a group of tree dwelling mammals arose called the (1) _____, the earliest of

which were the (2) _____. The (3) _____ include (4) _____, apes, and (5) _____. The

evolution from early prosimians to hominids required several adaptations. Excellent (6) _____ vision with (7)

_____ facing eyes allowed them to judge depth and distances better. This shift to greater visual abilities

lessened their dependence upon their sense of (8) _____. The (9) _____ and (10) _____

movements of the hands allowed hominids to (11) _____, which eventually led to the ability to make (12)

_____. The foramen magnum was adapted for (13) _____ while their (14) _____ and jaws evolved to a mixed diet. As the (15) _____ expanded, more complex (16) _____ occurred and this eventually lead to the development of (17) _____.

In central (18) _____, 6 or 7 million years ago, hominids were becoming distinct from the apes. During the Miocene and Pliocene, many different "southern apes" or (19) _____ evolved. The earliest forms of genus (20) _____ evolved in the East African Rift Valley. About (21) _____ million years ago, hominids began making crude stone tools. *H. erectus* had a larger (22) _____ and was a more advanced (23) _____ than its ancestors. This group eventually migrated out of Africa into both (24) _____ and (25) _____. By (26) _____ years ago, modern humans, (27) _____ were seen in East Africa and around (28) _____ years ago in the Middle East. As they spread, earlier lineages, like the massively built, cold-adapted (29) _____ of Europe, disappeared.

One debate in human evolution concerns the way in which *H. sapiens* arose. According to the (30) _____ model, modern humans arose in sub-Saharan Africa and migrated out, replacing (31) _____ populations from prior migrations. This is supported by the discovery of the oldest (32) _____ fossils in Africa. The (33) _____ model proposes that the *H. erectus* populations which migrated earlier, independently evolved into (34) _____.

Choice [pp.440-441]

Label each of the numbered primates with one of the following lettered groups.

 a. Prosimians   b. Anthropoids   c. Hominoids

35. _____ Humans
36. _____ Lemur
37. _____ Gibbon
38. _____ Baboon
39. _____ Gorillas
40. _____ Tarsier
41. _____ Orangutan

Labeling [p.440]

Match the following organisms to their appropriate place on the modern primate family tree.

42. _____ humans

43. _____ gorillas

44. _____ tarsiers

45. _____ lemurs, lorises, galagos

46. _____ Old World monkeys

47. _____ orangutans

48. _____ chimpanzees, bonobos

49. _____ New World monkeys

50. _____ gibbons

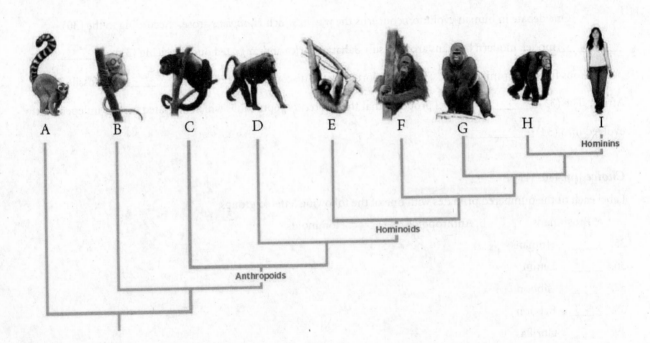

Labeling and Matching [p.443]

Label and match the following characteristics to either the gorilla (G) or the human (H). Use figure 26.5 as a guide.

51. _____ C-shaped curve in backbone

52. _____ flat foot with big opposable toe

53. _____ legs longer than arms, thighs angle inwards

54. _____ legs shorter than arms, thighs angle outwards

55. _____ spinal cord attaches at center of skull

56. _____ S-shaped curve in backbone

57. _____ spinal cord attaches at rear of skull

58. _____ food with arch, big toe aligned with others and non-opposable

Short Answer [pp.448-449]

59. Briefly explain the multiregional model of human evolution.

_____

_____

_____

60. Briefly explain the replacement model of human evolution.

_____

_____

_____

SELF-TEST

___ 1. Primitive primates generally live
    _____.
    a. in tropical forest canopies
    b. in temperate savanna and grassland habitats
    c. near rivers, lakes, and streams in the East African Rift Valley
    d. in caves where there are abundant supplies of insects

___ 2. A hominid in Europe and Near East that became extinct nearly 30,000 years ago was the _____.
    a. dryopith
    b. *Australopithecus*
    c. *Homo erectus*
    d. Neanderthal

___ 3. Which of the following is NOT a key trend that led to human traits?
    a. Enhanced daytime vision
    b. Upright walking
    c. Jaws and teeth adapted to eat meat
    d. Opposable thumbs

___ 4. The oldest, undisputed, bipedal hominid is
    _____.
    a. *Ardipithecus ramidus*
    b. paranthropus
    c. *Homo habilis*
    d. *Australopithecus afarensis*

___ 5. Which hominid was the 1st to migrate out of Africa?
    a. *Homo erectus*
    b. *Homo neanderthalensis*
    c. *Homo sapiens*
    d. *Homo floresiensis*

___ 6. Fossils of *Australopithecus* are believed to be in the lineage of humans. Where were their fossils found?
    a. Africa
    b. Asia
    c. North America
    d. Australia

___ 7. Which of these species is the oldest?
    a. *Homo erectus*
    b. *Ardipithecus kadabba*
    c. *Homo sapiens*
    d. *Australopithecus anamensis*

___ 8. Which of these species is most closely related to humans?
    a. gibbon
    b. bonobo
    c. lemur
    d. gorilla

9. About how long ago did humans first
   appear in the fossil record?
   a. 2.3 million years
   b. 200,000 years
   c. 1.8 million years
   d. 4 million years

10. How long ago did our most recent
    common ancestor live in Africa?
    a. 60,000 years
    b. 200,000 years
    c. 1.8 million years
    d. 4 million years

11. Which of the following is a difference
    between *Homo sapiens* and *Homo erectus*?
    a. higher, rounder skull
    b. protruding jaw
    c. smaller jawbones
    d. flatter face
    e. all of the above

12. The name *Homo habilis* is roughly
    translated as
    a. hairy ape
    b. caveman habitat
    c. handy man
    d. apelike human

13. Primates have
    a. touch-sensitive pads on their fingertips
    b. eyes set in front of skull
    c. shoulders with extensive range of
       motion
    d. grasping hands and feet
    e. all of the above.

14. The ancestor of humans and
    chimpanzees/bonobos diverged
    approximately _____ ago.
    a. 2.3 million years
    b. 8 million years
    c. 1.8 million years
    d. 4 million years

15. Humans differ from other primates in that
    they
    a. walk upright.
    b. are relatively hairless.
    c. cannot grasp with their feet.
    d. have inflexible hands.
    e. a, b, and c are all true.

Matching [pp.440-449]

Choose the one most appropriate answer for each term.

16. _____ anthropoids

17. _____ australopiths

18. _____ *Homo habilis*

19. _____ hominids

20. _____ hominoids

21. _____ *Homo erectus*

22. _____ primates

23. _____ prosimians

A. A group that includes apes and humans

B. Organisms in a suborder that includes New
   World and Old World monkeys, apes, and
   humans

C. One of the earliest members of genus Homo;
   found in Africa

D. A group that includes humans and human-like
   species

E. The first member of the genus Homo to leave
   Africa and migrate to the far corners of Eurasia

F. Organisms in a suborder that includes tarsiers,
   lemurs, and others

G. A group that includes prosimians and
   anthropoids

H. Bipedal organisms living from about 4 million to
   1 million years ago, petite with a narrow jaw and
   small teeth

CHAPTER OBJECTIVES / REVIEW QUESTIONS

1. List the five key adaptations of primate evolution. [pp.440-443]
2. State which anatomical features underwent the greatest changes along the evolutionary line from early anthropoids to humans. [pp.440-445]
3. Explain how you think *Homo sapiens* arose. Use the evidence presented throughout the chapter to support your explanation. [pp.444-449]

INTEGRATING AND APPLYING KEY CONCEPTS

1. Suppose someone told you that between 12 and 6 million years ago various populations of dryopiths were forced by predators to flee from their native forests and take up refuge along the banks of rivers, estuarine, and costal habitats. Over time and through various mutations they eventually lost their body hair, became bipedal, developed subcutaneous fat deposits as insulation, and developed a bridged nose that provided an advantage in a watery environment. The dryopith populations that did not flee to the safety of the watery habitats remained isolated and never developed these particular features. Over time the dryopiths that remained inland were killed by predators and their population became extinct. The water dwelling populations continued to adapt and successfully expand their population by learning to access the variety of available food sources (shellfish, fish, wild rice, oats, and various nuts, fruits, and tubers. In these aquatic habitats the first food gathering tools were developed. In order to be effective at food collecting some early form of communication was also established. How does this story fit with current speculations about the evolution of early humans? What evidence could be used to support or negate this story?

27

PLANT TISSUES

INTRODUCTION

This chapter examines simple and complex plant tissues and the specific types of cells that comprise those tissues. It also addresses the way these tissues are arranged to form plant organs, how the structure and function of these tissues and organs are interrelated, and how the organs combine to form a unified, functional plant body.

STUDY STRATEGIES

- Read all of chapter 27 with the goal of familiarizing yourself with the boldface terms
- Compare and contrast monocots and eudicots.
- Identify the major types of plant tissues including ground, dermal, meristem, and vascular.
- Recognize the following specialized tissues: parenchyma, collenchyma, sclerenchyma, epidermis, periderm, xylem, and phloem.
- Identify the anatomy of a primary shoot and how that translates into primary growth.
- Recognize the various types of leaves.
- Describe the primary leaf structure, including the roles of the epidermis, mesophyll, and veins.
- Identify the anatomy of a primary root and how that translates into primary growth.
- Compare and contrast primary and secondary growth, including where secondary growth takes place within the plant.
- Describe the structure of bark and its importance to woody plants.
- Compare and contrast heartwood versus softwood.
- Explain how dendochronologists can use tree rings to better understand climate.
- Describe some of the specializations of stems.

FOCAL POINTS

- Figure 27.2 *(animated)* [p.455] illustrates the basic structure of a flowering plant.
- Table 27.1 [p.454] compares structural differences between Eudicots and Monocots.
- Table 27.2 [p.456] summarizes the components of plant tissues and their functions.
- Figure 27.6 [p.458] compares the structure of monocot and eudicot stems.
- Figure 27.10 *(animated)* [p.461] is an excellent review of the tissue structure of the leaf and the movement of materials into and out of it.
- Figure 27.13 *(animated)* [p.462] details the production and organization of mature tissues in plant roots.
- Figure 27.14 *(animated)* [p.463] illustrates primary growth in a root.
- Figures 27.15 and 27.16 *(animated)* [pp.464-465] provide a comprehension of how lateral meristems function in the secondary growth of woody plants.

INTERACTIVE EXERCISES

27.1. SEQUESTERING CARBON [p.453]

27.2. THE PLANT BODY [pp.454-455]

Boldfaced Terms

ground tissue _____

_____

vascular tissue _____

_____

dermal tissue _____

_____

cotyledon _____

_____

meristem _____

_____

primary growth _____

_____

secondary growth _____

_____

Labeling [p.455]

Identify the numbered parts of the illustration. Choose from the following:

dermal tissues   roots   ground tissues   shoots   vascular tissues

1. _____

2. _____

3. _____

4. _____

5. _____

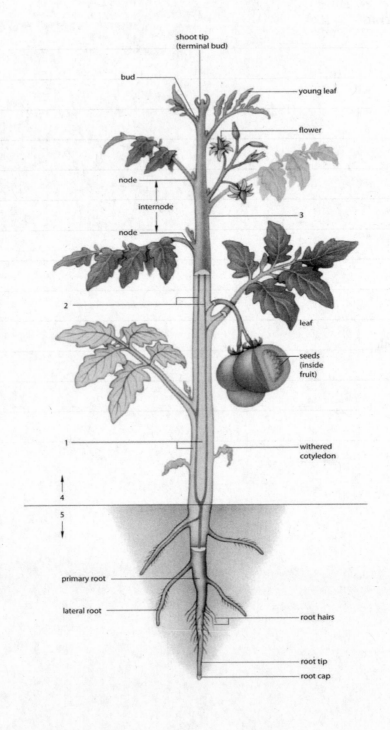

Matching [p.454]

Choose the most appropriate answer for each term.

6. _____ secondary growth

7. _____ cotyledons

8. _____ meristems

9. _____ apical meristems

10. _____ primary growth

11. _____ dermal tissue

12. _____ Monocot

13. _____ Eudicot

14. _____ ground tissue

15. _____ vascular tissue

16. _____ carbon sequestration

A. A lengthening of stems and roots originating from cell divisions at apical meristems

B. Localized regions of dividing cells

C. A subgroup of flowering plants; flower parts in 3s and multiples of 3; parallel leaf veins

D. Covers and protects plant surfaces

E. A thickening of stems and roots caused by activity of the lateral meristems

F. Located in the dome-shaped tips of all shoots and roots; responsible for lengthening those organs

G. A subgroup of flowering plants; flower parts in 4's and 5's or multiples thereof; net-veined leaves

H. "Seed leaves" that contain food for the growing embryo

I. Distributes water and nutrients throughout the plant body

J. Bulk of plant tissue including photosynthetic and storage tissues

K. Any process that removes CO_2 from the atmosphere

Labeling [p.455]

It is important to understand the terms that identify the thin sections (slices) of plant organs and tissues that are prepared for study. Label each of the following three diagrams. Choose from radial section (cut along the radius of the organ), transverse or cross section (cut perpendicular to the long axis of the organ), or tangential section (cut made at a right angle to the radius of the organ). Note: The dark area indicates the slice.

17._____ section   18. _____ section   19. _____ section

27.3. PLANT TISSUES [pp.456-457]

Boldfaced Terms

parenchyma _____

_____

mesophyll _____

_____

collenchyma _____

_____

sclerenchyma _____

_____

xylem _____

_____

tracheid _____

_____

vessel member _____

_____

phloem _____

_____

sieve-tube member _____

_____

companion cells _____

_____

epidermis _____

_____

Choice [p.456]

For questions 1 -3, state whether the tissue is *Simple* or *Complex.*

1. Parenchyma _____

2. Xylem _____

3. Epidermis _____

Choice [pp.456-457]

For questions 4-13, choose from the following Simple Tissues:

      a. parenchyma   b. collenchyma   c. sclerenchyma

4. _____ This tissue makes up the bulk of soft primary growth of plant organs

5. _____ The cells are dead at maturity and have thick, lignin-impregnated walls

6. _____ Cells are alive at maturity and can continue to divide

7. _____ Some types specialize in storage, secretion, and repair of wounds

8. _____ Cells are elongated, alive at maturity, and contain pectin in their cell walls

9. _____ Stretchable tissue, flexible support for rapidly growing plant parts

10. _____ Supports mature vascular tissues and strengthens seed coats

11. _____ Cells are thin-walled, pliable, and many-sided

12. _____ This type of cell, when specialized for photosynthesis, is called mesophyll

13. _____ Fibers and sclereids belong to this type of simple tissue

Choice [pp.456-457]

(a) Choose the cell type that matches the numbered description: companion cell, vessel member, tracheid, sieve tube member

(b) State whether the cell is **dead** or **alive** at maturity

(c) List the cell as belonging to the complex tissue of **Xylem** or **Phloem**

14. Water-conducting tube; cells tapered at ends, have pits in sides

(a) _____

(b) _____

(c) _____

15. Stacks of cells connect by perforated plates to move sugars throughout the plant

(a) _____

(b) _____

(c) _____

16. Parenchyma cells that load sugars into #15

(a) _____

(b) _____

(c) _____

17. Cylindrical cells stacked end to end, cell walls impregnated with lignin

(a) _____

(b) _____

(c) _____

27.4. PRIMARY SHOOT STRUCTURE [pp.458-459]

Boldfaced Terms

terminal bud _____

_____

lateral bud _____

_____

vascular bundle _____

_____

node _____

_____

Labeling [p.458]

Name the structures numbered in the following illustrations of the stem. Choose from these terms:

parenchyma, sclerenchyma   vessel   vascular bundle   sieve tube   pith   cortex   epidermis

1. _____

2. _____

3. _____

4. _____

5. _____

6. _____

7. _____

8. _____

9. _____ Is figure A a cross-section of a monocot or a eudicot stem?

10. _____ Is figure B a cross-section of a monocot or a eudicot stem?

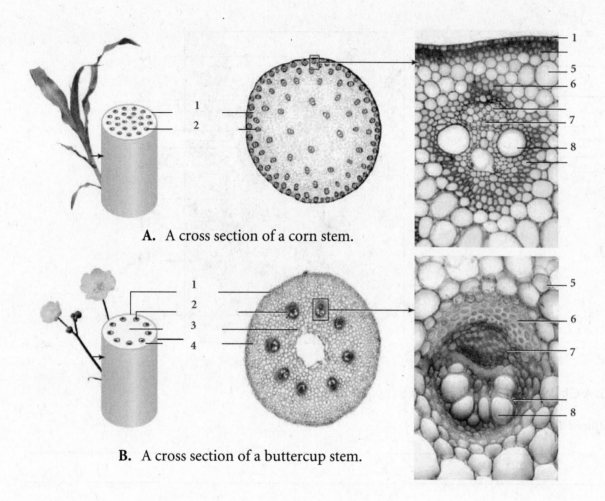

A. A cross section of a corn stem.

B. A cross section of a buttercup stem.

Name the structures numbered in the following illustration of the embryo. Choose from these terms:

procambium,   protoderm,   ground meristem,   immature leaf,   apical meristem

11. _____

12. _____

13. _____

14. _____

15. _____

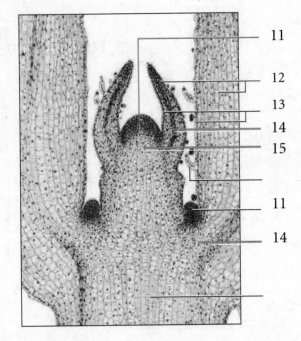

27.5. A CLOSER LOOK AT LEAVES [pp.460-461]

Boldfaced Terms

vein _____

_____

Labeling [p.460]

Name the structures numbered in the illustrations below.

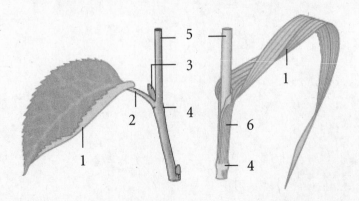

1. _____   3. _____   5. _____

2. _____   4. _____   6. _____

Dichotomous Choice [p.460]

Circle one of the two possible answers given between parentheses in each statement.

7. The leaf illustrated on the left in the preceding section is a (monocot/eudicot).

8. The leaf illustrated on the right in the preceding section is a (monocot/eudicot).

Labeling [p.461]

Identify each numbered part of the accompanying illustration.

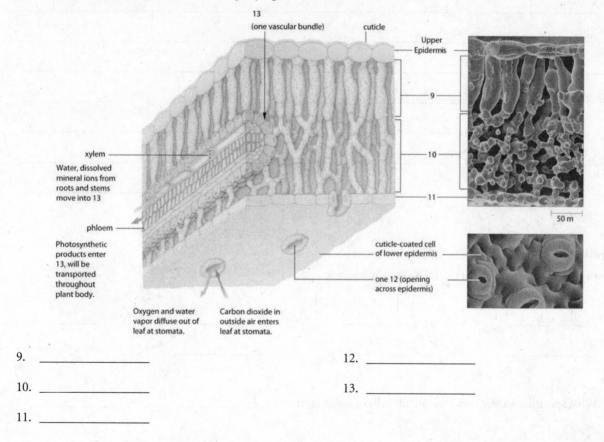

9. _____   12. _____

10. _____   13. _____

11. _____

27.6. PRIMARY ROOT STRUCTURE [pp.462-463]

Boldfaced Terms

root hairs _____

_____

vascular cylinder _____

_____

taproot system _____

_____

fibrous root system _____

_____

endodermis _____

_____

pericycle _____

_____

stele _____

_____

Labeling [p.462]

Identify each numbered part of the following illustrations.

1. _____   3. _____   5. _____   7. _____

2. _____   4. _____   6. _____

8.   What specialized structures separate the cells of tissue layer #6? _____

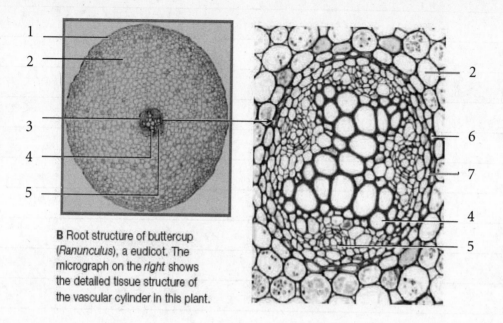

B Root structure of buttercup (*Ranunculus*), a eudicot. The micrograph on the *right* shows the detailed tissue structure of the vascular cylinder in this plant.

Fill-in-the-Blanks [pp.462-463]

When a seed germinates, the first structure to emerge is usually the (9) _____. At its tip, a protective layer, the (10) _____, forms. The (11) _____ of primary roots is the plant's absorptive surface. To increase absorption, some epidermal cells send out extensions called (12) _____, which greatly increase the (13) _____ available for absorption. Water, minerals, and food are transported through a large central (14) _____. The (15) _____ is a layer of cells that separates the stele from the cortex. Just inside that layer lies the (16) _____, whose cells divide perpendicular to the root axis to produce (17) _____ that branch off the original root. The root system of most eudicots is a(n) (18) _____ system having one major root with many branches. Monocots usually have a(n) (19) _____ root system consisting of many (20) _____ roots that sprout from the stem..

27.7. SECONDARY GROWTH [pp.464-465]

27.8. TREE RINGS AND OLD SECRETS [p.466]

Boldfaced Terms

vascular cambium _____

_____

wood _____

_____

cork cambium _____

_____

bark _____

_____

cork _____

_____

heartwood _____

_____

sapwood _____

_____

lateral meristem _____

_____

Matching [pp.464-466]

Choose the most appropriate answer for each term.

1. _____ tree rings
2. _____ heartwood
3. _____ sapwood
4. _____ early wood
5. _____ suberin
6. _____ rays

7. _____ bark
8. _____ cork
9. _____ late wood
10. _____ periderm
11. _____ sap
12. _____ vascular cambium

A. Produced by the cork cambium

B. Includes all tissues external to the vascular cambium

C. Alternating bands of early and late wood; reveal secondary growth patterns of xylem in response to changes in climate

D. Responsible for the production of secondary phloem and secondary xylem

E. The first xylem cells produced in response to moisture at the start of the growing season; tend to have large diameters and thin walls

F. Cells that radiate from the vascular cambium to the pith conducting water and dissolved solutes laterally

G. Dermal tissues consisting of parenchyma, cork cambium and cork

H. Formed in response to dry seasons or summer, xylem cells with smaller diameters and thicker walls

I. A fluid containing sugars that travels through secondary xylem from roots to buds in spring

J. Compressed, non-conducting xylem cells at the center of older stems and roots; collects metabolic wastes such as resins, oils, gums, and tannins; often dark in color

K. Secondary growth located between heartwood and the vascular cambium; wet, usually pale; contains conducting xylem

L. A fatty substance in the cell walls of cork cells

Short Answers [pp.464-465]

13. Compare and contrast hardwood and softwood trees.

_____

_____

14. Which tissues make up the bark of a tree?

_____

_____

15. Name the two lateral meristems and state their functions.

_____

_____

27.9. VARIATIONS ON A STEM [p.467]

Matching [p.467]

Choose the stem modification that matches each description.

1. _____ corm
2. _____ tuber
3. _____ bulb
4. _____ cladode
5. _____ stolon
6. _____ rhizome

A. A flattened photosynthetic stem that stores water
B. An underground stem with overlapping modified leaves called scales
C. A fleshy, horizontal underground stem with nodes that form aboveground structures
D. Runners that branch from a main stem; roots and shoots sprout from the nodes
E. Solid thickened underground stem; roots grow from a basal plate
F. Thickened portion of underground stolon; has nodes; used for nutrient storage

SELF-TEST

____ 1. _____ covers and protects the plant's surfaces.
   a. Ground tissue
   b. Dermal tissue
   c. Vascular tissue
   d. Pericycle

____ 2. Which of the following is not considered a type of simple tissue?
   a. Epidermis
   b. Parenchyma
   c. Collenchyma
   d. Sclerenchyma

___ 3. Of the following cell types, which one does not appear in vascular tissues?
   a. Vessel members
   b. Cork cells
   c. Tracheids
   d. Sieve-tube members
   e. Companion cells

___ 4. The _____ is a leaflike structure that is part of the embryo; monocot embryos have one, eudicot embryos have two.
   a. shoot tip
   b. root tip
   c. cotyledon
   d. apical meristem

___ 5. Each part of the stem where one or more leaves are attached is a(n) _____.
   a. node
   b. internode
   c. vascular bundle
   d. cotyledon

___ 6. Which of the following structures is not considered meristematic?
   a. Vascular cambium
   b. Apical meristem
   c. Cork cambium
   d. Endodermis

___ 7. New plants grow and older plant parts lengthen through cell divisions at _____ meristems present at root and shoot tips; older roots and stems of woody plants increase in diameter through cell divisions at _____ meristems.
   a. lateral; lateral
   b. lateral; apical
   c. apical; apical
   d. apical; lateral

___ 8. Vascular bundles called _____ form a network through a leaf blade.]
   a. xylem
   b. phloem
   c. veins
   d. cuticles
   e. vessels

___ 9. A primary root and its lateral branchings represent a(n) _____ system.
   a. lateral root
   b. adventitious root
   c. taproot
   d. prop root

___ 10. The _____ layer of a root divides to produce lateral roots.
   a. endodermis
   b. pericycle
   c. xylem
   d. cortex
   e. phloem

___ 11. Primary growth is _____ to the main axis of the plant body; secondary growth is _____ to the main axis of the plant body.
   a. parallel; parallel
   b. perpendicular; perpendicular
   c. parallel; perpendicular
   d. perpendicular; parallel

___ 12. Which of the following is not a Eudicot characteristic?
   a. 2 cotyledons
   b. Flower parts in 3's or multiples
   c. Netlike veins
   d. Vascular bundles organized in a ring in the stem

___ 13. Photosynthetic mesophyll cells are a type of
   a. sclerenchyma
   b. collenchyma
   c. epidermis
   d. parenchyma

___ 14. An onion, with its layers of scales overlapping an underground stem, is an example of a
   a. corm
   b. bulb
   c. rhizome
   d. tuber

CHAPTER OBJECTIVES/REVIEW QUESTIONS

1. Most of the aboveground parts of flowering plants are part of the _____ system; most of the plants belowground parts are part of the _____ system. [p.454]
2. Distinguish among ground tissues, vascular tissues, and dermal tissues . [p.455]
3. Plants grow at localized regions of self-perpetuating embryonic cells called _____. [p.459]
4. Visually identify and generally describe the simple tissues called parenchyma, collenchyma, and sclerenchyma. [pp.456-457]
5. _____ tissue conducts soil water and dissolved minerals, and it mechanically supports the plant. [p.457]
6. _____ tissue transports sugars and other solutes. [p.457]
7. Name and describe the functions of the conducting cells in xylem and phloem. [p.457]
8. What is the general function of the guard cells and stomata found within the epidermis of young stems and leaves? [p.460]
9. Visually distinguish between monocot stems and eudicot stems, as seen in cross section. [p.458]
10. How does the "simple" leaf type differ from "compound" leaves? [p.460]
11. Describe the structure (cells and layers) and major functions of leaf epidermis, mesophyll, and vein tissue. [pp.460-461]
12. How does a taproot system differ from a fibrous root system? [p.462]
13. Define the term adventitious root. [p.462]
14. Describe the origin and function of root hairs. [p.462]
15. Describe the relationship between apical and lateral meristems and primary and secondary growth. [pp.455, 458-459, 464]
16. Explain the origin of the annual growth layers (tree rings) seen in a cross section of a tree trunk. [p.465]
17. Describe how vascular cambium produces secondary xylem and secondary phloem. [p.464]
18. How is early wood different from late wood? [p.465]

INTEGRATING AND APPLYING KEY CONCEPTS

1. Many plants can be asexually reproduced through stem cuttings. What plant processes must occur for this to work?
2. Consider how the following plant pathogens could affect the structure and function of plant organs and tissues. (1) a fungus that enters the leaf through the stoma and grows in the leaf air spaces, (2) a worm that burrows into a root and lives there, (3) a bug that lays its eggs beneath the epidermis of a stem where the larvae hatch and grow.
3. Compare the various plant tissues used for support, transport of nutrients, protection and gas exchange with their animal counterparts.

28

PLANT NUTRITION AND TRANSPORT

INTRODUCTION

This chapter introduces plant physiology. Plants, like other organisms, must regulate the exchange of materials with the external environment as well as within the organism itself. In this chapter you will examine how cells regulate the movement of both water and organic materials. The chapter focuses on the structure and function of the vascular tissue of plants—the xylem and phloem. It also introduces some important mutualistic relationships that enhance the plant's ability to exchange materials within the roots. In addition, you will learn about the adaptations plants have evolved to prevent water loss, inorganic nutrients they require, and how the soil in which a plant grows affects its life functions.

STUDY STRATEGIES

- Read all of chapter 28 with the goal of familiarizing yourself with the boldface terms.
- Describe the major properties of soil.
- Discuss the major plant nutrients and their function.
- Identify major processes in the formation and breakdown of soil, including leaching and erosion.
- Explain regions and structures of root specialization.
- Describe how roots take up water and nutrients and transport them to the rest of the plant.
- Explain the cohesion-tension theory and identify the role of transpiration in the process.
- Discuss adaptations of stems and leaves to water conservation.
- Describe how organic compounds are transported from leaves to stems and roots and the specialized cells which assist in the process.

FOCAL POINTS

- Table 28.1 [p.472] provides a summary of plant nutrients and their functions.
- Figure 28.2 [p.472] illustrates soil horizons.
- Figure 28.4 [p.474] illustrates various root adaptations.
- Figure 28.5 [p.475] (animated) demonstrates how plants regulate water movement through root tissues and into the xylem.
- Figure 28.7 [p. 477] (animated) illustrates the cohesion-tension theory, essential to understanding the movement of water through the plant's xylem.
- Figures 28.8, 28.10 and 28.11 [pp.478-479] provide a visual representation of the relationship between the stomatal opening and the physiological condition of the guard cells.
- Figure 28.13 [p.481] (animated) demonstrates the pressure-flow theory of sugar movement through the phloem of a plant from a source (site of photosynthesis) to a sink (area of storage or usage).

INTERACTIVE EXERCISES

28.1. LEAFY CLEAN-UP CREWS [p.471]

28.2. PLANT NUTRIENTS AND AVAILABILITY IN SOIL [pp.472-473]

Boldfaced Terms

soil _____

_____

humus _____

_____

loam _____

_____

topsoil _____

_____

leaching _____

_____

soil erosion _____

_____

Matching [pp.471-473]

Choose the most appropriate statement for each term.

1. ____ soil
2. ____ humus
3. ____ loam
4. ____ phytoremediation
5. ____ topsoil
6. ____ nutrients
7. ____ macronutrients
8. ____ micronutrients
9. ____ leaching
10. ____ erosion

A. Element or molecule essential for an organism's growth and survival

B. The use of plants to clean contaminants from soil by degrading or concentrating them in their tissues

C. The decomposing organic material in soil

D. The loss of soil under the force of wind, running water, and ice

E. Chemical elements essential for plant growth necessary in only trace amounts

F. Consists of particles of minerals mixed with variable amounts of decomposing organic material

G. The removal of some of the nutrients in soil as water passes through it

H. Uppermost part of the soil, rich in organic matter (A horizon); zone of densest root growth

I. Soils having more or less equal proportions of sand, silt, and clay; best for plant growth

J.   Essential elements that are required in amounts   greater than 0.5% of the plant's dry weight

Short Answer [pp.472-473]

11. What role do clay, sand, and silt particles play in the nutrition of plants?

_____

_____

12. Compare and contrast the processes of leaching and erosion.

_____

_____

Choice [p.472]

Choose the soil layer that matches the description.

   a. A Horizon   b. B Horizon   c. C Horizon   d. O Horizon

13. _____   No organic material; partly weathered rocks and rock fragments; rests on bedrock

14. _____   Topsoil with decomposed organic material

15. _____   Leaf litter & other organic material on the surface of the soil

16. _____   Less organic material, more minerals and larger soil particles than #14

28.3. HOW DO ROOTS ABSORB WATER AND NUTRIENTS? [pp.474-475]

Boldfaced Terms

mycorrhiza (pl. mycorrhizae) _____

_____

nitrogen fixation_____

_____

root nodules _____

_____

Casparian strip _____

_____

Choice [p.474]

Choose the type of specialized root structure best described by each statement.

a. root hairs   b. mycorrhizae   c. root nodules

1. _____ Fungi that live in a mutualistic relationship with roots

2. _____ Penetrate the soil for only a few millimeters; live only a few days

3. _____ Swollen masses of bacteria-infected root cells

4. _____ Contain bacteria that live in a mutualistic relationship with roots

5. _____ Thin extensions of root epidermal cells that increase surface area

6. _____ Hyphae with a surface area larger than that of roots

7. _____ Organism that shares minerals it absorbs from the soil with plant roots

8. _____ Contain organisms that convert nitrogen gas to ammonia (nitrogen fixation)

Sequence [p.475]

Put the following steps of water absorption in correct sequential order by placing a 1 in the appropriate blank for the first step, a 2 for the second step, etc.

9. _____ Water passes through and between the parenchyma cells of the cortex

10. _____ Water passes from cell to cell via plasmodesmata

11. _____ Water diffuses across the epidermal cell layer

12. _____ Water diffuses across plasma membrane of endodermal cells

13. _____ Water is taken up by xylem cells

14. _____ Casparian strip prevents water from passing through cell walls

Short Answer [pp.475]

15. Describe the movement of mineral ions from soil to xylem.

_____

_____

_____

_____

28.4. WATER MOVEMENT INSIDE PLANTS [pp.476-477]

28.5. WATER-CONSERVING ADAPTATIONS OF STEMS AND LEAVES [pp.478-479]

Boldfaced Terms

cohesion–tension theory _____

_____

transpiration_____

_____

guard cells _____

_____

Fill-in-the-Blanks [pp.476-477]

In Angiosperms, xylem (1) _____ are responsible for most water conduction. (2) _____ in their side walls allow lateral water flow. (3) _____ separate stacked vessel (4) _____ in mature xylem and allow vertical movement of water from roots upwards.

By the (5) _____ theory, water inside the (6) _____ is pulled upward by air's drying power creating a continuous negative pressure called (7) _____which extends all the way from leaves to (8) _____. Most of the water a plant takes up is lost due to (9) _____ , the evaporation of water from aboveground plant parts, through (10) _____ on the plant's leaves and stems. This loss of water from leaves and stems causes a (11) _____ that exerts a pull on the continuous column of water in the xylem. The collective strength of many (12) _____ among water molecules imparts (13) _____ to liquid water. Thus, the negative pressure created by transpiration exerts (14) _____ on the entire column of water in a xylem tube extending from (15) _____ down through (16) _____ and on into young (17) _____ where water is pulled in from the (18) _____.

Matching [pp.478-479]

Select the term that best matches the lettered description.

19. ____ a hormone released by root cells to signal a scarcity of soil water
20. ____ an ion pumped into cells to cause water to enter the cell by osmosis
21. ____ water availability, CO_2 level, light intensity affect stomatal opening
22. ____ a gap between guard cells in the epidermis
23 ____ a water impermeable layer covering the epidermis
24. ____ turgid guard cells

25. ____ specialized epidermal cells that control size of stoma

A. stoma

B. cuticle

C. Guard cell

D. abscisic acid

E. environmental cues

F. potassium

G. condition resulting in stomatal opening

28.6. MOVEMENT OF ORGANIC COMPOUNDS IN PLANTS [pp.480-481]

Boldfaced Terms

translocation _____

_____

source_____

_____

sink_____

_____

pressure-flow theory _____

_____

sieve plate _____

_____

Matching [pp.480-481]

Choose the most appropriate statement for each term.

1. ____ translocation

2. ____ sieve tube

3. ____ companion cells

4. ____ source

5. ____ sink

6. ____ pressure-flow theory

7. ____ sieve plate

A. Any region where sugars are being produced

B. Cells adjacent to sieve-tube members that supply metabolic functions to them

C. Any region of the plant where sugars are being used or stored.

D. Process that move sucrose and organic compounds through phloem of vascular plants

E. Internal pressure builds up at the source end of a sieve-tube system and pushes the solute-rich solution toward a sink, where solutes and water are removed

F. Living cells with porous ends; found in phloem

G. Perforated ends of sieve tube members

Fill-in-the-Blanks [pp.480-481]

Phloem is a complex tissue consisting of (8) _____, _____ and _____. At maturity, a sieve-tube member cell has lost its (9) _____ and many other cytoplasmic structures. Adjacent to each member cell is a (10) _____ that supplies its metabolic needs. Sugars flow through mature phloem through a process known as (11) _____. These sugars flow from a (12) _____, where they are produced, to a (13) _____ where they are used or stored. The (14) _____ theory describes how this is accomplished. Companion cells aid sieve tube members by (15) _____ sugars into them. The sugar concentration in the sieve tubes increases making those cells (16) _____ relative to the surrounding cells. Through the process of (17) _____, water flows from the surrounding cells into the sieve tube cells. The resultant (18) _____ pressure pushes the sugars along a concentration gradient toward a (19) _____. Sugars are (20) _____ from the sieve tubes there.

SELF-TEST

_____ 1. The use of plant species to remove toxic chemicals from soils is called _____
   a. transpiration
   b. nitrogen fixation
   c. phytoremediation
   d. translocation

_____ 2. Plant macronutrients are the nine dissolved mineral ions that _____.
   a. are found in major concentrations in the earth's surface
   b. occur in only small traces in plant tissues
   c. are required in amounts above 0.5 percent of the plant's dry weight
   d. can function only without the presence of micronutrients

_____ 3. Soils that have the best oxygen and water penetration are called _____.
   a. topsoils
   b. humus
   c. clays
   d. loams

_____ 4. The greatest amount of decomposed organic matter is found in which of the following soil layers?
   a. O Horizon
   b. A Horizon
   c. B Horizon
   d. C Horizon

_____ 5. Gaseous nitrogen is converted to a plant-usable form by _____.
   a. root nodules
   b. mycorrhizae
   c. nitrogen-fixing bacteria
   d. sunlight and carbon dioxide

_____ 6. _____ prevent(s) inward-moving water from moving past the abutting walls of the root endodermal cells.
   a. Cytoplasm
   b. Plasma membranes
   c. Osmosis
   d. Casparian strips

_____ 7. Most of the water moving into a leaf is lost through _____.
   a. osmotic gradients being established
   b. transpiration
   c. pressure flow forces
   d. translocation

_____ 8. Mycorrhizae represent mutualistic relationships between roots and _____.
   a. bacteria
   b. fungi
   c. animals
   d. all of the above

9. Without _____, plants would rapidly wilt and die during hot, dry spells.
   a. a cuticle
   b. mycorrhizae
   c. phloem
   d. cotyledons

10. The _____ theory of water transport states that hydrogen bonding allows water molecules to maintain a continuous fluid column as water is pulled from roots to leaves.
    a. pressure-flow
    b. cohesion–tension
    c. evaporation
    d. nitrogen-fixation

11. In the pressure-flow theory, leaves are _____ regions while growing leaves, stems, fruits, seeds, and roots are _____ regions.
    a. source; source
    b. sink; source
    c. source; sink
    d. sink; sink

12. Which of the following is not involved in water conservation?
    a. Cuticle
    b. Guard cell
    c. Abscisic acid
    d. Companion cell

13. Which of the following is a plant micronutrient?
    a. Nitrogen
    b. Calcium
    c. Copper
    d. Sulfur

CHAPTER OBJECTIVES/REVIEW QUESTIONS

1. Explain the benefits of phytoremediation. [p.471]
2. Distinguish between macronutrients and micronutrients in relation to their role in plant nutrition. [p.472]
3. Define the terms soil, humus, and loam. [p.473]
4. Distinguish between leaching and erosion. [p.473]
5. Describe the roles of root nodules and mycorrhizae in plant nutrition. [p.474]
6. Define nitrogen fixation and explain what is involved in this process. [p.474]
7. Explain why root hairs are so valuable in root absorption. [p.474]
8. Explain the role of the Casparian strip in regulating water movement. [p.475]
9. Describe the passage of water from the soil to the xylem. [p.475]
10. Define transpiration. [p.477]
11. Use the cohesion–tension theory to describe how water moves in a plant. [pp.476-477]
12. Explain the role of the cuticle and stomata in regulating movements of materials into the plant. [pp.478-479]
13. Describe the role of phloem sieve-tube members and companion cells in the movement of organic compounds. [p.480]
14. Define the role of sinks and sources in translocation. [pp.480-481]
15. Explain the general principles of the pressure-flow theory. [pp.480-481]

INTEGRATING AND APPLYING KEY CONCEPTS

1. How do you think maple syrup is made from maple trees? Which specific systems of the plant are involved, and why are maple trees tapped only at certain times of the year?

2. What might be the result of stripping vegetation from the soil and exposing it to elements of climate? What role do you think ground cover (plants that are short with thick root systems) plays in preventing soil erosion?

3. As noted in the chapter, cuticles can actually prevent the movement of gases in the direction that the plant prefers. What balance must the plant obtain between the amount of cuticle and gas transport? What factors are at work? In what types of environment would you expect to find plants with thick cuticles?

4. What deleterious effect might a deficiency in nitrogen, magnesium and iron have on a plant?

5. Review the cohesion-tension theory and explain why the introduction of an air bubble into a cut rose stem will cause it to die quickly.

29

LIFE CYCLES OF FLOWERING PLANTS

INTRODUCTION

The ability of flowering plants to reproduce both asexually (vegetatively) and sexually has helped them become very successful inhabitants of this planet. Central to grasping the success of flowering plants is the understanding of floral and seed structure along with the various adaptations for pollination and seed dispersal that have coevolved in relation with members of the animal kingdom. This chapter follows the sexual reproductive process from alternation of gametophyte and sporophyte generations, through floral and pollinator co-evolution, pollination, the unique angiosperm process of double fertilization, to seed formation, fruiting, and fruit adaptations. It also discusses various types of vegetative reproduction.

STUDY STRATEGIES

- Read all of chapter 29 with the goal of familiarizing yourself with the boldface terms.
- Identify the major structures of the flower and their functions.
- Recognize the variation in flowers amongst angiosperms.
- Describe the various vectors and modes of pollination, including adaptations of flowers for pollination.
- Describe the alternation of generations in a flowering plant.
- Explain the process of sexual reproduction in flowering plants.
- Discuss the formation, germination, and development of seeds.
- Identify the structure and function of fruits.
- Explain the various modes of asexual reproduction in flowering plants.
- Discuss senescence in angiosperms.

FOCAL POINTS

- Sections 29.1 and 29.3 [pp. 485, 488-489], and Table 29.1 [p.489] focus on the important concept of coevolution of flowers and their pollinators.
- Figure 29.2 [p.486] (animated) summarizes the life cycle of flowering plants.
- Figure 29.3 [p.486] (animated) illustrates the structure of the reproductive shoot known as a flower.
- Figure 29.8 [pp.490-491] (animated) shows the life cycle of the cherry plant from the development of male and female gametophytes through pollination and fertilization.
- Figure 29.11 [p.493] (animated) demonstrates the transition of an ovule to a mature seed.
- Figure 29.14 [p.495] illustrates the structure and function of various fruit adaptations.
- Table 29.2 [p.494] provides a synopsis of the classification of fruit.
- Section 29.9 [p.498-499] describes the ability of flowering plants to populate their environment through asexual reproduction.
- Figures 29.16 and 29.17 illustrate germination and early growth of monocot and eudicot seeds.

INTERACTIVE EXERCISES

29.1. PLIGHT OF THE HONEYBEE [p.485]

29.2. REPRODUCTIVE STRUCTURES [pp.486-487]

Boldfaced Terms

anther _____

_____

calyx _____

_____

corolla _____

_____

stamens _____

_____

petal _____

_____

sepal _____

_____

stigma _____

_____

flower _____

_____

carpel _____

_____

ovary _____

_____

ovule _____

_____

Labeling [p.486]

Identify each numbered part of the accompanying illustration.

1. _____

2. _____

3. _____

4. _____

5. _____

6. _____

7. _____

8. _____

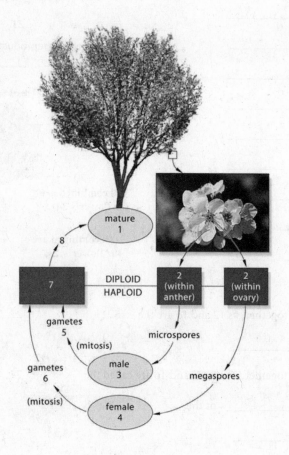

Short Answer [p.487]

9. Compare and contrast regular and irregular flowers.

_____

_____

10. Compare and contrast complete and incomplete flowers.

_____

_____

11. What is a perfect flower?

_____

_____

Labeling [p.486]

Identify each numbered part of the following diagram of the angiosperm flower.

12. _____

13. _____

14. _____

15. _____

16. _____

17. _____

18. _____

19. _____

20. _____

21 (male reproductive part)

22 (female reproductive part)

13

14

12

15

20

16

(all combined are the flower's 24)

(all combined are the flower's 23)

17 (forms within ovary)

18

21. Taken together, #s 12 and 13 are the

_____

22. Taken together, #s 14, 15, and 16 are called the

_____ or the

_____

23. A whorl of #19 is called the

_____

24. A whorl of #20 is called the

_____

29.3. FLOWERS AND THEIR POLLINATORS [pp.488-489]

Boldfaced Terms

pollination vector_____

_____

Fill-in-the-Blanks [pp.488-489]

In order to ensure sexual reproduction, Angiosperms depend on environmental agents known as (1)

_____ to transfer pollen from a(an) (2) _____ to the (3) _____ of a flower. Many flowers, like

grasses, release billions of pollen grains that are carried by the (4) _____ to a female flower. Other plants

require the help of animal (5) _____ to transfer pollen among individuals of the same species. (6)

_____, (7) _____, and (8) _____ are common pollination vectors. About 90% of flowering plants

have (9) _____ animal pollinators. Yellow, blue, and white pigments attract (10) _____, as do pigments that reflect (11) _____ light. Plants that release sweet or musky odors at night attract (12) _____ and

(13) _____, and the odors of rotting flesh and dung attract (14) _____ and (15) _____. Pollen

and (16) _____ are nutritious rewards that attract pollinators. Honeybees convert nectar to (17) _____.

Many flowers have specializations that exclude non-pollinators. An ample supply of nectar deeply hidden at the

bottom of a long (18) _____ may be reached only by specific pollinators such as (19) _____, (20)

_____, and (21) _____ that have evolved matching feeding devices.

29.4. A NEW GENERATION BEGINS [pp. 490-491]

29.5. FLOWER SEX [p.492]

Boldfaced Terms

microspore _____

_____

megaspore _____

_____

dormancy _____

_____

germination _____

_____

double fertilization _____

_____

endosperm _____

_____

Sequence [p.490]

Place the numbers 1-6 in the space before each statement to indicate its proper place in the life cycle of the flowering plant from production of the female gametophyte to seed formation.

1. _____   Germination of the pollen tube

2. _____   Megaspores form by meiosis

3. _____   Seed consists of maternal integuments, a diploid embryo, and triploid endosperm

4. _____   One sperm nucleus fertilizes the egg, a second fuses with the endosperm mother cell

5. _____   An ovule forms within the ovary of the sporophyte

6. _____   Repeated mitoses of the megaspore form the embryo sac, the female gametophyte

Dichotomous Choice [pp.490-491]

Circle one of the two possible answers between the parentheses to complete each statement.

7. The gametophyte stage is (haploid/diploid).

8. The sporophyte stage is (haploid/diploid).

9. The megaspore is (haploid/diploid).

10. The microspore is (haploid/diploid).

11. Nuclei within a pollen grain are (haploid/diploid).

12. The embryo is (haploid/diploid).

Sequence [p.490]

Place numbers 1-6 in the space before each statement to indicate its proper place in the life cycle of the male gametophyte of a flowering plant.

13. _____   A pollen grain lands on the stigma

14. _____   Meiosis produces four haploid microspores

15. _____   A pollen tube germinates from the pollen grain

16. _____   Pollen sacs form in the anther of the sporophyte flower

17. _____   Mitosis of a microspore followed by cellular differentiation forms a pollen grain

18. _____   A pollen grain is released from the anther

Fill-in-the-Blanks [p.492]

The outer layer of the coat of a pollen grain is made of (19) _____, a very durable substance. Sex in plants begins when (20) _____ on the epidermal cells of the stigma bind to molecules on the coat of the pollen grain. The pollen grain secretes lipids and proteins which bind to (21) _____ in the stigma cell membranes. Pollen is dry and its cells are (22) _____ until they receive nutrient-rich fluid from the stigma. The cells of the male gametophyte (pollen) then resume metabolism and grow a (23) _____ out of one of the (24) _____ or (25) _____ in the pollen's coat. Gradients of nutrients direct the growth of the pollen tube down through the (26) _____. The female gametophyte secretes (27) _____ to guide the pollen tube to the (28) _____. These signals are (29) _____ so that pollen tubes of different species may grow down the style but not be able to reach and (30) _____ the egg.

29.6. SEED FORMATION [p.493]

29.7. FRUITS [pp.494-495]

29.8. EARLY DEVELOPMENT [pp.496-497]

Boldfaced Terms

seed_____

_____

fruit _____

_____

coleoptile _____

_____

hypocotyl _____

_____

radicle _____

_____

plumule _____

_____

Labeling [p.493]

Identify each numbered part of the accompanying illustration of a developing seed.

1. _____

2. _____

3. _____

4. _____

5. _____

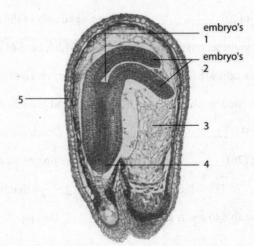

embryo's 1

embryo's 2

5

3

4

Choice [p.493]

Choose the part of the seed that contains each substance.

a. seed coat   b. embryo   c. endosperm

6. _____   7. _____   8. _____   9. _____   10. _____
   Vitamins   Minerals   Protein   Fiber   Starch

Short Answer [p.494]

11. Differentiate among simple fruits, aggregate fruits, and multiple fruits. Give an example of each.

_____

_____

_____

12. Differentiate between true fruits and accessory fruits. Give an example of each.

_____

_____

_____

Matching [p.494]

Match the fruit category with its description. Complete the exercise by inserting an example of each fruit type in the parentheses.

13. ____ dehiscent   (_____)

14. ____ indehiscent (_____)

15. ____ drupe   (_____)

16. ____ berry   (_____)

17. ____ pome   (_____)

A. Fleshy fruit with a hard pit surrounding the seed

B. Fleshy fruit; seeds in a core derived from the ovary; core is surrounded by fleshy tissue derived from the receptacle

C. Dry fruit wall splits along a seam to release the seeds

D. Fleshy fruit formed from one ovary; usually has many seeds; no pit

E. Dry fruit; wall does not split open; seeds are dispersed inside intact fruits

Choice [p.495]

Choose the type of seed dispersal that best fits the adaptation listed.

a. wind-dispersed fruits   b. fruits dispersed by animals

c. water-dispersed fruits   d. mechanical adaptations for dispersion

18. ____ Pods that pop open ejecting the seeds

19. ____ Thick-husked fruits of coconut palms

20. ____ Seed coats abraded by digestive enzymes assist in seed germination

21. ____ Wings of maple fruits

22. ____ Air bladders of sedges

23. ____ Red fleshy fruits of hawthorne

24. ____ Hairy modified sepals of dandelions, milkweed, and thistle

25. ____ Spines and hooks of cockleburs and bur clover

Sequence [p.496-497]

Place numbers 1-5 in the space before each statement to indicate its proper place in the process of germination.

26. _____ Active mitosis in meristems; embryo begins to grow.

27. _____ Enzymes in endosperm activate breaking starch into sugars.

28. _____ Radicle emerges from seed coat.

29. _____ Water enters seed.

30. _____ Seed coat ruptures.

29.9. ASEXUAL REPRODUCTION OF FLOWERING PLANTS [pp.498-499]

29.10. SENESCENCE [pp.500-501]

Boldfaced Terms

vegetative reproduction _____

_____

tissue culture propagation _____

_____

abscission _____

_____

senescence _____

_____

Matching [pp.498-499]

Match the following terms associated with asexual reproduction in flowering plants with their definitions.

1. _____ root suckers

2. _____ clone

3. _____ Tasmanian King's holly

4. _____ triploid

5. _____ sterile

6. _____ vegetative reproduction

7. _____ tissue culture propagation

8. _____ cutting

9. _____ grafting

10. _____ polyploidy

A. In this laboratory technique a single somatic cell may be induced to form an embryo and then an entire new plant

B. Any organism that cannot reproduce sexually

C. Any multiple of the "n" chromosome number beyond diploid (2n)

D. A group of plants that have arisen from the vegetative reproduction of one ancestor and share genetic identity with that ancestor

E. Shoots that sprout from the lateral roots of a plant

F. A stem fragment of a parent plant that has the capability of forming shoots as a new individual

G. Having the '3n' number of chromosomes in each cell

H. The oldest known living plant, more than 43,000 years old

I. New roots and shoots grow from extensions or fragments of a parent plant

J. Joining a cutting of one plant to the tissues of another plant, often a stem cutting to a rootstock

Matching [p.500-501]

Select the description that best fits the numbered term.

11. ____ abscission

12. ____ senescence

13. ____ ethylene

14. ____ abscission zone

15. ____ drought

A. Environmental condition that induces dormancy in trees

B. Aging of cells, individuals or communities

C. Area of a plant where cells are producing enzymes to digest their cell walls

D. Process by which plant parts are shed

E. Plant hormone responsible for abscission

SELF-TEST

____ 1. The joint evolution of flowers and their pollinators is known as ____.
   a. adaptation
   b. coevolution
   c. joint evolution
   d. covert evolution

____ 2. A stamen is _____.
   a. composed of a stigma
   b. the mature male gametophyte
   c. the site where microspores are produced
   d. part of the vegetative phase of an angiosperm

____ 3. The portion of the carpel that contains an ovule is the _____ .
   a. stigma
   b. anther
   c. style
   d. ovary

____ 4. The phase in the life cycle of plants that gives rise to spores is known as the _____.
   a. gametophyte
   b. embryo
   c. sporophyte
   d. seed

____ 5. A gametophyte is _____.
   a. a gamete-producing plant
   b. haploid
   c. both a and b
   d. the plant produced by the fusion of gametes

____ 6. A characteristic of a seed is that it _____.]
   a. contains an embryo sporophyte
   b. represents an arrested growth stage
   c. is covered by hardened and thickened integuments
   d. all of these

____ 7. Ovaries become _____ while ovules become _____ .
   a. fruit; megaspores
   b. fruits; seeds
   c. megaspores; fruits
   d. seeds; fruits

____ 8. In flowering plants, one sperm nucleus fuses with that of an egg, and a zygote forms that develops into an embryo. Another sperm fuses with _____.]
   a. a primary endosperm cell to produce three cells, each with one nucleus
   b. a primary endosperm cell to produce one cell with one diploid nucleus
   c. both nuclei of the endosperm mother cell, forming a primary endosperm cell with a single triploid nucleus
   d. one of the smaller megaspores to produce what will eventually become the seed coat

____ 9. "Simple, aggregate, multiple, and accessory" refer to types of ____.
   a. carpels
   b. seeds
   c. fruits
   d. ovaries

___ 10. Calyx is to corolla as _____ is (are) to _____.
a. ovary; ovule
b. stamens; pistil
c. petals; sepals
d. sepals; petals

___ 11. A tulip flower consisting of symmetric green sepals, pink petals, yellow stamens, and orange pistils is _____.
a. complete
b. perfect
c. regular
d. all of the above

___ 12. Which of the following is not considered a valuable pollinator?
a. Monarch butterfly
b. Hummingbird
c. Honeybee
d. Leaf-cutter ants

___ 13. Which of the following is not a type of asexual reproduction?
a. root suckers
b. grafting
c. germination
d. cutting

___ 14. Abscission involves ____.
a. leaf and fruit drop
b. ethylene
c. senescence
d. all of the above

CHAPTER OBJECTIVES /REVIEW QUESTIONS

1. Describe the role of a pollinator. [pp.485, 488-489]
2. Distinguish between sporophytes and gametophytes. [p.486
3. Identify the various parts of a typical flower and state their functions. [p.486]
4. Distinguish between a flower that is *perfect* and one that is *imperfect*. [p.487]
5. What structures represent the male gametophyte and the female gametophyte in flowering plants? List the contents of each. [pp.490-491]
6. Describe the double fertilization that occurs uniquely in the flowering plant life cycle. [pp.490-491]
7. Review seed structure and discuss how its design prepared it for survival in the harsh, dry, terrestrial climate. [p.496]
8. Review the general types of fruits produced by flowering plants. [pp.494-495]
9. Describe the various methods of seed dispersal by fruits. [p.495]
10. Discuss polyploidy in plants including the reasons for its existence and consequences to the plant. [p.498]
11. Distinguish between vegetative reproduction and tissue culture propagation, including a discussion of the various types of vegetative reproduction; cite an example for each. [pp.498-499]
12. Discuss the process of abscission and the reasons for it. [pp.500-501]

INTEGRATING AND APPLYING KEY CONCEPTS

1. In terms of botanical morphology, a flower is interpreted as "a shoot modified for reproductive functions." After studying floral structures in this chapter and shoot/leaf structure in text Chapter 29, can you think of any comparable structural evidence that might have led botanists to arrive at this conclusion?
2. This chapter revealed the intimate interrelationships between flowering plants and animals as pollinators and seed dispersal agents. Consider how circumstances like habitat destruction, the use of pesticides and animal diseases can have grave consequences on the continued success of flowering plants.

30
COMMUNICATION STRATEGIES IN PLANTS

INTRODUCTION

Like all living organisms, plants have mechanisms to address such essential life functions as energy requirements, reproduction, growth, development, and response to stimuli. Previous chapters have discussed energy requirements and reproduction. This chapter introduces the major mechanisms by which plants control and regulate their growth and development and respond to stimuli. Unlike many animals, plants do not have nervous tissue and so must rely solely on the use of hormone messengers and other chemical signaling. From germination of a seed and growth of its embryo to sexual reproduction and death, plants use chemical signaling to control all aspects of their life cycles. In order to gain insight into plant physiology, you should familiarize yourself with the major classes of hormones and their functions.

STUDY STRATEGIES

- Read all of chapter 30 with the goal of familiarizing yourself with the boldface terms.
- Explain the role of hormones in a plant.
- Identify major plant hormones and their functions in a plant.
- Describe auxin and its role as the master hormone in plants.
- Explain apical dominance.
- Discuss the roles of cytokinin, gibberellin, abscisic acid, ethylene as hormones in the plant body.
- Recognize the various types of tropisms and their importance as adaptations for plant fitness.
- Identify how circadian rhythms acts as a biological clock to measure days as well as seasons.
- Compare and contrast long-day versus short-day plants.
- Explain how vernalization acts as a biological clock.
- Discuss how plants cope with stress.

FOCAL POINTS

- Table 30.1 [p.507] summarizes the varied effects- of ten plant hormones.
- Figure 30.3 (animated) [p.508] demonstrates the effect of auxin on monocot seedling growth.
- Figure 30.6 [p.510] depicts the roles of auxin and cytokinin in apical dominance.
- Figure 30.8 [p.511] demonstrates the action of gibberellin in seed germination.
- Figure 30.11 [p.513] illustrates the role of ethylene in flower and fruit formation.
- Figure 30.13 (animated) [p.514] illustrates the function of statoliths in root tip response to gravity.
- Figure 30.14 (animated) [p.514] demonstrates the effect of auxins on stem growth in response to light.
- Figure 30.16 (animated) [p.516] demonstrates rhythmic movements in plants.
- Figure 30.18 (animated) [p.517] illustrates the role of photoperiodism in plant flowering.

INTERACTIVE EXERCISES

30.1. PRESCRIPTION: CHOCOLATE [p.505]

30.2. INTRODUCTION TO PLANT HORMONES [pp.506-507]

Boldfaced Terms

plant hormone _____

_____

Fill-in-the-Blanks [p.505-507]

Plants make many molecules that help them to regulate interactions with their (1) _____. When

we consume plants, these molecules may have effects on our physiology. The *Theobroma cacao* tree, from which we

get (2) _____ and (3) _____, produces many substances, among which are (4) _____, a type of

secondary metabolite. The seeds of the cacao tree are rich in (5) _____ produced to aid in plant (6)

_____. It is also beneficial to humans in the treatment of (7) _____, (8)_____ and (9)

_____. The consumption of large amounts of (10) _____is the most likely reason for the absence of

heart disease among the Kuna of South America.

Matching [p.507]

____ 11. Brassinosteroid   a. Enhances germination; involved in thigmotropism and stress defense

____ 12. Jasmonic Acid   b. Opposes auxin; involved in apical dominance

____ 13. Nitric oxide   c. Tissue defense; involved in development of anthers and pollen

____ 14. Salicylic Acid   d. Involved in thigmotropism; induces expression of defense genes

____ 15. Strigolactone   e. Activates systemic acquired resistance

30.3. AUXIN: THE MASTER GROWTH HORMONE [pp.508-509]

30.4. CYTOKININ [p.510]

30.5. GIBBERELLIN [p.511]

30.6. ABSCISIC ACID [p.512]

30.7. ETHYLENE [p.513]

Boldfaced Terms

gibberellin _____

_____

auxin_____

_____

apical dominance _____

_____

abscisic acid _____

_____

cytokinin _____

_____

ethylene _____

_____

Choice [pp.508-513]

For each of the following, choose from the following classes of plant hormones.

   a. auxins   b. gibberellins   c. cytokinins   d. abscisic acid (ABA)   e. ethylene

1. _____ Inhibits lateral root formation
2. _____ Influences levels of other plant hormones
3. _____ Stimulates fruit ripening
4. _____ Induce cell division and elongation between stem nodes
5. _____ Responsible for apical dominance by inhibiting cell division and elongation in lateral buds
6. _____ Closes stomata in times of stress
7. _____ Breaks dormancy in seeds and stimulates germination
8. _____ Most common type in nature is indole-3-acetic acid (IAA)
9. _____ In autumn, induces abscission of leaves
10. _____ Used to increase fruit size in seedless plants
11. _____ Used as a herbicide; causes uncontrolled cell division and death
12. _____ Synthesis begins in chloroplasts
13. _____ Allows fruit to be shipped green, then ripened at the store
14. _____ Promotes root formation in cuttings
15. _____ Induces cell division and differentiation of vascular tissues
16. _____ Stimulates differentiation in root apical meristem cells
17. _____ Stimulates lateral root growth

18. ____ Transported by efflux carriers

19. ____ Mobilizes food reserves in germinating seeds

20. ____ Inhibits germination

Fill-in-the-Blanks [pp.505-507]

The five major types of plant hormones are (21) _____, (22) _____, (23) _____, (24) _____, and (25) _____. During germination, water absorbed by a seed causes the release of (26) _____ which induces transcription of the gene for (27) _____. Stored starch in the (28) _____ is then broken down into sugars that are used to make ATP to fuel rapid cell division in the meristem cells. (29) _____ is a member of the auxin family of hormones that has many effects on plant development. It causes cells to expand by increasing the activity of (30) _____ that pump hydrogen ions into the cell wall, thus softening it. (31) _____ pressure allows the cell wall to stretch irreversibly. IAA is made mainly in (32) _____ and (33) _____ where it is most highly concentrated, and is transported by the (34) _____ over long distances from the site of its production. Active transport proteins called (35) _____ direct the flow of auxin through cells. (36) _____ produced in roots has a dampening effect on auxin transport by affecting the functioning of efflux carriers. In the process of (37) _____, auxin inhibits lateral bud growth. When the terminal bud is removed, (38) _____ stimulates growth of lateral buds. (39) _____ prevents a seed from germinating too soon and is responsible for closing stomata in times of stress. (40) _____ plays an important role in abscission, fruit ripening and stress responses.

30.8. TROPISMS [pp.514-515]

Boldfaced Terms

tropism _____

_____

gravitropism _____

_____

statolith _____

_____

phototropism _____

_____

thigmotropism _____

_____

heliotropism_____

_____

Choice [pp.514-515]

For each description, choose the most appropriate tropic response.

    a. phototropism   b. gravitropism   c. thigmotropism   d. mechanical stress   e. heliotropism

1. _____ Growth response made possible by statoliths in plant cells
2. _____ Plant parts change position in response to changing angles of the sun
3. _____ Growth response to the direction of light
4. _____ Auxin migrates to one side of a stem causing greater cell elongation on that side
5. _____ Growth response to the Earth's gravitational forces
6. _____ Growth response to contact with a solid object
7. _____ Growth response to factors such as prevailing winds and grazing animals
8. _____ Allows roots to grow around rocks in the soil

Fill-in-the-Blanks [pp.514-515]

A (9) _____ is a growth response of a plant to an environmental (10) _____. Tropisms are mediated by (11) _____. No matter how a seed is positioned in the soil, when it germinates, the primary root always grows (12) _____ and the primary shoot always grows (13) _____ due to a growth response known as (14) _____. Root cap cells contain starch-filled amyloplasts called (15) _____ that respond to gravity by occupying the side of the cell closest to gravity. Their placement at the bottom of cells causes a redistribution of (16) _____ which results in downward growth of root tips. The effects of (17) _____ cause stems to bend toward the source of light. The mechanism responsible for this involves the hormone (18) _____. Pigments called (19) _____ absorb (20) _____ and use its energy in a cascade of intracellular signals which ultimately redistributes auxin to the (21) _____ side of a shoot. The result is that cells on that side of the stem (22) _____ faster than the cells on the illuminated side. When a vine's (23) _____ touch an object, they begin to curl around it due to the growth response called (24) _____. This is due to a decrease in cell (25) _____ at the point of contact while the cells on the opposite side of the shoot continue to elongate. (26) _____, such as shaking, wind shear, or the action of grazing animals, also inhibits stem elongation at the point of stress.

30.9. SENSING RECURRING ENVIRONMENTAL CHANGES [pp.516-517]

30.10. RESPONSES TO STRESS [pp.518-519]

Boldfaced Terms

circadian rhythm _____

_____

phytochrome _____

_____

photoperiodism _____

_____

vernalization _____

_____

systemic acquired resistance _____

_____

Matching [p.516-517]

1. _____ "clock"
2. _____ circadian rhythm
3. _____ CO gene
4. _____ phytochrome
5. _____ photoperiodism
6. _____ short-day plant
7. _____ long-day plant
8. _____ day-neutral plant
9. _____ vernalization
10. _____ cryptochrome
11. _____ FT gene
12. _____ bolting

A. "Flowering locus T"; found in companion cells

B. A cycle of activity that has an approximate duration of 24 hours

C. Plants that flower only when the hours of darkness exceed a critical value

D. Blue green photoreceptor pigments sensitive to red and far-red light

E. The return to warmth following cold winter temperatures; necessary for flowering in some plants

F. Plants that flower regardless of night length

G. An internal mechanism that governs the timing of rhythmic cycles of activity

H. When a plant switches from leaf production to flower production

I. Expression is affected by phytochrome and cryptochrome

J. Photoreceptor that absorbs blue-green and UV light

K. The plant's response to changes in the length of night relative to the length of day

L. Plants that flower only when the hours of darkness fall below a critical value

Fill-in-the-Blanks [pp.518-519]

Plant stressors can be (13) _____, from nonliving environmental conditions, or (14) _____, from pathogens or herbivores. The synthesis of (15) _____ can be triggered by temperature extremes and a lack of water. While (16) _____ is involved in the above responses, it is also activated during certain biotic responses. A bacterial invasion triggers the synthesis of (17) _____, which in turn leads to an ABA-mediated nitric oxide response that closes (18) _____ to prevent further bacterial invasion. In a (19) _____ response, plant cells in an infected area commit suicide to prevent spread of the infection to other plant parts, often killing the pathogens also. This is ineffective against those pathogens that kill plant host tissue to obtain nutrients. To combat this, plants use a whole-body, long-term defense called (20) _____. An infected tissue signals other parts of the plant to produce (21) _____ which permits transcription of a whole host of molecules involved in pathogen resistance including (22) _____ produced by the cacao tree.

Tissue damage caused by a caterpillar chewing on a leaf triggers the synthesis of ABA, hydrogen peroxide, ethylene and (23) _____. The latter activates certain genes resulting in the release of volatile chemicals into the (24) _____. These secondary metabolites attract parasitic wasps to the injured plant where they deposit an (25) _____ inside the caterpillar that will hatch and consume its host.

SELF-TEST

For questions 1-4, choose from the following answers:
   a. gibberellins
   b. ethylene
   c. abscisic acid
   d. auxins

____ 1. Promoting fruit ripening and abscission of leaves, flowers, and fruits is a function of _____.

____ 2. The effects of apical dominance are caused by _____.

____ 3. Stem lengthening is a result of cell division and elongation stimulated by _____.

____ 4. Closing of stomates is stimulated by _____.

____ 5. _____ is demonstrated by a germinating seed whose primary root always curves down while its primary shoot always curves up.
   a. Phototropism
   b. Photoperiodism
   c. Gravitropism
   d. Thigmotropism

____ 6. Light in the _____ part of the visible spectrum is the main stimulus for phototropism. [p.531]
   a. blue
   b. yellow
   c. red
   d. green

___ 7. Plants whose flowers are closed during the day but open at night are exhibiting a _____.
   a. growth movement
   b. circadian rhythm
   c. biological clock
   d. both b and c are correct

___ 8. Vernalization is _____.
   a. the death of plant parts in response to cold
   b. daily changes in plant orientation toward gravity
   c. flowering in response to warmth following a period of cold
   d. a protein-related disease

___ 9. Cytokinin _____.
   a. aids in fruit ripening
   b. activates systemic acquired resistance
   c. mediates tropisms
   d. inhibits lateral root formation

___ 10. Which of the following inhibits the recycling of auxin efflux carriers?
   a. ethylene
   b. cytokinin
   c. gibberellins
   d. strigolactone

___ 11. Fruit ripening is induced by _____.
   a. ethylene
   b. auxins
   c. gibberellins
   d. cytokinins

___ 12. Plants that do not require a specific amount of darkness to flower are called day-_____ plants.
   a. long
   b. short
   c. neutral
   d. indifferent

___ 13. Salicylic acid _____.
   a. aids in fruit ripening
   b. activates systemic acquired resistance
   c. mediates tropisms
   d. inhibits lateral root formation

___ 14. Statoliths are _____.
   a. found in root cap cells
   b. amyloplasts
   c. found in chloroplasts
   d. both a and b

CHAPTER OBJECTIVES/REVIEW QUESTIONS

1. Define the term *plant hormone*. [p.506]
2. Describe the role of epicatechin in plant physiology and its effects on human health. [p.505]
3. Understand negative and positive feedback mechanisms in hormonal control. [p.506]
4. Describe the general role of each class of plant hormone. [p.507]
5. Describe the process of apical dominance. [p.509]
6. Understand the general action of auxins. [pp.508-509]
7. Understand the functions of cytokinin. [p.510]
8. What is the role of gibberellin in germination? [p.511]
9. Describe the various functions of Abscisic acid and ethylene. [pp.512-513]
10. Define *phototropism*, *gravitropism*, and *thigmotropism*, and *heliotropism* and give examples of each. [pp.514-515]
11. Define the role of statoliths in gravitropism. [p.514]
12. Explain the role of phototropins in phototropism. [p.515]
13. Give an example of how mechanical stress can affect plants. [p.515]
14. Describe the process of photoperiodism as it relates to circadian cycles and biological clocks. [pp.516-517]
15. Describe the action of phytochromes. [p.516]
16. Describe the photoperiodic responses of "long-day," "short-day," and "day-neutral" plants. [p.517]
17. Define *vernalization*. [p.517]
18. Explain the cascade of events that begin when bacteria infect a plant. [pp.518-519]

INTEGRATING AND APPLYING KEY CONCEPTS

1. An oak tree has grown up in the middle of a forest. A lumber company has just cut down all the surrounding trees except for a narrow strip of woods that includes the oak. How will the oak likely adjust to its changed environment? To what new stresses will it be exposed? Which hormones will most probably be involved in the adjustment?

2. You have been hired by a company in Costa Rica to raise a rare breed of northern plant for distribution. This plant requires 12 hours of continuous dark over a three-week period to flower and form seeds. Explain what facilities you would need to make this possible.

3. As an agricultural consultant, your job is to maximize profits by growing plants as quickly as possible and reducing the amount of loss due to damage or rotting on the way to market. Describe the hormones and other chemicals that you would most likely use in your work and when they would be applied to the plant.

4. You just purchased a property with several greenhouses and want to supply florists with many flowers such as roses, chrysanthemums, carnations, daisies, irises, tulips, daffodils, sunflowers, poinsettias, and lilies. Considering the problem of photoperiodism and flowering, how would you go about designing special set-ups for short-day, long-day, and day-neutral plants? Which plants could be grouped together in each greenhouse?

5. Consider what you have learned about flavonoids and other secondary metabolites concerning their purposes in plant physiology and their uses in human health. What are the consequences, for both plants and humans, of destruction or degradation of plant habitats for plant species, and the introduction of exotic plant pathogens?

31

ANIMAL TISSUES AND ORGAN SYSTEMS

INTRODUCTION

Chapter 31 starts with a description of how cells are organized and connected to form organs, organ systems, and multicellular animals. The four types of animal tissues are described: epithelial, connective, muscle, and nerve. Vertebrate organ systems are introduced along with a more detailed description of how tissues are integrated into a specific organ system—the integumentary system.

STUDY STRATEGIES

- Read all of Chapter 31 with the goal of familiarizing yourself with the boldface terms.
- Recognize the organization of multicellular organisms from cells through organ systems.
- Explain the role of epithelial tissues and their basic structure.
- Describe the roles of connective tissues and identify the various types.
- Discuss the differences between the three types of muscular tissue.
- Explain the role of nervous tissue.
- Identify the different types of organ systems in humans.
- Discuss the role of integument, identify the various parts of skin, and the type of cancers that can affect skin.

FOCAL POINTS

- Figure 31.2 [p.524] illustrates how cells are attached to each other to form coherent tissues.
- Figure 31.5 [p.526] shows various types of epithelium.
- Figure 31.7 [pp.528-529] illustrate connective tissues.
- Figure 31.8 [p.530] has images of muscle tissues.
- Figure 31.9 [p.531] shows a typical neuron.
- Figure 31.12 [pp.532-533] describe anatomical terms and outline the major human organ systems.
- Figure 31.14 [p.534] diagrams human skin structure.
- Figure 31.17 [p.536] illustrates a typical negative feedback mechanism in homeostasis.

INTERACTIVE EXERCISES

31.1. STEM CELLS – IT'S ALL ABOUT POTENTIAL [p.523]

31.2. ORGANIZATION OF ANIMAL BODIES [pp.524-525]

Boldfaced, Page-Referenced Terms

stem cells_____

_____

extracellular fluid _____

_____

interstitial fluid _____

_____

Fill-in-the-Blanks [pp.523-525]

(1) _____ are cells that are undifferentiated and have the capability to produce any type of cell. You started as a single call and all of the cells of your body came from (2) _____ stem cells. Repair and replacement of cells in your body today starts from (3) _____ stem cells. (4) _____ is about how an animal is put together while (5) _____ is about how animals function. Four types of tissue occur in all (6) _____ bodies. (7) _____ tissues cover body surfaces and line internal cavities. (8) _____ tissues hold body parts together and provide structural support. (9) _____ tissues move the body or its parts. (10) _____ tissues detect stimuli and relay information.

(11) _____ is the environment in which the body cells lives. It bathes the cells and provides nutrients and other substances necessary to stay alive. It is also the location in which (12) _____ are deposited. In vertebrates, this fluid is termed (13) _____ and (14) _____.

31.3. EPITHELIAL TISSUE [pp. 526-527]

Boldfaced, Page-Referenced Terms

epithelial tissue _____

_____

basement membrane_____

_____

microvilli_____

_____

gland cells _____

_____

exocrine glands _____

_____

endocrine glands _____

_____

Fill-in-the-Blanks [pp.526-527]

(1) _____ tissue has a free surface, which faces either a body fluid or the outside environment. The

opposite surface is attached to a (2) _____ which helps to attach the tissue to underlying structures. This

tissue may be (3) _____ or (4) _____ layers.

(5) _____ has a single layer of cells and functions as a lining for body cavities, ducts, and tubes. (6)

_____ has two or more layers and typically functions in protection, as it does in the skin. (7) _____

glands secrete mucus, saliva, earwax, milk, oil, digestive enzymes, and other cell products, These products are usually

released onto a free (8) _____ surface through ducts or tubes. (9) _____ glands lack ducts; their

products are (10) _____, which are secreted directly into the fluid bathing the gland. Typically, the (11)

_____ picks up the hormone molecules and distributes them to target cells elsewhere in the body. (12)

_____ junctions only occur in epithelial tissue and prevent cells from leaking fluids between the cells. (13)

_____ junctions are those that act as anchoring points but do not form a seal.

True/False [pp.526-527]

If the statement is true, write a "T" in the blank. If the statement is false, make it correct by changing the underlined word(s) and writing the correct word(s) in the answer blank.

14. _____ Epithelial tissues are named based on the shape of the cells and the number of <u>nuclei</u> in each cell.

15. _____ Simple epithelial tissues have <u>one</u> cell layer.

16. _____ Because individual cells in stratified squamous epithelium are thin, the tissue is best suited for <u>diffusion and osmosis</u>.

17. _____ Columnar cells that are involved in absorption have microvilli on the free edge to increase <u>surface area</u>.

18. _____ Endocrine glands deliver their secretions through <u>ducts</u>.

31.4. CONNECTIVE TISSUES [pp.528-529]

31.5. MUSCLE TISSUES [pp.530-531]

31.6. NERVOUS TISSUES [p.531]

Boldfaced, Page-Referenced Terms

connective tissues _____

_____

cartilage_____

_____

adipose tissue _____

_____

bone tissue _____

_____

blood_____

_____

Choice [pp.528-529]

For questions 1-8, choose from the following types of soft connective tissue.

     a. loose   b. fibrous, irregular   c. fibrous, regular

1. _____ Contains many fibers, mostly collagen-containing ones, in no particular orientation, and a few fibroblasts

2. _____ Rows of fibroblasts often intervene between the bundles of fibers

3. _____ Has its fibers and cells loosely arranged in a semifluid ground substance

4. _____ Has parallel bundles of many collagen fibers and resists being torn apart

5. _____ Forms protective capsules around organs that do not stretch much

6. _____ Often serves as a support framework for epithelium

7. _____ Found in tendons, which attach skeletal muscle to bones

8. _____ Found in elastic ligaments, which attach bones to each other

Matching [pp.528-529]

Match each of the specialized connective tissues to its definition.

9. _____ adipose tissue

10. _____ blood

11. _____ bone

12. _____ cartilage

A. Chock-full of large gat cells; stores excess energy as fats; richly supplied with blood

B. Intercellular material, solid yet pliable, resists compression; structural models for vertebrate embryo bones; maintains shape of nose, outer ear, and other body parts; cushions joints

C. Derived mainly from connective tissue, has transport functions; circulating within plasma are a great many red blood cells, white blood cells, and platelets

D. The weight-bearing tissue of vertebrate skeletons, which support or protect softer tissues and organs; mineral-hardened with calcium-salt-laden collagen fibers and ground substance; interact with skeletal muscles attached to them

Dichotomous Choice [pp.530-532]

Circle one of the two possible answers given between parentheses in each statement.

13. Contractile cells of (skeletal / smooth) muscle tissue taper at both ends.

14. Walls of the stomach and intestine contain (smooth / skeletal) muscle tissue.

15. The only muscle tissue attached to bones is (skeletal / smooth).

16. (Smooth / Skeletal) muscle cells are bundled together in parallel.

17. "Voluntary" muscle action is associated with (smooth / skeletal) muscle tissue.

18. The term *striated* means (bundled / striped).

19. The function of smooth muscle tissue is to (pump blood / cause movement in the internal organs).

20. Cell junctions fuse together the plasma membranes of (smooth / cardiac) muscle cells.

21. Excitable cells are the (neuroglia / neurons).

22. (Neuroglia / Muscle) cells protect the neurons and support them structurally and metabolically.

23. Different types of (neuroglia / neurons) detect specific stimuli, integrate information, and issue or relay commands for response.

Labeling and Matching [pp.526-531] Label each of the following illustrations on the next page with one of the following terms for tissue types: connective, epithelial, muscle, nervous

Complete the exercise by writing all appropriate letters from each of the following groups in the parentheses after each label.

24. _____ (   )

25. _____ (   )

26. _____ (   )

27. _____ (   )

28. _____ (   )

29. _____ (   )

30. _____ (   )

31. _____ (   )

32. _____ (   )

33. _____ (   )

34. _____ (   )

35. _____ (   )

A. Adipose tissue
B. Bone
C. Cardiac muscle
D. Fibrous, regular connective tissue
E. Loose connective tissue
F. Simple columnar epithelium
G. Simple cubodial epithelium
H. Simple squamous epithelium
I. Smooth muscle
J. Skeletal muscle
K. Blood
L. Neurons

a. Absorption
b. Communication by means of electrical signals
c. Energy reserve
d. Contraction for voluntary movements
e. Diffusion
f. Padding
g. Contracts to propel substances along internal passageways; not striated
h. Attaches muscle to bone and bone to bone
i. In vertebrates, provides the strongest internal framework of the organism
j. Elasticity
k. Secretion
l. Pumps circulatory fluid; striated
m. Insulation
n. Transport of nutrients and waste products to and from body cells

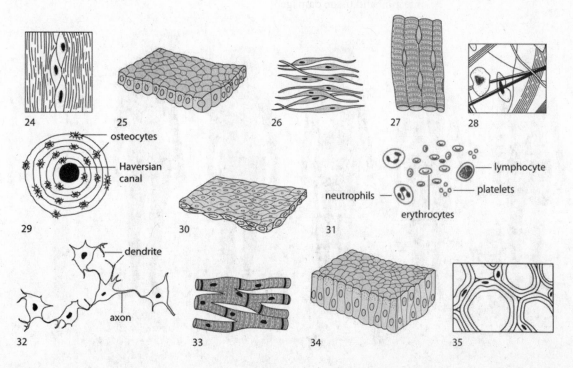

24   25   26   27   28

osteocytes

Haversian canal

lymphocyte

neutrophils

platelets

erythrocytes

29   30   31

dendrite

axon

32   33   34   35

31.7. ORGAN SYSTEMS [pp.532-533]

Labeling and Matching [pp.532-533]
Write the name of each organ system described. Complete the exercise by entering the proper letter form the following illustration in the parentheses after each label.

1. _____ (   ) Rapidly transports many materials to and from cells; helps stabilize internal pH and temperature

2. _____ (   ) Rapidly delivers oxygen to the tissue fluid that bathes all living cells; removes carbon dioxide wastes of cells; helps regulate pH

3. _____ (   ) Maintains the volume and composition of internal environment; excretes excess fluid and blood-borne wastes

4. _____ (   ) Supports and protects body parts; provides muscle attachment sites; produces red blood cells; stores calcium, phosphorus

5. _____ (   ) Hormonally controls body function; works with nervous system to integrate short-term and long-term activities

6. _____ (   ) *Female:* produces eggs; after fertilization, affords a protected, nutritive environment for the development of new individual. *Male:* produces and transfers sperm to the female. Hormones of both systems also influence other organ systems

7. _____ (   ) Ingests food and water; mechanically and chemically breaks down food and absorbs small molecules into internal environment; eliminates food residues

8. _____ (   ) Moves body and its internal parts; maintains posture; generates heat (by increases in metabolic activity)

9. _____ (   ) Detects both external and internal stimuli; controls and coordinates responses to stimuli; integrates all organ system activities

10. _____ (   ) Protects body from injury, dehydration, and some pathogens; controls its temperature; excretes some wastes; receives some external stimuli

11. _____ (   ) Collects and returns some tissue fluid to the bloodstream; defends the body against infection and tissue damage

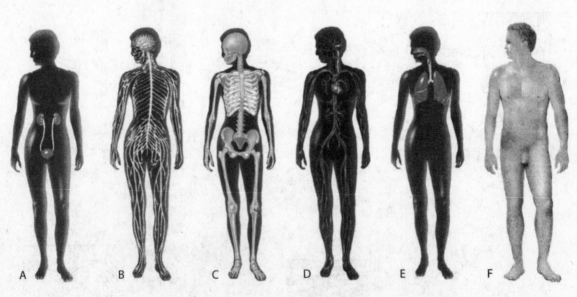

A   B   C   D   E   F

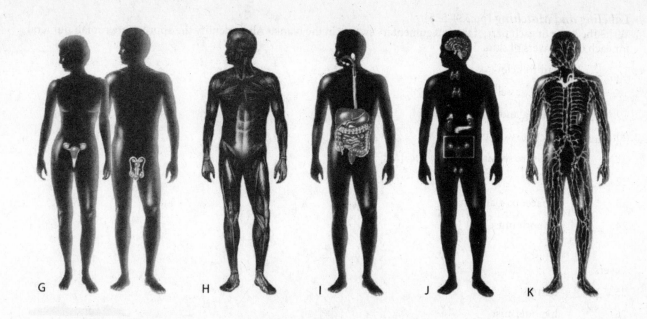

G   H   I   J   K

31.8. HUMAN INTEGUMENTARY SYSTEM [pp.534-535]

31.9. NEGATIVE FEEDBACK IN HOMEOSTASIS [p.536]

Boldfaced, Page-Reference Terms

epidermis _____

_____

dermis _____

_____

Fill-in-the-Blanks [pp.534-535]

The human organ with the largest surface area is the (1) _____. Its upper layer, the (2) _____,

is a(n) (3) _____ epithelium, and its lower region, the (4) _____, is primarily dense (fibrous) (5)

_____. (6) secrete a tough, water-resistant protein called (7) _____ that also makes up most of the

structure of (8) _____. (9) _____ secrete the brownish-black pigment, (10) _____ that helps

prevent (11) _____ radiation from damaging the skin. However, enough of this radiation must get in to

stimulate (12) _____ to produce (13) _____, which helps the body absorb calcium. As we age, skin

becomes less (14) _____. Since adults make new skin cells every day of their lives, that can be (15) _____

for medical uses. These tissues can be used for chronic wounds or deep (16) _____ where all skin cells have

been destroyed.

Labeling and Matching [pp.534-535]

Write the letter of each part of the integumentary system in the blank. Also, identify the appropriate roman numeral for each of the layers of skin.

17. _____ hair follicle

18. _____ blood vessel

19. _____ sweat gland

20. _____ duct of sweat gland

21. _____ pressure sensitive sensory receptor

22. _____ hair

23. _____ sebaceous gland

24. _____ smooth muscle

Layers:

25. _____ dermis

26. _____ hypodermis

27. _____ epidermis

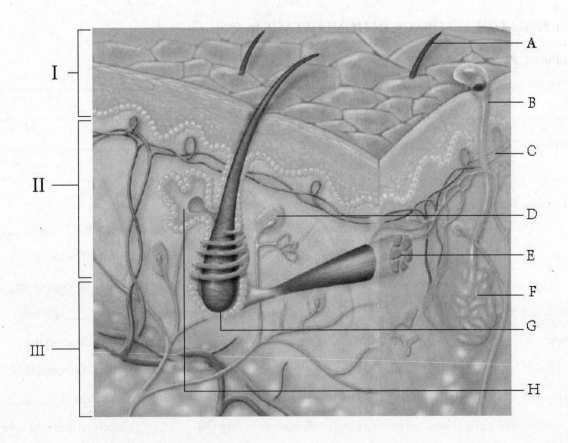

S

Short Answer [pp.534-535]

28. Describe how the skin serves as the first line of defense of the immune system. _____

_____

_____

_____

29. Skin tissues grown in a culture lack what structures normally found in the skin? _____

_____

_____

_____

30. Explain what is meant by a negative feedback loop. _____

_____

_____

_____

Matching [pp.536]

Match each of the actions of negative feedback control in thermoregulation with the appropriate type of control tissue or organ. Some controls may be used more than once.

31. _____ exertion on a hot day; body temperature rises

32. _____ skeletal muscles in the chest wall contract more frequently, increasing the rate of breathing.

33. _____ endocrine glands that affect general activity levels slow secretion of hormones that stimulate activity.

34. _____ monitors internal temperature and sends signal when it increases

35. _____ sweat glands secrete more sweat, which cools the body as it evaporates.

36. _____ receives signals from sensory receptors and signals muscles and glands.

37. _____ smooth muscle in blood vessels supplying the skin relax and widen; more blood flows to skin, and more heat radiates to surrounding air.

38. _____ body temperature declines

A. stimulus

B. sensory receptors

C. brain

D. muscles and glands

E. response

SELF-TEST

___ 1. Which of the following is not one of the connective tissues?
  - a. bone
  - b. blood
  - c. cartilage
  - d. skeletal muscle

___ 2. Gland cells are contained in _____ tissues.
  - a. muscular
  - b. epithelial
  - c. connective
  - d. nervous

___ 3. Blood is considered to be a(n) _____ tissue.
  - a. epithelial
  - b. muscular
  - c. connective
  - d. none of these

___ 4. _____ are abundant in cardiac muscles where they promote diffusion of ions and small molecules from cell to cell.
  - a. Adhesion junctions
  - b. Filter junctions
  - c. Gap junctions
  - d. Tight junctions

___ 5. Muscle that is not striped and is involuntary is _____.
  - a. cardiac
  - b. skeletal
  - c. striated
  - d. smooth

___ 6. A(n) _____ is a group of cells and intercellular substances, all interacting in one or more tasks.
  - a. organ
  - b. organ system
  - c. tissue
  - d. cuticle

___ 7. A graduate student in developmental biology accidentally stabbed a fish embryo. Later, the embryo developed into a creature that could not move and had no supportive or circulatory systems. Which embryonic tissue had suffered the damage?
  - a. ectoderm
  - b. endoderm
  - c. mesoderm
  - d. protoderm

___ 8. A tissue whose cells are striated and fused at the ends be cell junctions so that the cells hold together during forceful contraction is called _____ tissue.
  - a. smooth muscle
  - b. dense fibrous connective
  - c. supportive connective
  - d. cardiac muscle

___ 9. The secretion of tears, milk, sweat, and oil are functions of _____ tissues.
  - a. epithelial
  - b. loose connective
  - c. lymphoid
  - d. nervous

___ 10. Memory, decision making, and issuing commands to effectors are functions of _____ tissue.
  - a. connective
  - b. epithelial
  - c. muscle
  - d. nervous

___ 11. _____ are only present in epithelial tissues.
  - a. Adhesion junctions
  - b. Filter junctions
  - c. Gap junctions
  - d. Tight junctions

___ 12. This layer of the integumentary system contains the blood vessels, nervous tissue, and hair follicles.
  - a. Adhesion junctions
  - b. Filter junctions
  - c. Gap junctions
  - d. Tight junctions

___ 13. Which of the following muscle tissues works with the skeletal system to move a vertebrate?
  - a. smooth
  - b. skeletal
  - c. cardiac
  - d. none of the above

___ 14. A _____ is a supportive cell of the nervous system.
  - a. neuron
  - b. adipose
  - c. neuroglia
  - d. squamous

15. _____ glands do not have ducts.
    a. exocrine
    b. endocrine
    c. taste
    d. epithelial

16. This tissue is responsible for rapid exchange of gasses and regulates pH.
    a. urinary
    b. digestive
    c. respiratory
    d. reproductive

17. This tissue is maintains the volume and composition of the internal environment.
    a. urinary
    b. digestive
    c. respiratory
    d. reproductive

18. This tissue chemically and mechanically breaks down food molecules and eliminates solid waste products.
    a. urinary

    b. digestive
    c. respiratory
    d. reproductive

19. This type of system is the primary way that organisms maintain homeostasis through feedback loops.
    a. positive feedback
    b. negative feedback
    c. baroreceptors
    d. all of the above

20. Which is the correct order for control over homeostasis?
    a. muscles/glands, receptors, response, stimulus, brain
    b. receptors, brain, muscles/glands, response, stimulus
    c. stimulus, receptors, brain, muscles/glands, response
    d. brain, receptors, stimulus, muscles/glands, response

Matching

Choose the most appropriate answer for each term.

21. _____ circulatory system

22. _____ digestive system

23. _____ endocrine system

24. _____ integumentary system

25. _____ muscular system

26. _____ nervous system

27. _____ reproductive system

28. _____ respiratory system

29. _____ skeletal system

30. _____ urinary system

A. Picks up nutrients absorbed from gut and transports them to cells throughout the body

B. Helps cells use nutrients by supplying them with oxygen and relieving them of CO_2 wastes

C. Helps maintain the volume and composition of body fluids that bathe the body's cells

D. Provided basic framework for the animal and supports other organs of the body

E. Uses chemical messengers to control and guide body functions

F. Produces younger, temporarily smaller versions of the animal

G. Breaks down larger food molecules into smaller nutrient molecules that can be absorbed by body fluids and transported to body cells

H. Consists of contractile parts that move the body through the environment and propel substances about in the animal

I. Serves as an electrochemical communications system in the animal's body

J. In the meerkat, serves as a heat catcher in the morning and protective insulation at night

CHAPTER OBJECTIVES/REVIEW QUESTIONS

1. Distinguish simple epithelium from stratified epithelium. [pp.526-527]
2. Name and describe the various types of epithelial tissues as well as their location and general functions. [pp.526-527]
3. Describe the glands that usually secrete their products onto a free epithelial surface through ducts or tubes; cite examples of their products. [p.527]
4. Name the glands that lack ducts; describe their products which are secreted directly into the fluid bathing the gland. [p.527]
5. Distinguish between loose connective tissue; fibrous, irregular connective tissue; and fibrous, regular connective tissue on the basis of their structures and functions. [p.528]
6. Cartilage, bone, adipose tissue, and blood are known as the specialized connective tissues; describe their structures and various functions. [pp.528-529]
7. Distinguish among skeletal, smooth, and cardiac muscle tissues in terms of location, structure, and function. [p.530]
8. Explain the differences between neurons and neuroglia. [p.531]
9. List each of the eleven principal organ systems in humans and list the main task of each. [pp.532-533]

INTEGRATING AND APPLYING KEY CONCEPTS

1. What are some of the uses for adult and embryonic stem cells? Are both equally good for these uses?
2. Explain why, of all places in the body, marrow is located on the interior of long bones. Explain why your bones are remodeled after you reach maturity. Why does your body not keep the same mature skeleton throughout life?
3. You observe a tissue under the microscope and see multiple layers of cells along a free edge (outside environment). What other observations would you need to know to know exactly what type of tissue this is?
4. Explain why dense, regular connective tissue is the best type of connective tissue for ligaments.
5. Select any three organ systems and explain how they interact with each other.

32

NEURAL CONTROL

INTRODUCTION

This chapter provides an overview of how nervous systems, specifically vertebrate nervous systems, function. It describes how neurons work by developing a membrane and then transmitting action potentials. The chapter builds on this information by discussing the role of the major neurotransmitters and how signal integration occurs. The chapter then explores the anatomy and physiology of the human nervous system. It concludes with a discussion of how drugs influence the activity of the nervous system. One challenge in this chapter is that each section presents a number of new terms you need to understand before proceeding to the next section. The other complex process is how the neuron functions on a cell/molecular level—pay attention to the placement and movement of ions. Finally, focus on how the different parts of the nervous system interact with each other.

FOCAL POINTS

- Figure 32.4 [p.543] diagrams the overall structure of vertebrate nervous systems and the interaction of the various divisions.
- Sections 32.3 and 32.4 [pp.544-547] describe the role of a neuron and how it conducts messages using action potentials. Understanding this information is necessary in order to understand the remainder of the chapter.
- Figures 32.5 and 32.6 [p.544] illustrate the three types of neurons and their functional zones.

INTERACTIVE EXERCISES

32.1. IN PURSUIT OF ECSATSY [p.541]

32.2. EVOLUTION OF NERVOUS SYSTEMS [pp.542-543]

Boldfaced, Page-Referenced Terms

nerve net _____

_____

cephalization_____

_____

sensory neurons _____

_____

interneurons _____

_____

motor neurons _____

_____

ganglion _____

_____

nerve cords _____

_____

central nervous system _____

_____

peripheral nervous system _____

_____

nerve _____

_____

Matching [pp.542-543]

Match each of the following terms with its correct definition.

1. _____ nerve net
2. _____ ganglion
3. _____ cephalization
4. _____ motor neuron
5. _____ interneurons
6. _____ sensory neurons
7. _____ nerves
8. _____ neuroglia

A. An asymmetrical mesh of neurons found in organisms with radial symmetry

B. Detect information about stimuli

C. Cells that structurally and metabolically support neurons

D. Accept and process sensory input

E. Relay signals to effectors that carry out responses

F. A cluster of nerve cell bodies

G. Having many sense organs concentrated at one end of the animal

H. Long-distance cable of the nervous system

Labeling [pp.542-543]

Provide the missing label for each numbered item in the accompanying figure.

9. _____

10. _____

11. _____

12. _____

13. _____

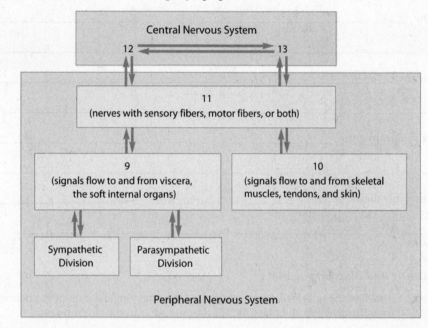

Labeling and Matching [pp.542-543]

Match each nerve or nervous system organ with its appropriate label from the figure.

14. _____ cranial nerves

15. _____ sacral nerves

16. _____ ulnar nerve

17. _____ brain

18. _____ cervical nerves

19. _____ thoracic nerves

20. _____ sciatic nerve

21. _____ spinal cord

22. _____ lumbar nerves

23. _____ coccygeal nerves

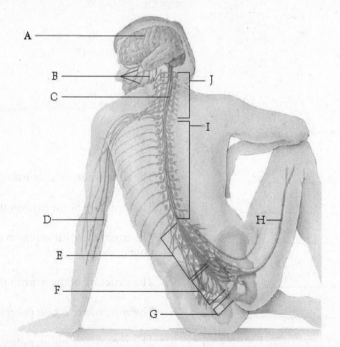

32.3. NEURONS—THE GREAT COMMUNICATORS [pp.544-545]

Boldfaced, Page-Referenced Terms

axon _____

_____

dendrites _____

_____

membrane potential _____

_____

resting membrane potential _____

_____

Labeling and Matching [p.544]

First, label each of the indicated structures in the accompanying diagram of the neuron. Then choose the correct function for each structure and place the appropriate letter in the parentheses.

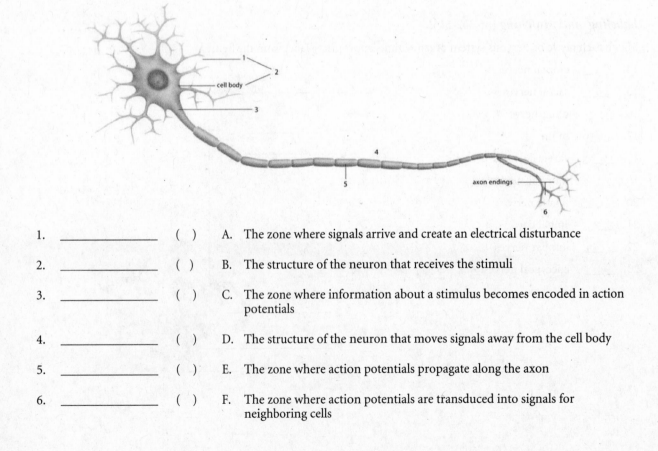

1. _____ ( )   A.   The zone where signals arrive and create an electrical disturbance

2. _____ ( )   B.   The structure of the neuron that receives the stimuli

3. _____ ( )   C.   The zone where information about a stimulus becomes encoded in action potentials

4. _____ ( )   D.   The structure of the neuron that moves signals away from the cell body

5. _____ ( )   E.   The zone where action potentials propagate along the axon

6. _____ ( )   F.   The zone where action potentials are transduced into signals for neighboring cells

Labeling and Matching [p.544]

Match each part of the neuron with its appropriate structure. The structures may be used more than once. Then, label each of the three types of neurons for 19-21.

7. _____   B.   axon

8. _____   C.   cell body

9. _____   D.   axon terminal

10. _____   E.   dendrites

11. _____   F.   peripheral axon

12. _____

13. _____

14. _____

15. _____

16. _____

17. _____

18. _____

19. _____

20. _____

21. _____

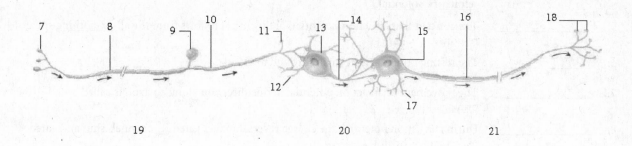

A.   receptor endings

Short Answer [pp.544-545]

22. Describe how neuron structure and function are different from cells in other tissues.

_____

_____

23. Describe the difference in function of dendrites and axons.

_____

_____

32.4. THE ACTION POTENTIALS [pp.546-547]

Boldfaced, Page-Referenced Terms

action potential _____

_____

threshold potential _____

_____

positive feedback _____

_____

True/False [pp.546-547]

If the statement is true, write a "T" in the blank. If the statement is false, make it correct by changing the underlined word(s) and writing the correct word(s) in the answer blank.

1. _____ The ions important in establishing a resting potential are potassium, sodium, and <u>hydrogen</u>.

2. _____ The membrane of a neuron leaks sodium ions. The <u>sodium-potassium pump</u> maintains the resting potential.

3. _____ As an action potential is initiated, <u>potassium</u> channels are the first to open..

4. _____ The minimum amount of stimulus required to start an action potential is called the <u>refractory potential</u>.

5. _____ The concept that all action potentials are identical follows from the <u>all-or-nothing</u> spike response.

6. _____ The resting potential is normally around a <u>+55 mV</u>.

7. _____ The movement of an action potential in one direction along an axon is called <u>propagation</u>.

8. _____ During an action potential, the charge reversal makes gated <u>K^+</u> channels shut and gated <u>Na^+</u> channels open.

Labeling and Matching [p.544]

Match each of the steps of the action potential with the appropriate diagram.

9. _____ Na+ diffuses down its concentration gradient into the neuron.

10. _____ K diffuses down its concentration gradient out of the neuron, making neuron cytoplasm once again more negative than the interstitial fluid.

11. _____ Threshold is reached and Na+ channels open.

12. _____ High membrane potential makes Na+ channels close and K channels open.

13. _____ The Na+ influx increases membrane potential, making the neuron cytoplasm even more positive than the interstitial fluid.

14. _____ Neuron at rest.

15. _____ Once closed, Na channels are briefly inactivated, preventing action potentials from moving "backward."

16. _____ High Na+ in interstitial fluid; high K+inside neuron.

17. _____ Interstitial fluid has more positively charged ions than neuron cytoplasm does, making the membrane potential negative.

18. _____ K channels close.

19. _____ Diffusion of Na to regions farther along the axon propagates the action potential.

20. _____ Gated channels for Na+ or K+ are closed.

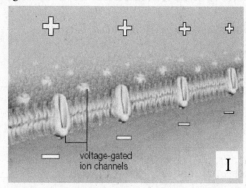

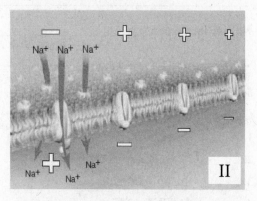

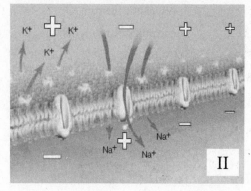

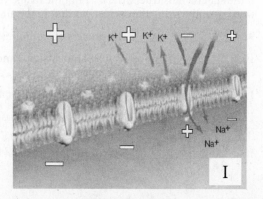

Neural Control   **379**

32.5. HOW NEURONS SEND MESSAGES TO OTHER CELLS [pp.548-549]

32.6. A SMORGASBORD OF SIGNALS [pp.550-551]

Boldfaced, Page-Referenced Terms

synapse _____

_____

neuromuscular junction _____

_____

neurotransmitter _____

_____

synaptic integration _____

_____

Matching [pp.548-549]

Choose the most appropriate statement for each of the following terms.

1. _____ excitatory effect
2. _____ chemical synapse
3. _____ neurotransmitter
4. _____ inhibitory effect
5. _____ presynaptic neuron
6. _____ neuromuscular junction
7. _____ postsynaptic cell
8. _____ acetylcholine

A. Contains gated channels for calcium ions

B. Contains receptors for neurotransmitters

C. Moving a membrane away from the threshold of an action potential

D. A type of neurotransmitter used in neuromuscular junctions

E. A form of synapse between a motor neuron and skeletal muscle

F. Driving a membrane toward the threshold of an action potential

G. Signaling molecules made by neurons

H. A functional bridge between a neuron and another cell

True/False [pp.548-549]

If the statement is true, write a "T" in the blank. If the statement is false, make it correct by changing the underlined word(s) and writing the correct word(s) in the answer blank.

9. _____ The process of summing all the inputs to a neuron is called neuron differentiation.

10. _____ If an inhibitory signal and an excitatory signal arrive at the same time, they normally cancel each other.

11. _____ The action potential is passed to the next neuron through the output zone.

12. _____ Neurotransmitters diffuse across the synapse and attach to sodium ions on the postsynaptic neuron.

13. _____ Neurotransmitters must be cleaned out of the synapse after the signal transfer is complete.

Matching [pp.550-551]

Match each of the following neurotransmitters and neuropeptides to the correct statement.

14. _____ serotonin

15. _____ GABA

16. _____ neuromodulators

17. _____ enkephalins and endorphins

18. _____ norepinephrine and epinephrine

19. _____ dopamine

20. _____ substance P

A. The major neurotransmitter inhibitor in the brain; derived from glutamate

B. Influences mood and memory; derived from tryptophan

C. Prime the body to respond to stress

D. A neuromodulator that enhances pain reception

E. Natural painkillers

F. Influences fine motor control and pleasure-seeking behaviors

G. A class of chemicals that magnify or reduce the effect of a neurotransmitter

Complete the Table [p.551]

21. Drugs disrupt signaling at synapses. Complete the following table, which summarizes information about these molecules.

Drug	Category	Description/Function
a. nicotine		
b. cocaine		
c. amphetamine		
d. alcohol		
e. barbituates		
f. morphine		
g. oxycodone		
h. LSD		
i. marijuana		

Short Answer [p.551]

22. Describe what drug addiction is. _____

_____

_____

23. Explain five of the warning signs of drug addiction. _____

_____

_____

32.7. THE PERIPHERAL NERVOUS SYSTEM [pp.552-553]

Boldfaced, Page-Referenced Terms

myelin sheath _____

_____

somatic nervous system _____

_____

autonomic nervous system _____

_____

sympathetic neurons _____

_____

parasympathetic neurons _____

_____

fight-flight response _____

Matching [pp.552-553]

Match each of the following terms with the appropriate statement.

1. _____ myelin sheath
2. _____ Schwann cell
3. _____ Autonomic Nervous System
4. _____ Peripheral Nervous System
5. _____ Fight-Flight Response

A. An electric insulator formed by Schwann cells

B. When nerve signals put you in a state of intense arousal

C. Contains all nerves outside of the central nervous system

D. A part of the nervous system concerned with signals to and from internal organs

E. A glial cell that produces the myelin around peripheral nerves

Short Answer [pp.552-553]

6. Explain the benefit of having a myelin sheath on a peripheral nerve.

   _____

   _____

7. Explain the interaction between the sympathetic and parasympathetic divisions of the nervous system.

   _____

   _____

Labeling [p.553]

Label each numbered part of the accompanying illustration.

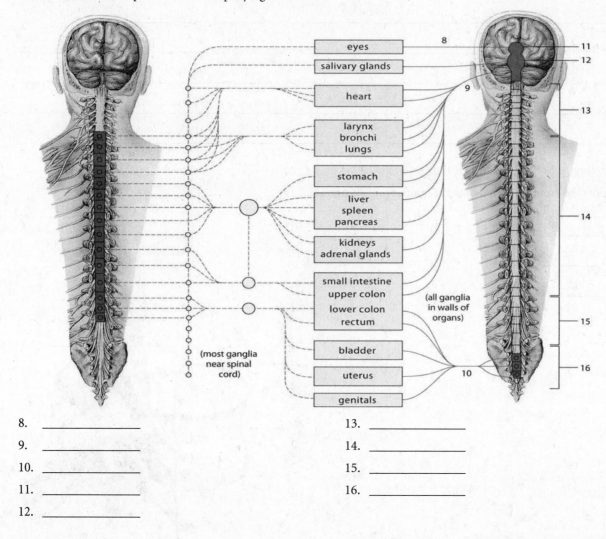

8. _____

9. _____

10. _____

11. _____

12. _____

13. _____

14. _____

15. _____

16. _____

32.8. THE SPINAL CORD [pp. 554-555]

Boldfaced, Page-Referenced Terms

spinal cord _____

_____

meninges _____

_____

cerebrospinal fluid _____

_____

white matter _____

_____

gray matter _____

_____

reflex _____

_____

Labeling [p.554]

Identify the numbered parts of the accompanying illustration.

1. _____

2. _____

3. _____

4. _____

5. _____

Matching [p.555]

Match each of the following statements to the appropriate letter in the accompanying diagram.

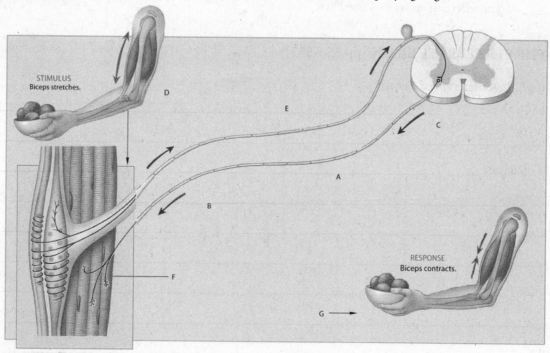

6. _____ Axons of the motor neurons synapse with muscle cells

7. _____ Neurotransmitter is released from the sensory neuron and stimulates the motor neuron

8. _____ Neurotransmitter released from the motor neuron stimulates the plasma membrane of muscle cells

9. _____ The action potential is propagated along the axon of the motor neuron

10. _____ The load is placed on the muscle tissue

11. _____ The stretching of the muscle tissue in response to the load stimulates receptors, generating an action potential

12. _____ The muscle is stimulated and contracts

Choice [pp.552-555]

For each of the numbered statements, choose the most appropriate group of nerves from the lettered list:

    a. peripheral—somatic nerves   b. peripheral—autonomic sympathetic nerves

    c. peripheral—autonomic parasympathetic nerves   d. spinal cord nerves

13. _____ Dominate when the body is in a state of relaxation.

14. _____ Dominate in times of stress, excitement, and danger.

15. _____ Can be attacked by meningitis.

16. _____ The sensory part of these nerves delivers information from receptors in the skin, skeletal muscles, and tendons to the central nervous system.

17. _____ Tend to slow down the body overall and divert energy to basic "housekeeping" tasks, such as digestion.

18. _____ The meninges belong to this division.

19. _____ An expressway for signals between the peripheral nervous system and the brain.

20. _____ The fight–flight response is directed by this division.

21. _____ Some reflex actions are processed here.

32.9. THE VERTEBRATE BRAIN [pp.556-557]

Boldfaced, Page-Referenced Terms

blood-brain barrier _____

_____

medulla oblongata _____

_____

pons _____

_____

cerebellum _____

_____

cerebrum _____

_____

thalamus _____

_____

hypothalamus _____

_____

Choose the most appropriate statement for each term.

1. _____ medulla oblongata
2. _____ cerebellum
3. _____ pons
4. _____ brain stem
5. _____ cerebrum
6. _____ thalamus
7. _____ hypothalamus
8. _____ blood–brain barrier

A. Main center for homeostatic control of the internal environment

B. Center for sorting out sensory information and sending it to the cerebrum

C. In vertebrates, site where olfactory information is integrated and responded to

D. Houses reflex centers for respiration and circulation

E. Protects the spinal cord and brain from harmful substances

F. Uses sensory input to control motor skills and posture

G. Controls signal traffic between the cerebellum and forebrain

H. The most ancient nervous tissue; continuous with the spinal cord

Choice [pp.556-557]

Indicate the area of the brain to which each structure belongs.

    a. forebrain   b. hindbrain   c. midbrain

9. _____ cerebrum
10. _____ thalamus
11. _____ hypothalamus
12. _____ cerebellum
13. _____ medulla oblongata
14. _____ pons
15. _____ tectum

32.10. A CLOSER LOOK AT THE HUMAN CEREBRUM [pp.558-559]

32.11. EMOTION AND MEMORY [p.559]

Boldfaced, Page-Referenced Terms

cerebral cortex _____

_____

primary motor cortex _____

_____

primary somatosensory cortex _____

_____

limbic system _____

_____

Choice [pp.558-559]

For each statement, choose the appropriate area of the brain from the list provided. Some answers may be used more than once.

1. _____ Controls emotions and functions in memory

2. _____ Promotes chemical changes that affect states of consciousness

3. _____ Receiving center for information from skin and joints

4. _____ Governs learned patterns of motor skills

5. _____ Controls and coordinates movements of skeletal muscles

6. _____ Involved in interpreting social cues

7. _____ Contains the primary visual cortex

8. _____ Translates thoughts into speech

9. _____ Coordinates organ behavior for self-gratifying behaviors

10. _____ Perceptions of sound and odor arise here

11. _____ Personality and intellect start here

A. Broca's area

B. primary somato sensory cortex

C. temporal lobe

D. primary motor cortex

E. premotor cortex

F. occipital lobe

G. prefrontal cortex

H. limbic system

I. reticular formation

Short Answer [p.559]

12. In an effort to relieve the frequent seizures of severe epilepsy, neural surgeon Roger Sperry cut the neural bridge of the corpus callosum of several of these patients. The seizures did subside in frequency and intensity. Summarize the subsequent findings of Sperry regarding the function of the corpus callosum.

_____

_____

_____

Fill-in-the-Blanks [p.559]

(13) _____ term memory lasts just a few seconds or hours. This stage holds a few bits of

information—a set of numbers, the words of a sentence, and so on. In (14) _____ -term memory, a seemingly

unlimited quantity of larger bits is stored more or less (15) _____. Different forms of input are stored and

called up by different mechanisms. Retention is best for (16) _____ memories, which are created when you

consciously repeat an activity over and over. As the skill is being learned, the (17) _____ cortex signals (18)

_____ areas of the cortex. Signals flow into the sensory cortex, the (19) _____, and the corpus

striatum—a part of the (20) _____. Once the skill is mastered, the corpus striatum is able to call for

appropriate movements, which frees you from having to consciously think about how to make the movements. (21)

_____ memory allows you to remember how a lemon smells. It starts with signals from the (22) _____

cortex to the amygdala, which acts as the (23) _____ for those memories. The amygdala connects to the (24)

_____, which serves as a(n) (25) _____ center. Signals must loop repeatedly through the hippocampus,

cortex (26) _____, and (27) _____ for a memory to be retained.

32.12. NEUROGLIA—THE NEURON'S SUPPORT STAFF [pp.560-561]

Complete the Table [pp.560-561]

1. There are four types of neuroglial cells. Complete the following table, which summarizes information about
   these cells.

RNA Molecule	*Description/Function*
a. Ependymal cell	
b. Microglia	
c. Astrocyte	
d. Oligodendrocyte	

SELF-TEST

___ 1. Which of the following is not true of an action potential?
   a. It is a short-range message that can vary in size
   b. It is an all-or-none brief reversal in membrane potential
   c. It doesn't decay with distance
   d. It is self-propagating

___ 2. The conducting zone of a neuron is the _____.
   a. axon
   b. axon endings
   c. cell body
   d. dendrite

___ 3. The output zone of a neuron is the _____.
   a. axon
   b. axon endings
   c. cell body
   d. dendrite

___ 4. An action potential is brought about by _____.
   a. a sudden membrane impermeability
   b. the movement of negatively charged proteins through the neuronal membrane
   c. the movement of lipoproteins to the outer membrane
   d. a local change in membrane permeability caused by a greater-than-threshold stimulus

___ 5. The resting membrane potential _____.
   a. exists as long as a voltage difference sufficient to do work exists across a membrane
   b. occurs because there are more potassium ions outside the neuronal membrane than inside
   c. occurs because of the unique distribution of receptor proteins located on the dendrite exterior
   d. is brought about by a local change in membrane permeability caused by a greater-than-threshold stimulus

___ 6. The phrase "all-or-nothing" used in conjunction with discussion about an action potential means that _____.
   a. a resting membrane potential has been received by the cell
   b. an impulse does not diminish or dissipate as it travels away from the trigger zone
   c. the membrane either achieves total equilibrium or remains as far from equilibrium as possible
   d. propagation along the neuron is much faster than in other neurons

___ 7. An action potential passes from neuron to neuron across a synaptic cleft by _____.
   a. myelin bridges
   b. the resting membrane potential
   c. neurotransmitter substances
   d. neuromodulator substances

___ 8. _____ nerves dominate when the body is not receiving much outside stimulation.
   a. Ganglia
   b. Pacemaker
   c. Sympathetic
   d. Parasympathetic
   e. All of the above

___ 9. Oligodendrocytes and astrocytes are examples of _____.
   a. neurons
   b. neuroglial cells
   c. nerves
   d. neuropeptides

___ 10. The _____ are the protective coverings of the brain and spinal cord.
   a. ventricles
   b. meninges
   c. tectums
   d. olfactory bulbs
   e. pineal gland

11. The _____ regulates body temperature, thirst, and hunger and serves as an endocrine gland.
    a. medulla
    b. pons
    c. thalamus
    d. hypothalamus

12. The part of the brain that controls the basic responses necessary to maintain life processes (respiration, blood circulation) is the _____.
    a. cerebral cortex
    b. cerebellum
    c. corpus callosum
    d. medulla oblongata

13. The _____ integrates sensory input from the eyes, ears, and muscle spindles with motor signal from the fore-brain; it also helps control motor dexterity.
    a. cerebrum
    b. pons
    c. cerebellum
    d. hypothalamus
    e. thalamus

14. The _____ evolved as a coordinating center for sensory input and as a relay station for signals to the cerebrum.
    a. medulla
    b. pons
    c. reticular formation
    d. hypothalamus
    e. thalamus

15. Which neurotransmitter affects fine motor control and pleasure-seeking behaviors?
    a. Epinephrine
    b. Dopamine
    c. GABA
    d. Serotonin

16. Nerve nets are the rudimentary nervous systems of which of the following?
    a. Sea anemones
    b. jelly fish
    c. corals
    d. all of the above

17. Flatworms have a pair of _____ which often serves as their integrating center.
    a. nerves
    b. ganglia
    c. brains
    d. spinal cords

18. The input zone of a neuron is the _____.
    a. axon
    b. axon endings
    c. cell body
    d. dendrite

19. The two ions involved in the action potential of a nerve are
    a. oxygen and potassium
    b. calcium and magnesium
    c. sodium and potassium
    d. sodium and magnesium

20. Schwann cells produce _____ that ensures quick firing of neurons.
    a. adipose
    b. actin
    c. myosin
    d. myelin

21. This system is directly involved in emotion and memory
    a. sympathetic
    b. parasympathetic
    c. limbic
    d. olfactory

22. This portion of the brain controls reflexes such as breathing, swallowing, coughing, and sneezing.
    a. corpus callosum
    b. medulla oblongata
    c. cerebellum
    d. pons

23. This disease is caused by disorders in the dopamine-secreting neurons.
    a. Alzheimer's
    b. Parkinson's
    c. Multiple sclerosis
    d. Muscular dystrophy

24. Ritalin is a medication used in the treatment of _____.
    a. depression
    b. migraines
    c. ADHD
    d. Alzheimer's

25. The ventricles are critical to the circulation of this fluid.
    a. cerebrospinal fluid
    b. blood
    c. urine
    d. interstitial fluid

CHAPTER OBJECTIVES/REVIEW QUESTIONS

1. Define the terms *sensory neuron*, *interneurons*, *motor neurons*, and *ganglia*. [p.542]
2. Explain the major stages in the evolution of nervous systems. [pp.542-543]
3. Draw a neuron and label it according to its three general zones, its specific structures, and the specific function(s) of each structure. [p.544]
4. Define *resting membrane potential*; explain what establishes it and how it is used by the cell neuron. [pp.544-545]
5. Define *action potential* and state how sodium and potassium ions are used to generate an action potential. [pp.546-547]
6. Understand the importance of the sodium–potassium pump in maintaining the resting membrane potential. [pp.547-548]
7. Explain how graded signals differ from action potentials. [pp.547-548]
8. Explain the importance of a threshold level in generating an action potential. [pp.547-548]
9. Understand the relationship between neurotransmitters and chemical synapses. [pp.548-549]
10. Understand the difference between pre- and postsynaptic cells. [pp.548-549]
11. Understand the process of synaptic integration. [p.549]
12. Understand the three mechanisms by which neurotransmitters are removed from the synaptic cleft. [pp.548-549]
13. Recognize the general role of the major neurotransmitters and neuropeptides. [pp.550-551]
14. List the major classes of psychoactive drugs and provide an example of each class. [p.551]
15. Explain what the stretch reflex is and tell how it helps an animal survive. [pp.554-555]
16. Describe the function of each major division of the peripheral nervous system. [pp.554-555]
17. Describe the structure and function of the spinal cord. [pp.554-555]
18. List the parts of the brain found in the hindbrain, midbrain, and forebrain, and tell the basic functions of each. [pp.556-557]
19. In terms of structure and function, explain how the mechanism called the blood–brain barrier protects the brain and spinal cord. [pp.556-557]
20. Recognize the general function of each area of the cerebrum. [p.558]
21. Explain the function of the limbic system. [p.559]
22. Distinguish between short-term and long-term information storage. [p.559]
23. Explain the roles of the various types of neuroglial cells. [pp.560-561]

INTEGRATING AND APPLYING KEY CONCEPTS

1. What do you think might happen to human behavior if inhibitory postsynaptic potentials did not exist and if the threshold stimulus necessary to provoke an action potential were much higher?
2. Suppose that anger is eventually determined to be caused by excessive amounts of specific transmitter substances in the brains of angry people. Also suppose that an inexpensive antidote to anger that neutralizes these anger-producing transmitter substances is readily available. Can violent murderers now argue that they have been wrongfully punished because they were victimized by their brain's transmitter substances and could not have acted in any other way? Suppose an antidote is prescribed to curb violent tempers in an easily angered person. Suppose also that the person forgets to take the pill and subsequently murders a family member. Can the murderer still claim to be victimized by transmitter substances?

33

SENSORY PERCEPTION

INTRODUCTION

This chapter describes the different kinds of sensory receptors and shows how they are used in the special senses. The five major senses of vertebrates are looked at in detail.

STUDY STRATEGIES

- Read all of Chapter 33 with the goal of familiarizing yourself with the boldface terms.
- Identify the various types of sensory receptors and pathways.
- Explain sensory perception and adaptation.
- Compare and contrast somatic and visceral sensations.
- Describe the somatosensory cortex, the sense of touch, muscle sense, and pain.
- Explain the phenomena of referred pain.
- Discuss the chemical senses of olfaction and taste.
- Describe the sense of vision, including the various types of eyes in the animal kingdom.
- Recognize the structures of the human eye and the function of each.
- Explain visual processing and some common visual disorders.
- Describe the senses of hearing and equilibrium.

FOCAL POINTS

- Figure 33.7 [p.569] illustrates site of referred pain
- Figures 33.8 and 33.9 [pp.570-571] depict the structures involved in the chemical senses.
- Figure 33.22 [p.578] shows the anatomy of the human ear.
- Figure 34.33.15 [p.574] diagrams the human eye.
- Figures 33.18 [p.576] look at rods and cones and how they are arranged in the retina.
- Section 33.8 [p.577] describes visual disorders and eye diseases.

INTERACTIVE EXERCISES

33.1. A WHALE OF A DILEMMA [p.565]

33.2. OVERVIEW OF SENSORY PATHWAYS [pp.566-567]

Boldfaced, Page-Referenced Terms

stimulus _____

mechanoreceptors _____

_____

chemoreceptors _____

_____

pain receptors _____

_____

thermoreceptors _____

_____

photoreceptors _____

_____

osmoreceptors _____

_____

sensation _____

_____

sensory adaptation _____

_____

sensory perception _____

_____

True/False [pp.565-566]

If the statement is true, write a "T" in the blank. If the statement is false, make it correct by changing the underlined word(s) and writing the correct word(s) in the answer blank.

1. _____ <u>Perception</u> is the conscious awareness of change in internal or external conditions.

2. _____ A generic term for any energy that activates a specific receptor is a <u>stimulus</u>.

3. _____ <u>Nociceptors</u> detect chemicals dissolved in the fluid around them.

4. _____ Changes in concentration of solutes in the body are detected by <u>osmoreceptors</u>.

5. _____ Signals coming over the optic nerve are always interpreted as <u>vision</u> by the brain regardless of the original stimulus.

6. _____ A weak stimulus recruits the <u>same number of</u> sensory receptors as a stronger stimulus.

7. _____ The fact that you don't feel a hat on your head a few minutes after putting it on is an example of <u>sensory integration</u>.

8. _____ Sensory receptors in the skin and skeletal muscles and near joints produce <u>visceral</u> sensations.

Select the best term for each of the following descriptions.

   a. chemoreceptors   b. mechanoreceptors   c. nociceptors   d. photoreceptors   e. thermoreceptors

9. _____ Associated with vision

10. _____ Associated with pain

11. _____ Detect odors

12. _____ Detect sounds

13. _____ Detect CO_2 concentration in the blood

14. _____ Detect environmental temperature

15. _____ Detect internal body temperature

16. _____ Detect touch

17. _____ Rods and cones

18. _____ Hair cells in the ear's organ of Corti

19. _____ Pacinian corpuscles in the skin

20. _____ Olfactory receptors in nose

21. _____ Associated with the movement of fluid in the inner ear

Labeling and Matching[p.565-566]

Match the letter from the figure to each of the mechanoreceptors. Then, place the appropriate type of sensation in the parentheses.

22. _____ (   ) Free nerve endings

23. _____ (   ) Meissner's corpuscles

24. _____ (   ) Ruffini endings

25. _____ (   ) bulb of Krause

26. _____ (   ) Pacinian corpuscles

a.   detect light touches.
b.   respond to more pressure
c.   adapt more slowly than other pressure sensors.
d.   responds to touch and cold.
e.   can detect hair movements, temperature changes, or tissue damage

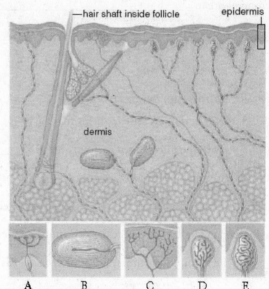

33.3. SOMATIC AND VISCERAL SENSATIONS [pp.568-569]

33.4. CHEMICAL SENSES [pp.570-571]

Boldfaced, Page-Referenced Terms

somatic sensations _____

_____

visceral sensations _____

_____

pain _____

_____

olfaction _____

_____

taste receptors _____

_____

pheromone _____

_____

vomeronasal organ _____

_____

Fill-in-the-Blanks [pp.568-571]

Somatic sensations arise in the primary (1) _____. The two largest parts of this brain area correspond to body parts with the most sensory activity, the (2) _____ and the (3) _____. This region responds to input from receptors of various types. (4) _____ are simple, usually unmyelinated receptors that respond to pressure, temperature, and (5) _____. Encapsulated receptors, like the (6) _____ abundant in the lips, fingertips, and genitalia, also detect sensory input. Tissue injury triggers the release of chemicals that stimulate (7) _____ receptors. Sometimes, this stimulation leads to the release of the natural opiates (8) _____ and (9) _____. Chemical receptors detect molecules that become dissolved in fluid next to them. Receptors are the modified endings of (10) _____ neurons. Animals smell substances by means of (11) _____ receptors, such as the ones in your (12) _____; a human has about (13) _____ million of these. Sensory nerve pathways—the (14) _____ bulb and nerve tract—lead from the nasal cavity to the region of the brain where odors are identified and associated with their sources. (15) _____ are signaling molecules secreted by one individual that influence the behavior of another. These molecules also target (16) _____ receptors in the (17)

_____ organ. In the case of taste, receptors are located on animal tongues, often as part of sensory organs called (18) _____, which are enclosed by circular papillae.

33.5. DIVERSITY OF VISUAL SYSTEMS [pp.572-573]

33.6. A CLOSER LOOK AT THE HUMAN EYE [pp.574-575]

33.7. LIGHT RECEPTION AND VISUAL PROCESSING [pp.576-577]

33.8. VISUAL DISORDERS [p.577]

Boldfaced, Page-Referenced Terms

vision _____

_____

eyes _____

_____

lens _____

_____

compound eyes _____

_____

camera eyes _____

_____

retina _____

_____

conjunctiva _____

_____

cornea _____

_____

sclera _____

_____

choroid _____

_____

iris _____

_____

pupil _____

_____

visual accomodation _____

_____

ciliary muscle _____

_____

Fill-in-the-Blanks [pp.572-577]

Complex eyes have a(n) (1) _____ to focus light rays onto the retina. Even more complex eyes have

a(n) (2) _____ for additional focusing and a ring of contractile tissue, the (3) _____, which helps to

regulate the amount of light entering the eye. The (4) _____, a dense fibrous layer, protects the eyeball. In

humans, muscles can change the shape of the (5) _____ in a process called (6) _____. Two types of

receptors are found in the retina, (7) _____, which respond to dim light, and (8) _____, which respond

to bright light and color. (8) _____ are most concentrated in a retinal area called the (9) _____. Rods

contain (10) _____, which is stimulated by photons of (11) _____ wavelengths. Sensory input from the

receptors is carried to the brain via the (12) _____. The region where the optic nerve exits the eye has no rods

or cones and is called the (13) _____.

Labeling and Matching [p.574]

Match each indicated part to the appropriate letter of the accompanying illustration.   Then, place the function of each of the structures in the parentheses.

14. _____ ( ) retina

15. _____ ( ) iris

16. _____ ( ) optic nerve

17. _____ ( ) ciliary body

18. _____ ( ) sclera

19. _____ ( ) lens

20. _____ ( ) vitreous humor

21. _____ ( ) choroid

22. _____ ( ) aqueous humor

23. _____ ( ) fovea

24. _____ ( ) pupil

25. _____ ( ) cornea

26. _____ ( ) optic disk (blind spot)

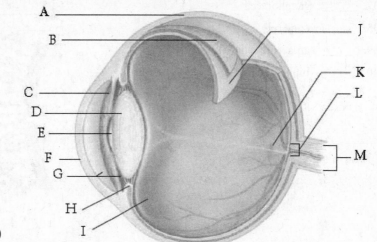

a. Protects eyeball
b. Focuses light
c. Serves as entrance for light
d. Adjusts diameter of pupil
e. Its muscles control the lens shape and hold lens in place
f. Its blood vessels nutritionally support wall cells; its pigments stop light scattering
g. Carries signals to brain
h. Absorbs, transduces light energy
i. Focuses light on photoreceptors
j. Transmits light, maintains fluid pressure
k. Transmits light, supports lens and eyeball

True/False [p.577]

If the statement is true, write a "T" in the blank. If the statement is false, make it correct by changing the underlined word(s) and writing the correct word(s) in the answer blank.

27. _____ <u>Nearsightedness</u> occurs when images are focused behind the retina.

28. _____ Macular degeneration occurs when there a loss of cones in the <u>fovea</u>.

29. _____ Increased pressure of the aqueous humor is called <u>glaucoma</u>.

30. _____ Cataracts are defects in the <u>cornea</u> that cloud vision.

31. _____ Astigmatism is caused by defects in the <u>lens</u> that prevent light from being focused properly.

32. _____ Nutritional blindness can be caused by a lack of <u>vitamin A</u> in the diet.

33. _____ The most common form of color blindness is the inability to distinguish <u>blue</u> from green.

34. _____ Onochocerciasis is a form of blindness caused by the <u>bacterium</u> *Chlamydia trachomatis*.

Labeling and Matching [p.576]

Match each indicated retinal cell to the appropriate letter of the accompanying illustration.

35. _____ rod cell

36. _____ horizontal cell

37. _____ ganglion cell

38. _____ cone cell

39. _____ amacrine cell

40. _____ bipolar cell

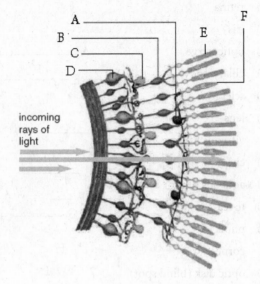

33.9. SENSE OF HEARING [pp.578-579]

33.10. ORGANS OF EQUILIBRIUM [p.580]

Boldfaced, Page-Referenced Terms

hearing _____

_____

outer ear _____

_____

middle ear _____

_____

ear drum _____

_____

inner ear _____

_____

cochlea _____

_____

organ of Corti _____

_____

organs of equilibrium _____

_____

vestibular apparatus _____

_____

True/False [pp.578-580]

If the statement is true, write a "T" in the blank. If the statement is false, make it correct by changing the underlined word(s) and writing the correct word(s) in the answer blank.

1. _____ Semicircular canals sense changes in rotation when hair cells are deformed by the movement of <u>calcium grains</u>.

2. _____ <u>Vertigo</u> is the sensation that the world is spinning around you.

3. _____ The saccule and utricle sense <u>dynamic equilibrium</u>.

4. _____ The <u>eardrum</u> transforms sound waves into vibrations of a thin membrane.

5. _____ The <u>middle ear bones</u> amplify the sound waves and transfer them to the fluid of the inner ear.

6. _____ The sound waves are converted to action potentials when movement of the <u>tectorial membrane</u> causes bending of hair cells.

7. _____ The inner ear contains both the vestibular apparatus and the <u>hammer, anvil, and stirrup</u>.

8. _____ Differences in pitch of a sound are determined by which part of the <u>basilar membrane</u> vibrates.

Labeling [pp.578-579]

Match each indicated structure in the human ear to the appropriate letter of the accompanying illustration.

9. _____ rod cell

10. _____ cochlea

11. _____ round window

12. _____ auditory nerve

13. _____ eardrum

14. _____ stirrup

15. _____ oval window

16. _____ hammer

17. _____ anvil

18. _____ auditory canal

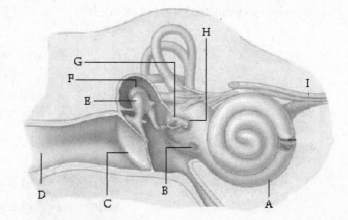

Labeling [pp. 579]

Match each indicated structure of the inner ear to the appropriate letter of the accompanying illustration.

19. _____ cochlear duct
20. _____ basilar membrane
21. _____ Organ of Corti
22. _____ vestibular duct

23. _____ hair cells
24. _____ tympanic duct
25. _____ tectonic membrane

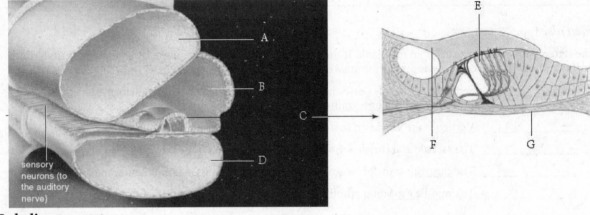

Labeling [pp. 580]

Match each indicated structure of the structures of equilibrium to the appropriate letter of the accompanying illustration.

26. _____ saccule
27. _____ sensory neurons
28. _____ semicircular canals
29. _____ hair cells
30. _____ utricle
31. _____ vestibular nerve
32. _____ gelatinous membrane

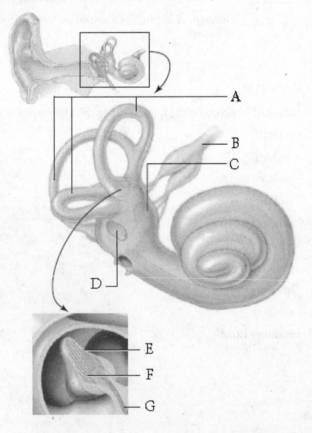

33. If consistent exposure to loud sounds can damage the sensory hairs of the inner ear, what could be the long term result and why?

_____

_____

_____

SELF-TEST

__ 1. Mechanoreceptors are important to
    _____.
    a. hearing
    b. regulating blood pressure
    c. both a and b
    d. none of the above

__ 2. Chemoreceptors are important to
    _____.
    a. taste
    b. smell
    c. both a and b
    d. none of the above

__ 3. Which part of the middle ear directly transfers vibrations to the oval window?
    a. Anvil
    b. Eardrum
    c. Hammer
    d. Stirrup

For questions 4–8, choose from the following answers:
    a. fovea
    b. cornea
    c. iris
    d. retina
    e. sclera

__ 4. The white protective fibrous tissue of the eye is the _____.

__ 5. Rods and cones are located in the _____.

__ 6. The highest concentration of cones is in the _____.

__ 7. The adjustable ring of contractile and connective tissues that controls the amount of light entering the eye is the _____.

__ 8. The outer transparent protective covering of part of the eyeball is the _____.

__ 9. Visual accommodation in mammals involves the ability to _____.
    a. change the sensitivity of the rods and cones by means of transmitters
    b. change the thickness of the lens by relaxing or contracting certain muscles
    c. change the curvature of the cornea
    d. adapt to large changes in light intensity
    e. all of the above

__ 10. Nearsightedness is caused by _____.
    a. eye structure that focuses an image in front of the retina
    b. uneven curvature of the lens
    c. eye structure that focuses an image posterior to the retina
    d. uneven curvature of the cornea
    e. none of the above

__ 11. Astigmatism is caused by _____.
    a. eye structure that focuses an image in front of the retina
    b. uneven curvature of the lens
    c. eye structure that focuses an image posterior to the retina
    d. uneven curvature of the cornea
    e. none of the above

__ 12. When visceral pain is felt in an area of the body different than where the injury/damaged tissues are, that is called _____ pain.
    a. phantom
    b. continuous
    c. referred
    d. chronic

For questions 13–17, choose from the following
answers:
  a. eardrum
  b. auditory canal
  c. ear bones
  d. cochlea
  e. semicircular canals

___ 13. Pressure waves in this structure trigger hair
     cells that conduct a nerve signal.

___ 14. These vibrate against each other and the
     oval window, converting mechanical energy
     into fluid waves.

___ 15. Otherwise known as the tympanic
     membrane.

___ 16. Used to monitor balance and equilibrium.

___ 17. Funnels sound waves from the atmosphere
     towards the middle ear.

___ 18. "Car sickness" is caused by mismatched
     signals between _____.
     a. eyes and inner ear
     b. smell and eyes
     c. eyes and taste
     d. inner ear and touch

___ 19. Nocturnal animals often have _____
     that are larger than similar diurnal animals.
     a. eyes
     b. noses
     c. ears
     d. tongues

___ 20. Insects have _____.
     a. complex eyes
     b. camera eyes
     c. compound eyes
     d. eye spots

___ 21. Pheromones are useful for _____ in
     animals.
     a. defense
     b. feeding
     c. birth
     d. communication

___ 22. These substances block pain receptors and
     are often available over the counter.
     a. substance P
     b. analgesics
     c. neurotransmitters
     d. pheromones

CHAPTER OBJECTIVES/REVIEW QUESTIONS

1. Define and distinguish among chemoreceptors, mechanoreceptors, photoreceptors, and thermoreceptors.
   Name at least one example of each type that appears in an animal. [pp.566-567]
2. Explain how a taste bud works. [p.571]
3. Contrast the structure of compound eyes with the structures of invertebrate ommatidia and of the human eye.
   [pp.572-575]
4. Describe how the human eye perceives color and black-and-white. [p.576]
5. Define *nearsightedness* and *farsightedness* and relate each to eyeball structure. [p.577]
6. Indicate the causes of the following disorders of the human eye: (a) red-green color blindness, (b)
   astigmatism, (c) cataracts, (d) glaucoma, and (e) retinal detachment. [p.577]
7. Follow a sound wave from the pinna to the organ of Corti; mention the name of each structure it passes and
   state where the sound wave is amplified and where the pattern of pressure waves is translated into
   electrochemical impulses. [pp.578-579]
8. Explain how the three semicircular canals of the human ear detect changes of position and acceleration in a
   variety of directions. [p.580]

INTEGRATING AND APPLYING KEY CONCEPTS

1. Explain why adaptation is a positive process for animals.
2. Discuss the function of pain in an animal's behavior.
3. How might human behavior be changed if human eyes were compound eyes composed of ommatidia and if humans perceived only vibrations—as fish do—rather than sounds?
4. Human and squid eyes are quite similar. Is it more likely that humans and squid shared a common ancestor with camera-type eyes or that these eyes are a product of convergent evolution?

34

ENDOCRINE CONTROL

INTRODUCTION

This chapter looks at the glands that make up the endocrine system, the hormones they produce, and the effects of those hormones. It will also focus on the important interaction between the endocrine and nervous systems that essentially allows them to operate as a single system. Types of hormones will be described as well as how the endocrine glands are controlled to ensure the appropriate level of each hormone is produced and circulated. Disorders caused by incorrect amounts of some hormones are also discussed.

STUDY STRATEGIES

- Read all of Chapter 34 with the goal of familiarizing yourself with the boldface terms.
- Identify the role of the endocrine system in vertebrates and the major endocrine glands.
- Recognize the two types of hormones and how each interact with their target cells.
- Describe the connection between the hypothalamus and pituitary glands.
- Discuss the hormones of the posterior and anterior pituitary glands and their actions.
- Explain the role of growth hormones and the disorders that can result from too little or too much GH.
- Identify other vertebrate hormones and glands including thyroid, parathyroid, adrenal, thymus, pancreas, pineal, and gonads.
- Discuss endocrine disorders as they relate to the endocrine glands and hormones.
- Describe invertebrate hormones and the role they play in the life cycle.

FOCAL POINT

- Figure 34.2 [p.587] shows the main endocrine glands and describes what they do.
- Figure 34.3 [p.589] diagrams the differences between the mechanisms of steroid and protein hormone action.
- Figures 34.5and .34.6[p.591] show the function of the pituitary and its relationship to other endocrine glands.
- Figure 34.12 [p.596] describes the function of the pancreas in blood sugar regulation.
- Tables 35.2 and 35.3 [pp.590-593] review the functions of the endocrine glands.

INTERACTIVE EXERCISES

34.1. HORMONES IN THE BALANCE [p.585]

34.2. THE VERTEBRATE ENDOCRINE SYSTEM [pp.586-587]

Boldfaced, Page-Referenced Terms

endocrine disruptor _____

_____

local signaling molecules_____

_____

animal hormones _____

_____

endocrine system _____

_____

Fill-in-the-Blanks [p.585]

(1) _____ is an insecticide that is effective and widely used in the U.S. Unfortunately, studies have shown that

it is also a(n) (2) _____. Male tadpoles exposed to (1) in the lab often became (3) _____. Field-collected

male frogs in agricultural areas where (1) was in use showed (4) _____ sex organs. Humans may also be

affected. Two pesticides banned in the U.S., (5) _____ and (6) _____, may also act as endocrine

disruptors.

Matching [pp.586-587]

Choose the most appropriate answer for each term.

7. _____ hormones

8. _____ neurotransmitters

9. _____ Gap Junctions

10. _____ target cells

11. _____ local signaling molecules

12. _____ pheromones

A. Signaling molecules released from axon endings of neurons; act swiftly on target cells molecule

B. Cell connections that allow signals to move directly from cytoplasm of one cell to another

C. Released by many types of body cells; alter conditions within localized regions of tissues

D. Signaling molecules released by one animal that act on cells of other animals of the same species and help integrate social behavior

E. Secretions from endocrine glands, endocrine cells, and some neurons; distributed by the bloodstream to nonadjacent target cells

F. Cells that have receptors for a given type of signaling

Complete the Table [p.587]

13. Complete the following table. First, identify the numbered components of the endocrine system shown in the accompanying illustration. Then list the hormones produced by each.

Gland Name	Number	Hormones Produced
a. Hypothalamus		
b. Pituitary, Anterior lobe		
c. Pituitary, posterior lobe		
d. Adrenal gland (cortex)		
e. Adrenal gland (medulla)		
f. Ovaries		
g. Testes		
h. Pineal		
i. Thyroid		
j. Parathyroid		
k. Thymus		
l. Pancreas		

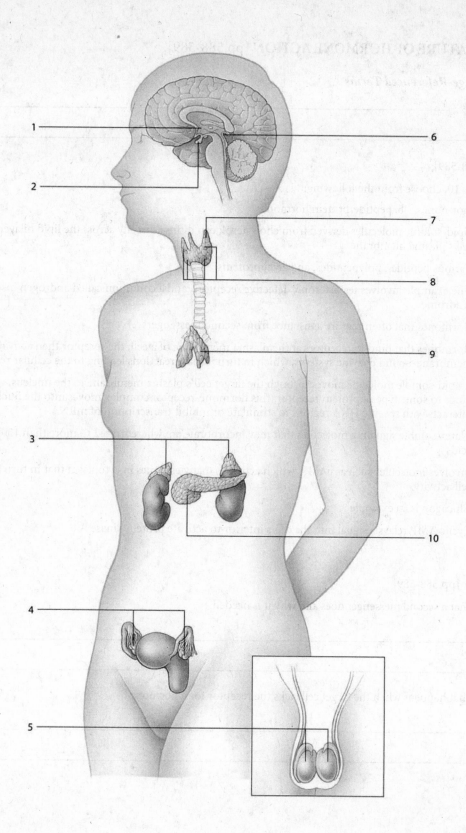

34.3. THE NATURE OF HORMONE ACTION [pp.588-589]

Boldfaced, Page-Referenced Terms

second messenger _____

_____

Choice [pp.588-589]

For questions 1-10, choose from the following:

    a. steroid hormones   b. peptide/protein hormones

1. _____ Lipid-soluble molecules derived from cholesterol; can diffuse directly across the lipid bilayer of a target cell's plasma membrane

2. _____ Various peptides, polypeptides, and glycoproteins

3. _____ One example involves testosterone, defective receptors, and a condition called androgen insensitivity syndrome

4. _____ Hormones that often require assistance from second messengers

5. _____ Hormones that bind to receptors at the plasma membrane of a cell; the receptor then activates specific membrane-bound enzyme systems, which in turn initiate reactions leading to the cellular response

6. _____ A lipid-soluble molecule moves through the target cell's plasma membrane to the nucleus, where it binds to some type of protein receptor; this hormone-receptor complex moves into the nucleus and interacts with specific DNA regions to stimulate or inhibit transcription of mRNA

7. _____ Water-soluble signaling molecules that may incorporate anywhere from 2 to more than 150 amino acids

8. _____ Involves molecules such as cAMP, which activates many enzymes in cytoplasm that in turn alter some cell activity.

9. _____ Glucagon is an example

10. _____ Cyclic AMP relays a signal into the cell's interior to activate protein kinase A

Short Answer [pp.588-589]

11. Explain what a second messenger does and why it is needed.

_____

_____

12. Explain what happens when the target cell lacks the receptor for a hormone.

_____

_____

Ordering [pp.588-589]

13. Place the following steps of action of a steroid hormone in order from 1-5.

A. _____ The hormone–receptor complex triggers transcription of a specific gene.

B. _____ mRNA moves into the cytoplasm and is transcribed into a protein.

C. _____ Being lipid soluble, the hormone easily diffuses across the cell's plasma membrane.

D. _____ A steroid hormone molecule is moved from blood into interstitial fluid bathing a target cell.

E. _____ The hormone diffuses through the cytoplasm and nuclear envelope. It binds with its receptor in the nucleus.

Ordering [pp.588-589]

14. Place the following steps of action of a peptide hormone in order from 1-5.

A. _____ Cyclic AMP activates another enzyme in the cell.

B. _____ A peptide hormone molecule, glucagon, diffuses from blood into interstitial fluid bathing the plasma membrane of a liver cell.

C. _____ The enzyme activated by cyclic AMP also inhibits glycogen synthesis.

D. _____ Glucagon binds with a receptor. Binding activates an enzyme that catalyzes the formation of cyclic AMP from ATP inside the cell.

E. _____ The enzyme activated by cyclic AMP activates another enzyme, which in turn activates another kind that catalyzes the breakdown of glycogen to its glucose monomers.

34.4. THE HYPOTHALAMUS AND PITUITARY GLAND [pp.590-591]

Boldfaced, Page-Referenced Terms

hypothalamus _____

_____

pituitary gland _____

_____

releasing hormones_____

_____

inhibiting hormones_____

_____

Labeling and Matching [p.587; also see summary table on p.590]

Label each hormone listed below with an A if it is secreted by the anterior lobe of the pituitary, a P if it is released from the posterior pituitary. Complete the exercise by entering the letter of the corresponding action in the parentheses after each label.

1. _____ ( ) ACTH
2. _____ ( ) ADH
3. _____ ( ) FSH
4. _____ ( ) STH (GH)
5. _____ ( ) LH
6. _____ ( ) MSH
7. _____ ( ) oxytocin (OCT)
8. _____ ( ) PRL
9. _____ ( ) TSH

A. Stimulates egg and sperm formation in ovaries and testes

B. Targets pigmented cells in skin and other surface coverings; induces color changes in response to external stimuli and affects some behaviors

C. Stimulates and sustains milk production in mammary glands

D. Stimulates progesterone secretion, ovulation, and corpus luteum formation in females; promotes testosterone secretion and sperm release in males

E. Induces uterine contractions and milk movement into secretory ducts of the mammary glands

F. Stimulates release of thyroid hormones from the thyroid gland

G. Acts on the kidneys to conserve water; required in control of extracellular fluid volume

H. Stimulates release of adrenal steroid hormones from the adrenal cortex

I. Promotes growth in young; induces protein synthesis and cell division; roles in adult glucose and protein metabolism

True/False [pp.590-591]

If the statement is true, write a "T" in the blank. If the statement is false, make it correct by changing the underlined word(s) and writing the correct word(s) in the answer blank.

10. _____ The posterior pituitary gland contains nerve endings that originate in the hypothalamus and secrete hormones.

11. _____ The anterior pituitary secretions are regulated by the posterior pituitary.

12. _____ Releasers are hormones released from the anterior pituitary.

13. _____ Anterior pituitary hormones are produced in the hypothalamus.

14. _____ Oxytocin and antidiuetic hormones are synthesized in the hypothalamus.

15. _____ The adrenocorticotropic hormone and thyroid stimulating hormone are synthesized in the hypothalamus.

16. _____ The anterior pituitary consists of nerve tissue and the posterior pituitary consists of glandular tissue.

17. _____ The production and subsequent release of milk require two hormones: prolactin and luteinizing hormone.

34.5. GROWTH HORMONE FUNCTION AND DISORDERS [p.592]

34.6. SOURCES AND EFFECTS OF OTHER VERTEBRATE HORMONES [p.593]

Matching [pp.592-593]

Match each of the following terms with its correct definition/hormone.

1. _____ Pituitary dwarfism   A. Excess growth hormone as an adult

2. _____ Acromegaly   B. Erythropoeitin

3. _____ Pituitary gigantism   C. Secretin

4. _____ Small intestine   D. Atrial natriuretic peptide

5. _____ Kidney   E. Excess growth hormone as a child

6. _____ Heart   F. Lack of growth hormone as a child

Short Answer [pp.592-593]

7. What is one of the treatments for children with lower than normal rhGH? What are the ethical concerns with this treatment?

_____

_____

_____

8. Briefly explain the general interaction between hormones and organs.

_____

_____

_____

34.7. THYROID AND PARATHYROID GLANDS [pp.594-595]

Boldfaced, Page-Referenced Terms

thyroid gland _____

_____

parathyroid glands _____

_____

Fill-in-the-Blanks [pp.594-595]

In a(n) (1) _____ loop, an increase in a hormone's level can inhibit the secretion of the hormone. This type of loop is in effect when TRH from the (2) _____ prevents the pituitary from secreting (3) _____, which in turn stops the (4) _____ from secreting its hormones. The mineral (5) _____ is required for the production of thyroid hormone. Deficiencies of (5) in the diet can lead to (6) _____. Excess thyroid hormone, or (7) _____, can cause many symptoms including anxiety and tremors. The four (8) _____ glands, located on the posterior surface of the thyroid, control the levels of (9) _____ in the blood. PTH targets cells in the (10) _____ and (11) _____. Low levels of (9) increase PTH secretion and can lead to (12) _____ in young children. Thyroid hormone is also important in the (13) _____ of a tadpole to a frog. Some pesticides are (14) _____ and can lead to developmental defects including lack of (13) in frogs living in contaminated ponds.

Short Answer [pp.596-597]

15. Why is iodized salt such a staple in the developed world and what does it work to prevent?

_____

_____

_____

16. What are symptoms of hyperthyroidism and how can it be treated?

_____

_____

_____

34.8. PANCREATIC HORMONES [pp.596-577]

34.9. THE ADRENAL GLANDS [pp.598-599]

34.10. THE GONADS [p.600]

34.11. THE PINEAL GLAND [p.601]

Boldfaced, Page-Referenced Terms

pancreas _____

_____

insulin _____

_____

glucagon _____

_____

adrenal glands _____

_____

adrenal cortex _____

_____

adrenal medulla _____

_____

cortisol _____

_____

gonads _____

_____

sex hormones _____

_____

puberty _____

_____

secondary sexual traits _____

_____

pineal gland _____

_____

melatonin _____

_____

circadian rhythm _____

_____

thymus _____

_____

ecdysone _____

_____

Matching [pp.596-602]

First match the gland/organ or cell group to the hormone(s) it produces. Then choose the best description of the action of the hormone(s).

a. glucagon b. testosterone c. cortisol d. estrogens and progesterone

e. epinephrine f. melatonin g. insulin h. thymulin

1. ___,___ adrenal cortex

2. ___,___ adrenal medulla

3. ___,___ testes

4. ___,___ ovaries

5. ___,___ pancreas (alpha cells)

6. ___,___ pancreas (beta cells)

7. ___,___ pineal

A. Required in egg maturation and release, and in preparation of uterine lining for pregnancy and its maintenance in pregnancy; influence growth, development, and female genital development; maintains sexual traits

B. Promotes protein and fatty acid breakdown to provide fuel to most cells other than in the brain

C. Increases glucose uptake in liver, muscle, and adipose tissue, resulting in lower blood sugar levels

D. Required in sperm formation, male genital development and maintenance of sexual traits; influences growth and development

E. Influences daily biorhythms, gonad development, and reproductive cycles

F. Raises metabolism; increases heart rate and force of contraction

G. Raises blood sugar level, an effect opposite to insulin

Matching [pp.596-602]

Match each of the following diseases/disorders with the hormone and gland involved. Some letters may be used more than once.

a. too little insulin b. too little testosterone c. too much cortisol d. too much insulin e. too much melatonin

8. ___,___ diabetes mellitus

9. ___,___ hypoglycemia

10. ___,___ Cushing's syndrome

11. ___,___ seasonal affective disorder

12. ___,___ decreased libido

A. Pancreas

B. Adrenal cortex

C. Pineal gland

D. Gonads

True/False [pp.596-600]

If the statement is true, write a "T" in the blank. If the statement is false, make it correct by changing the underlined word(s) and writing the correct word(s) in the answer blank.

13. _____ Type I diabetes usually begins when the person is an <u>adult</u>.

14. _____ Hypoglycemia occurs when blood sugar levels are too <u>high</u>.

15. _____ Cortisol levels are controlled by a <u>negative</u> feedback loop.

16. _____ Cushing's disease is characterized by <u>depressed</u> levels of cortisol.

17. _____ Hypercotisolism results in <u>lowered</u> white blood cell counts.

18. _____ The adrenal cortex hormones are involved in body responses to <u>short-term</u> stress.

Ordering [pp.596]

19. Place the following steps of the regulation of blood sugar in order from 1-8.

A. _____ Pancreatic cells stop secreting glucagon and are stimulated to secrete insulin.

B. _____ Blood level of glucose declines to its normal level.

C. _____ Blood glucose increases to the normal level.

D. _____ Decrease of blood glucose between meals encourages glucagon secretion and slows insulin secretion

E. _____ After a meal, glucose enters blood faster than cells can take it up, so blood glucose increases.

F. _____ Adipose and muscle cells take up and store glucose; cells in the liver and muscle make more glycogen.

G. _____ Blood glucose declines as cells take it up and use it for metabolism.

H. _____ Glucagon causes cells to break glycogen down into glucose, which enters the blood.

Short Answer [pp.596-597]

20. Explain three of the complications caused by diabetes.

_____

_____

_____

21. What are contributions to the rapid increase in frequency of Type II diabetes?

_____

_____

_____

22. What health impacts might result from chronic stress?

_____

_____

_____

23. Explain the disorder of seasonal affective disorder (SAD).

_____

_____

_____

23. Low thymus levels might contribute to which two diseases later on in life?

_____

_____

_____

34.13. INVERTEBRATE HORMONES [pp.602-603]

Boldfaced, Page-Referenced Terms

ecdysone _____

_____

Fill-in-the-Blanks [pp.602-603]

Many invertebrate organisms (1) _____, periodically shedding their body covering as they grow. This process is controlled by hormones, such as the arthropod hormone (2) _____, which is structurally a(n) (3) _____ hormone. In crustaceans, the (4) _____ organ at the base of the (5) _____ inhibits the secretion of (2). The process is similar in (6) _____, but they do not have a(n) (7) _____ hormone.

SELF-TEST

1. The _____ region of the forebrain monitors internal organs, influences certain forms of behavior, and secretes some hormones.
   a. hypothalamus
   b. pancreas
   c. thyroid
   d. pituitary
   e. thalamus

2. If you were lost in the desert and had no fresh water to drink, the level of _____ in your blood would increase as a means to conserve water.
   a. insulin
   b. corticotropin
   c. oxytocin
   d. antidiuretic hormone
   e. salt

For questions 3-5, choose from the following answers:
   a. estrogen
   b. PTH
   c. FSH
   d. growth hormone (GH)
   e. prolactin

3. Stimulates bone cells to release calcium and phosphate and the kidneys to conserve it.

4. Stimulates and sustains milk production in mammary glands.

5. Is the hormone associated with pituitary dwarfism, gigantism, and acromegaly.

For questions 6-8, choose from the following answers:
   a. adrenal medulla
   b. adrenal cortex
   c. thyroid
   d. anterior pituitary
   e. posterior pituitary

6. Produces cortisol and other glucocorticoids that help increase the level of glucose in blood.

7. The gland that is most closely associated with emergency situations is the _____.

8. The overall metabolic rates of warm-blooded animals, including humans, depend on hormones secreted by the _____ gland.

Matching

Choose the most appropriate answer for each term.

9. _____ ADH
10. _____ ACTH and TSH
11. _____ FSH and LH
12. _____ GH
13. _____ cortisol
14. _____ epinephrine and norepinephrine
15. _____ thyroxine and triiodothyronine
16. _____ estrogens and progesterone
17. _____ PTH
18. _____ glucagon
19. _____ testosterone
20. _____ insulin
21. _____ pheromones
22. _____ melatonin
23. _____ ecdysone
24. _____ aldosterone
25. _____ thymosins
26. _____ calcitonin
27. _____ oxytocin
28. _____ erythropoietin
29. _____ atrial natriuretic peptide
30. _____ leptin
31. _____ secretin

A. In times of excitement or stress, these adrenal medulla hormones help adjust blood circulation and fat and carbohydrate metabolism

B. Secreted by the ovaries; influence sexual traits

C. Secreted by beta pancreatic cells; lowers blood glucose level

D. Abnormal amount of this anterior pituitary lobe hormone has different effects on human growth during childhood and adulthood

E. Secreted by anterior pituitary lobe; orchestrate secretions from the adrenal gland and thyroid gland, respectively

F. Secreted by adrenal cortex; helps increase the level of glucose in blood

G. Secreted by alpha pancreatic cells; raises the blood glucose level

H. Largely controls molting in insects and crustaceans

I. Major thyroid hormones having widespread effects such as controlling the overall metabolic rates of warm-blooded animals

J. Secreted by the testes; influences sexual traits

K. Secreted by the pineal gland; influences the growth and development of gonads

L. Antidiuretic hormone released by the posterior pituitary lobe; promotes water reabsorption when the body must conserve water

M. Hormone-like secretions of certain exocrine glands; they diffuse through water or air to cellular targets outside the animal body

N. Anterior pituitary lobe hormones. Act through the gonads to influence gamete formation; secretion of the sex hormones required in sexual reproduction

O. Hormone secreted by the parathyroid glands in response to low blood calcium levels

P. Hormone that lowers blood calcium level

Q. Hormones involved with T lymphocytes

R. Stimulates the kidneys to excrete water and salt.

S. Stimulates maturation and production of oxygen-transporting red blood cells

T. Acts in the brain and suppresses appetite.

U. Stimulates the kidneys to reabsorb salt.

V. Acts on the pancreas

W. Induces milk movement into secretory ducts and uterine contractions during childbirth

CHAPTER OBJECTIVES/REVIEW QUESTIONS

1. Define the terms *neurotransmitters, local signaling molecules,* and *endocrine disruptors.* [p.586]
2. Collectively, the body's sources of hormones came to be called the _____ system. [p.586]
3. Contrast the proposed mechanisms of hormonal action on target cell activities by (a) steroid hormones and (b) peptide and protein hormones. [pp.588-587]
4. Outline the major human hormone sources, their secretions, main targets, and primary actions as shown in text Tables 34.1, 34.2 and 34.3. [pp.587, 590, 593].
5. The _____ and the pituitary gland interact closely as a major neural-endocrine control center. [pp.590-591]
6. Identify the hormones released from the posterior lobe of the pituitary, and state their target tissues. [pp.590-591]
7. Identify the hormones produced by the anterior lobe of the pituitary, and tell which target tissues or organs are affected. [pp.590-591]
8. Pituitary dwarfism, gigantism, and acromegaly are all associated with abnormal secretion of _____ by the pituitary gland. [p.592]
9. Describe the characteristics of hypothyroidism and hyperthyroidism. [pp.594-595]
10. Name the glands that secrete PTH, and state the function of this hormone. [p.595]
11. Name the hormones secreted by alpha and beta pancreatic cells; list the effect of each. [pp.596-597]
12. Describe the symptoms of diabetes mellitus, and distinguish between type 1 and type 2 diabetes. [p.597]
13. Describe the role of cortisol in a stress response. [p.598]
14. List the features of the fight–flight response. [pp.598-599]
15. The pineal gland secretes the hormone _____; relate two examples of the action of this hormone. [p.601]
16. The thymus is responsible for regulating _____ and _____. [p.602]
17. Gonads include ovaries and gonads and produce _____.[p.600]
18. Identify the various invertebrate hormones.[pp.602-603]

INTEGRATING AND APPLYING KEY CONCEPTS

1. Suppose you suddenly quadruple your already high daily consumption of calcium. State which body organs would be affected, and tell how they would be affected. Name two hormones whose levels would most probably be affected, and tell whether your body's production of them would increase or decrease. Suppose you continue this high rate of calcium consumption for 10 years. Predict which organs would be subject to the most stress as a result.
2. What would happen to hormonal control if the nerves between the hypothalamus and pituitary were severed?
3. Explain how events that are perceived through the nervous system can affect hormone levels in the endocrine system. Give one example of this occurring in humans.

35

STRUCTURAL SUPPORT AND MOVEMENT

INTRODUCTION

This chapter explores how muscles and bones interact to allow movement. It starts with a discussion of skeletal systems in the invertebrates and introduces the evolutionary reasons for the development of an endoskeleton. However, the majority of the chapter details how muscles work at the cellular level. A key point is the sliding-filament model of how muscle contractions are powered.

STUDY STRATEGIES

- Read all of Chapter 35 with the goal of familiarizing yourself with the boldface terms.
- Identify the role of skeletal and muscular systems in vertebrates.
- Recognize the three forms of skeletal systems in animals.
- Identify the major components of the vertebrate endoskeleton.
- Describe internal bone structure and function.
- Discuss how long bones are formed.
- Explain the structure and function of the various types of joints, with special emphasis on synovial joints, and common joint disorders.
- Describe the interaction of skeletal and muscular systems in locomotion.
- Recognize the major muscles of the human body.
- Describe the action of muscle contraction and the process of nervous control.
- Explain muscle metabolism and the role of exercise in maintaining a healthy skeletomuscular system.

FOCAL POINTS

- Figure 35.7 [p.610] gives an overview of the human skeleton; it is the framework the muscles are attached to.
- Figure 35.15 [p.617] shows the major muscles of the human body.
- Figures 35.16 and 35.17 [pp.618-619] illustrate the sliding-filament model of muscle contraction and focuses on the interaction of the proteins myosin and actin.
- Figure 35.12 [p.614] shows a typical synovial joint

INTERACTIVE EXERCISES

35.1 MUSCLES AND MYOSTATIN [p.607]

35.2. INVERTEBRATE SKELETONS [pp.608-609]

35.3. THE VERTEBRATE ENDOSKELETON [pp.610-611]

Boldfaced, Page-Referenced Terms

hydrostatic skeleton _____

_____

exoskeleton _____

_____

endoskeleton _____

_____

vertebral column _____

_____

vertebrae _____

_____

intervertebral disks _____

_____

axial skeleton _____

_____

appendicular skelton _____

_____

Choice [pp.608-611]

For each of the following statements, choose the most appropriate category of skeleton.

　　a. hydrostatic skeleton　　　b. endoskeleton　　　c. exoskeleton

1. _____ Uses external body parts to receive the applied force of muscle contractions
2. _____ Earthworms and soft-bodied invertebrates use this form of skeleton
3. _____ Echinoderms use this type of skeleton
4. _____ Uses internal body parts to receive the applied force of muscle contractions
5. _____ In this case muscles work against an internal body fluid
6. _____ This skeleton type is typical of the arthropods

Matching [pp.610-611]

Match each of the following structures with its correct definition/description.

7. _____ axial skeleton

8. _____ appendicular skeleton

9. _____ pectoral girdle

10. _____ pelvic girdle

11. _____ vertebrae

A. Consists of two sets of fused bones that support weight of upper body when standing
B. The central supporting column of the trunk and head
C. Bony segments that are stacked to make up the backbone
D. Consists of the pectoral girdle, pelvic girdle and limbs
E. The set of bones in the upper trunk to which the arms are attached

Labeling [p.611]

Identify each indicated part of the accompanying illustration.

12. _____

13. _____

14. _____

15. _____

16. _____

17. _____

18. _____

19. _____

20. _____

21. _____

22. _____

23. _____

24. _____

25. _____

26. _____

27. _____

28. _____

29. _____

30. _____

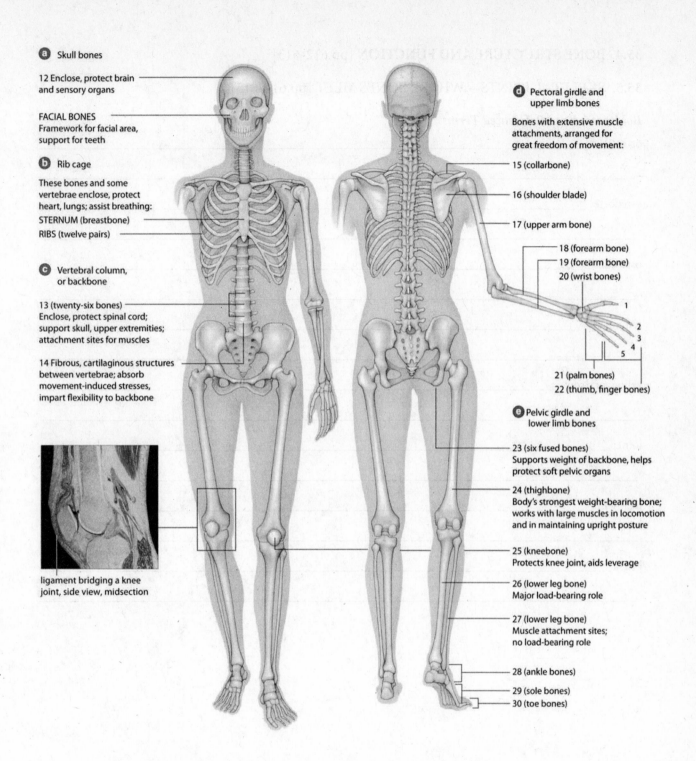

a Skull bones

12 Enclose, protect brain
and sensory organs

FACIAL BONES
Framework for facial area,
support for teeth

b Rib cage

These bones and some
vertebrae enclose, protect
heart, lungs; assist breathing:
STERNUM (breastbone)
RIBS (twelve pairs)

c Vertebral column,
or backbone

13 (twenty-six bones)
Enclose, protect spinal cord;
support skull, upper extremities;
attachment sites for muscles

14 Fibrous, cartilaginous structures
between vertebrae; absorb
movement-induced stresses,
impart flexibility to backbone

ligament bridging a knee
joint, side view, midsection

d Pectoral girdle and
upper limb bones

Bones with extensive muscle
attachments, arranged for
great freedom of movement:

15 (collarbone)

16 (shoulder blade)

17 (upper arm bone)

18 (forearm bone)
19 (forearm bone)
20 (wrist bones)

1

2
3
4
5

21 (palm bones)
22 (thumb, finger bones)

e Pelvic girdle and
lower limb bones

23 (six fused bones)
Supports weight of backbone, helps
protect soft pelvic organs

24 (thighbone)
Body's strongest weight-bearing bone;
works with large muscles in locomotion
and in maintaining upright posture

25 (kneebone)
Protects knee joint, aids leverage

26 (lower leg bone)
Major load-bearing role

27 (lower leg bone)
Muscle attachment sites;
no load-bearing role

28 (ankle bones)
29 (sole bones)
30 (toe bones)

Structural Support and Movement   **425**

35.4. BONE STRUCTURE AND FUNCTION [pp.612-613]

35.5. SKELETAL JOINTS—WHERE BONES MEET [pp.614-615]

Boldfaced, Page-Referenced Terms

osteoblast _____

_____

osteocyte _____

_____

osteoclast _____

_____

red marrow _____

_____

yellow marrow _____

_____

joint _____

_____

ligaments_____

_____

bursa_____

_____

Matching [pp.612-615]

Choose the most appropriate statement for each of the following terms.

1. _____ yellow marrow
2. _____ osteoclasts
3. _____ osteocytes
4. _____ ligaments
5. _____ joints
6. _____ red marrow
7. _____ osteoblasts
8. _____ bone remodeling

A. Bone-forming cells
B. An ongoing task between osteoblasts and osteoclasts
C. The cells that break down bone using acids and enzymes
D. Straps of dense connective tissue
E. Mature bone cells
F. The fatty region of most mature bones in adults
G. The location, within bone, of red blood cell formation
H. Areas of contact, or near-contact, between bones

Choice [pp.614-615]

For each of the following statements, choose the most appropriate type of joint from the list below.

a. fibrous joint   b. cartilaginous joint   c. synovial joint

9. _____ Contains fluid to help lubricate joint
10. _____ Connect vertebrae to each other
11. _____ Joint between pelvic girdle and femur
12. _____ Bones held securely together—little flex
13. _____ Contain ligaments
14. _____ Pads or disks connect bones
15. _____ Has widest range of movement
16. _____ Hold teeth in sockets

Labeling and Matching [pp.614-615]

Match the correct name for each structure with the letter from the figure of the synovial joint below.

17. _____ Cartilage
18. _____ Femur
19. _____ Tibia
20. _____ Fibula
21. _____ Menisci
22. _____ Patella
23. _____ Cruciate ligaments

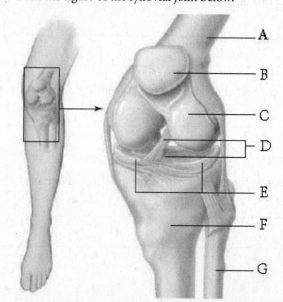

Labeling and Matching [pp.614-615]

Match the correct name for each structure with the letter from the figure of structure of bone below. Two of the structures will have two matches.

24. _____ nutrient canal

25. _____ , _____ spongy bone

26. _____ blood vessel

27. _____ yellow marrow

28. _____ dense connective tissue

29. _____ , _____ compact bone

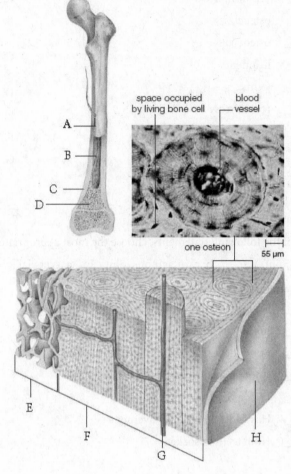

space occupied by living bone cell

blood vessel

A

B

C

D

one osteon

55 μm

E

F

G

H

Short Answer [pp.614-615]

30. Describe in general what happens in a sprain. _____

_____

_____

31. Describe what happens when a cruciate ligament tears. Include what joint this would be in. _____

_____

_____

32. Describe the difference between arthritis and bursitis. _____

_____

_____

35.6. SKELETAL-MUSCULAR SYSTEMS [pp.616-617]

35.7. HOW DOES SKELETAL MUSCLE CONTRACT? [pp.618-619]

Boldfaced, Page-Referenced Terms

tendon_____

_____

muscle fiber _____

_____

myofibrils _____

_____

sarcomere _____

_____

sliding filament model _____

_____

Fill-in-the-Blanks [pp.616-619]

Skeletal muscle cells are not your typical cells. Groups of them fuse together into one multinucleated (1)

_____. A cordlike or strap-like (2) _____ attaches to (3) _____. They act as a(n) (4) _____

system, in which a rigid rod is attached to a(n) (5) _____ point and moves about it. Muscles connect to bones

near a(n) (6) _____. When they contract, they transmit (7) _____ that makes the bones move. Bear in

mind, only (8) _____ muscle is the functional partner of bone. Often, two muscles work in (9) _____ to

each other; the action of one muscle (10) _____ the action of the other. (11) _____ muscle is mainly a

component of soft internal organs, such as the (12) _____. Cardiac muscle forms only in the (13) _____

wall. A strap-like (14) _____ connects skeletal muscles to bones.

Labeling and Matching [pp.616-617]

On the next page, provide the correct name for each indicated muscle in this diagram. Then, in the parentheses, match the muscle to its correct function.

15. _____( )

16. _____( )

17. _____( )

18. _____( )

19. _____( )

20. _____( )

21. _____( )

22. _____( )

23. _____( )

24. _____( )

25. _____( )

26. _____( )

27. _____( )

28. _____( )

29. _____( )

30. _____( )

A. Flexes the foot toward the shin

B. Bends the lower leg at the knee while walking

C. Straightens the forearm at the elbow

D. Flexes and draws the thigh toward the body

E. Draws the thigh backward and bends the knee

F. Flexes the thigh at the hips; extends the leg at the knee

G. Draws the arm forward and toward the body

H. Extends and rotates the thigh outward when walking and running

I. Draws the shoulder blade forward; helps raise the arm

J. Rotates and draws the arm backward and toward the body

K. Depresses the thoracic cavity

L. Bends the thigh at the hip; bends the lower leg at the knee

M. Compresses the abdomen; assists in lateral rotation of the torso

N. Raises the arm

O. Bends the forearm at the elbow

P. Lifts the shoulder blade; draws the head back

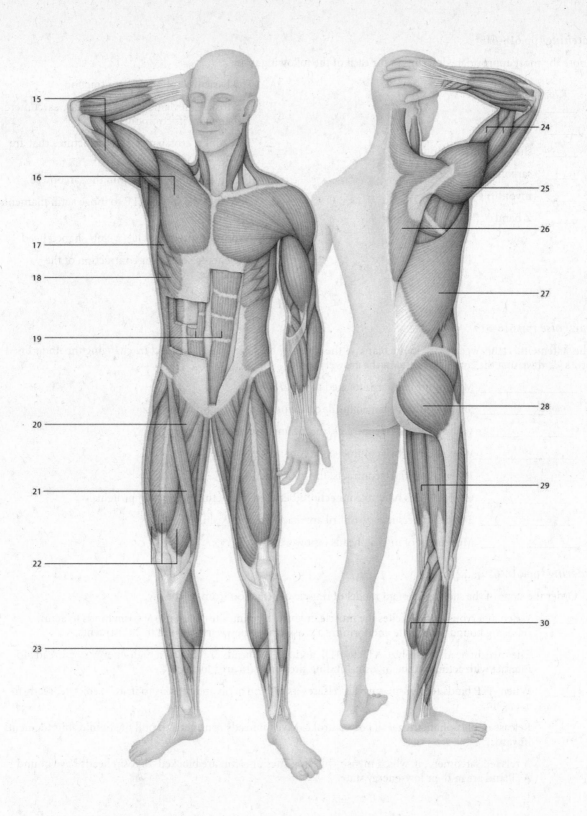

Matching [pp.618-619]

Choose the most appropriate description for each of the following terms.

31. _____ myosin

32. _____ actin

33. _____ ATP

34. _____ sliding-filament model

35. _____ sarcomeres

36. _____ myofibrils

37. _____ Z band

A. The basic unit of muscle contraction

B. A cytoskeletal element that flanks the sarcomere and anchors its components

C. Threadlike, crossbanded cell structures that are arranged in parallel

D. Thin filaments found within the sarcomere

E. Describes the use of ATP to move actin filaments past myosin

F. A motor protein that has a club-shaped head

G. The energy source for contraction of the sarcomere

True/False [pp.618-619]

If the statement is true, write a "T" in the blank. If the statement is false, make it correct by changing the underlined word(s) and writing the correct word(s) in the answer blank.

38. _____ <u>Myosin</u> proteins are attached to the Z band.

39. _____ Small, golf club shaped heads on myosin bind to <u>actin</u> filaments.

40. _____ When a myofibril contracts, the Z bands move <u>farther apart</u>.

41. _____ When a myofibril contracts, the A band <u>gets smaller</u>.

42. _____ When a myofibril contracts, the I band gets <u>larger</u> .

43. _____ Cross bridges refer to connections between the <u>actin and myosin</u> proteins.

44. _____ The thin fibers in a myofibril are made up of <u>myosin</u>.

45. _____ Movement of myosin heads is powered by <u>ATP</u>.

Ordering [pp.618-619]

46. Order the steps of the sliding-filament model of muscle contraction from 1-5 below.

A. _____ When a nervous signal excites the muscle, myosin-binding sites open up. Myosin binds to actin, releasing bound phosphate and forming a cross-bridge between thin and thick filaments.

B. _____ The myosin heads hydrolyze ATP to ADP and Pi. The heads are now in a high-energy state, ready to interact with actin, but the myosin-binding sites on actin are blocked.

C. _____ When ATP binds to myosin, myosin releases its grip on actin and returns to its relaxed state, ready to act again.

D. _____ Release of phosphate triggers a power stroke. Myosin heads contract and pull the bound thin filaments inward.

E. _____ A relaxed sarcomere, in which myosin-binding sites on actin are blocked. Myosin heads have bound ATP and are in their low-energy state.

35.8. NERVOUS CONTROL OF MUSCLE CONTRACTION [pp.620-621]

35.9. MUSCLE METABOLISM [pp.622-623]

Boldfaced, Page-Referenced Terms

sarcoplasmic reticulum _____

_____

motor unit _____

_____

muscle tension _____

_____

myoglobin_____

_____

Matching [pp.620-621]

Choose the most appropriate statement for each of the following terms.

1. _____ muscle cramp

2. _____ muscle fatigue

3. _____ muscular dystrophies

4. _____ tetanus (disease)

5. _____ botulism

6. _____ tetanus (muscle state)

7. _____ motor unit

8. _____ muscle twitch

9. _____ creatine phosphate

10. _____ excitable cells

11. _____ sarcoplasmic reticulum

12. _____ action potential

A. A disease in which over-stimulated muscles stiffen and contract; sometimes called lockjaw

B. A disease caused by bacteria that affects the neurons that synapse with muscle cells

C. The result of a sustained state of contraction due to high-frequency stimulation

D. A decrease in a muscle's capacity to generate force

E. Involuntary, painful contractions of muscles, probably the result of dehydration

F. Genetic disorders in which the muscles progressively weaken and degenerate

G. The brief interval in which a motor unit contracts

H. A motor neuron and all of the muscle cells that form junctions with its endings

I. Restores muscle fuel for about the rest 15 seconds of muscle contraction

J. The cellular structure that takes up and stores calcium ions

K. A reversal of the voltage difference across a membrane

L. Have the ability to reverse the voltage difference across their membrane as a result of stimulation

Short Answer [pp.620-621]

13. Explain the roles of troponin and tropomyosin in muscle contraction. _____

_____

_____

14. Explain the difference between isotonically and isometrically contracting muscles. _____

_____

_____

15. What are the benefits of exercise on the skeletal-muscular systems? _____

_____

_____

16. How does *Clostridium tetani* affect the muscular system? _____

_____

_____

_____

17. How does polio affect the muscular system? _____

_____

_____

_____

sELF-TEST

___ 1. Insects have what form of skeleton?
   a. Endoskeleton
   b. Exoskeleton
   c. Hydroskeleton
   d. No skeleton

___ 2. The major site of blood cell formation in the human body is the _____ .
   a. ligaments
   b. osteoclasts
   c. yellow marrow
   d. red marrow

___ 3. Bone-forming cells are called _____ .
   a. osteocytes
   b. osteoclasts
   c. osteoblasts
   d. none of the above

___ 4. Which of the following is made from dense connective tissue?
   a. Sarcomeres
   b. Joints
   c. Cartilage
   d. Ligaments

___ 5. The protein responsible for contraction in skeletal muscle is called _____ .
   a. actin
   b. myosin
   c. Z band
   d. myofibril

Chapter Thirty-Five

6. The sarcoplasmic reticulum does which of the following?
   a. Is involved in the sliding-filament model
   b. Restores voltage differences following an action potential
   c. Stores and releases calcium ions
   d. Restores ATP levels following contraction
   e. None of the above

7. _____ is a decrease in a muscle's capacity to generate force.
   a. Tetanus
   b. Muscle fatigue
   c. Muscle tension
   d. Muscle twitch

8. Which of the following has a genetic component?
   a. muscular dystrophies
   b. muscle fatigue
   c. muscle cramps
   d. tetanus
   e. botulism

9. Many soft-bodied animals have what form of skeleton?
   a. Endoskeleton
   b. Exoskeleton
   c. Hydroskeleton
   d. No skeleton

10. Which of the following bones would be included in the appendicular skeleton?
    a. skull
    b. rib
    c. radius
    d. vertebrae

11. Which of the following bones would be included in the axial skeleton?
    a. skull
    b. ulna
    c. radius
    d. tibia

12. Bones of the leg would include all except
    a. tibia
    b. ulna
    c. fibula
    d. femur

13. The element critical to the formation of bone is
    a. calcium
    b. sodium
    c. magnesium
    d. oxygen

14. Which of the following is not a function of bone?
    a. movement
    b. red blood cell formation
    c. protection
    d. support
    e. all of the above

15. Which hormone slows bone turnover?
    a. estrogen
    b. testosterone
    c. cortisol
    d. epinephrine

16. The most common type of inflammation of a joint is
    a. osteoarthritis
    b. rheumatoid arthritis
    c. bursitis
    d. dislocation

17. This type of muscle interacts directly with bone
    a. smooth
    b. skeletal
    c. cardiac
    d. none of the above

18. The unit of muscle contraction is the
    a. muscle fiber
    b. Z band
    c. myofibril
    d. sarcomere

19. This muscle rotates the arm
    a. trapezius
    b. hamstring
    c. lattisimus dorsi
    d. pectoralis major

20. This muscle is the longest in the body
    a. rectus abdominus
    b. sartorius
    c. gluteus maximus
    d. gastrocnemius

CHAPTER OBJECTIVES/REVIEW QUESTIONS

1. Define and give examples of the three types of skeletons found in animals. [pp.608-609]
2. Give examples of how the vertebrate skeleton allows life on land. [p.610-611]
3. Distinguish between the axial and appendicular skeleton. [pp.610-611]
4. Explain the importance of intervertebral disks for upright locomotion. [p.610]
5. Recognize the major bones of the human body and their roles. [p.611]
6. Explain the various roles of osteoblasts, osteoclasts, osteocytes, red marrow, and yellow marrow. [pp.612-613]
7. Distinguish between joints and ligaments. [pp.614-615]
8. Understand the relationship between the human skeleton and sprains, osteoarthritis, rheumatoid arthritis, and osteoporosis. [pp.614-615]
9. Understand the interaction of skeletal muscle and bones. [p.616]
10. Recognize the locations and functions of the major muscles. [p.617]
11. Describe the fine structure of a muscle fiber; use terms such as *myofibril*, *sarcomere*, *motor unit*, *actin*, and *myosin*. [pp.618-622]
12. Understand the sliding-filament model for muscle contraction. [pp.618-619]
13. Understand the importance of calcium in muscle contraction. [pp.620-621]
14. Identify the major sources of energy for muscle contraction. [pp.620-621]
15. Recognize how the terms *tetanus*, *muscle fatigue*, *muscle tension*, and *muscle twitch* relate to the properties of whole muscles. [pp.620-621]
16. Identify how the diseases muscular dystrophy, tetanus, and botulism interact with the muscle system. [pp.620-621]

INTEGRATING AND APPLYING KEY CONCEPTS

1. If humans had an exoskeleton rather than an endoskeleton, would they move differently from the way they do now? Name any advantages or disadvantages that having an exoskeleton instead of an endoskeleton would present in human locomotion.
2. Why should pregnant mothers increase their calcium intake? Where would the fetus get its calcium supply from naturally? Why should mothers continue increased calcium consumption following pregnancy?
3. What would be the effect of a chemical that prevented the interaction of myosin and actin?
4. What are the benefits of hydrostatic skeletons and exoskeletons to invertebrates? Why are endoskeletons so rare in these animals?

36
CIRCULATION

INTRODUCTION

This chapter looks at the structure and function of the circulatory system and its components, concentrating on humans. You will learn about the evolutionary background of our circulatory system as well as its parts—blood, blood vessels, and the heart. The interaction of these parts with the rest of the body will be covered.

STUDY STRATEGIES

- Read all of Chapter 36 with the goal of familiarizing yourself with the boldface terms.
- Compare and contrast open and closed circulatory systems.
- Discuss the evolution of closed circulatory systems in vertebrates.
- Identify the major components of the human circulatory system.
- Compare and contrast the pulmonary and systemic circuits of the human circulatory systems and identify the structures involved in each.
- Discuss the functions of blood and recognize its components.
- Describe the development of red blood cells, white blood cells, and platelets from stem cells.
- Explain hemostasis.
- Describe the ABO and Rh blood typing systems.
- Identify the structures of the human heart, the nervous control of the heartbeat, and how blood flows through the various chambers.
- Compare and contrast the various blood vessels.
- Discuss blood pressure and capillary exchange.
- Describe blood and cardiovascular disorders.
- Recognize the connection between the cardiovascular and lymphatic systems.

FOCAL POINTS

- Figure 36.2 [p.628] compares open and closed systems.
- Figure 36.3 [p.629] compares the circulatory systems of fish, amphibians, and birds and mammals.
- Figure 36.4 [p.630] shows the major vessels of the human circulatory system.
- Figure 36.5 [p.631] compares the systemic and pulmonary circuits.
- Figure 36.6 [p.632] analyzes the components of blood.
- Figure 36.7 [p.633] shows how the various blood cells are derived.
- Figure 36.8 [p.635] illustrates blood types and cross reactions between them.
- Figure 36.11 [p.636] illustrates the human heart.
- Figure 36.14 [p.638] compares various blood vessels.
- Figure 36.23 [p.644] diagrams the human lymphatic system.

INTERACTIVE EXERCISES

36.1 A SHOCKING SAVE [p.627]

36.2. THE NATURE OF BLOOD CIRCULATION [pp.628-629]

Boldfaced, Page-Referenced Terms

circulatory systems _____

_____

hearts _____

_____

open circulatory system _____

_____

hemolymph_____

_____

closed circulatory system _____

_____

pulmonary circuit _____

_____

systemic circuit _____

_____

Fill-in-the-Blanks [pp.627-629]

(1) _____ occurs when the heart abruptly stops beating. If oxygen is not delivered to the brain,

irreversible damage occurs.(2) _____ is used to keep blood and oxygen flowing until the heart can be shocked

into beating with a(n) (3) _____. In humans and many other animals, substances move rapidly to and from

living cells by way of a(n) (4) _____ circulatory system. (5) _____, a fluid connective tissue within the

(6) _____ and blood vessels, is the transport medium. Most of the cells of animals are bathed in a(n) (7)

_____; blood is constantly delivering nutrients and removing wastes from that fluid. The (8) _____

generates the pressure that keeps blood flowing. Blood flows (9) _____ [choose one] (rapidly, slowly) through

large-diameter vessels to and from the heart, but where the exchange of nutrients and wastes occurs, in the (10)

_____ beds, the blood is divided up into vast numbers of smaller-diameter vessels with tremendous surface

area that enables the exchange to occur by diffusion. The speed of the blood (11) _____ in these vessels. Fish

have a(n) (12) _____ circuit of blood flow, whereas in birds and mammals, blood flows through the (13)

_____ circuit to the lungs and through the (14) _____ circuit to the rest of the body. One of the advantages of having two circuits is that blood (15) _____ can be regulated independently in each.

Labeling and Short Answer [p.628]

16. Label the numbered parts in the following illustrations.

A. _____

B. _____

C. _____

D. _____

E. _____

F. _____

Creature #1

Creature #2

Describe the kind of circulatory system in:

17. Creature #1 _____

_____

18. Creature #2 _____

_____

36.3 HUMAN CARDIOVASCULAR SYSTEM [pp.630-631]

36.4. COMPONENTS AND FUNCTIONS OF BLOOD [pp.632-633]

36.5. HEMOSTASIS [p.634]

36.6. BLOOD TYPING [pp.634-635]

Boldfaced, Page-Referenced Terms

arteries _____

_____

arterioles _____

_____

capillaries _____

_____

venule _____

_____

veins _____

_____

aorta _____

_____

plasma_____

_____

red blood cells _____

_____

cell counts _____

_____

white blood cells _____

_____

platelets_____

_____

hemostasis_____

_____

agglutination_____

_____

ABO blood typing_____

_____

Rh blood typing _____

_____

True/False [pp.630-631]

If the statement is true, write a "T" in the blank. If the statement is false, make it correct by changing the underlined word(s) and writing the correct word(s) in the answer blank.

1. _____ The systemic circuit moves blood to the <u>lungs</u> and back to the heart.

2. _____ When blood moves from one capillary bed through a vein to a second capillary bed before returning to the heart, the vessel between the two capillary beds is called a <u>transfer</u> vein.

3. _____ The <u>pulmonary</u> circuit takes blood to the lungs to pick up oxygen and get rid of carbon dioxide.

4. _____ Arteries always contain blood <u>that is oxygenated</u>.

5. _____ The <u>aorta</u> is the largest artery in the body.

Complete the Table [p.632]

6. Fill in items a–g in the following table, which describes the components of blood.

Components	Relative Amounts	Functions
Plasma Portion (50%-60% of total volume)		
Water	91%-92% of plasma volume	a.
b.	7%-8%	Defense, clotting, lipid transport, roles in extracellular fluid volume, and so forth
Ions, sugars, lipids, amino acids, hormones, vitamins, dissolved gasses	1%-2%	Roles in extracellular fluid volume, pH, and so on
Cellular Portion (40%-50% of total volume)		
c.	4,800,000-5,400,000 per microliter	O2, CO2 transport
White Blood Cells		
d.	3,000-6,750	Phagocytosis
e.	1,000-2,700	Immunity
Monocytes (macrophages)	150-720	Phagocytosis
Eosinophils	100-360	f.
Basophils	25-90	Roles in inflammatory response, fat removal
g.	250,000-300,000	Roles in clotting

Short Answer [p.632-633]

7. Describe the functions of blood. _____

_____

_____

8. What is a possible complication if a mother who is Rh⁻ becomes pregnant with a second child who is Rh⁺? _____

_____

_____

True/False [pp.632-633]

If the statement is true, write a "T" in the blank. If the statement is false, make it correct by changing the underlined word(s) and writing the correct word(s) in the answer blank.

9. _____ Blood is categorized as a <u>connective</u> tissue.

10. _____ <u>Leukocytes</u> are white blood cells.

11. _____ By volume, 50-60 percent of the blood is made up of actual <u>cells</u>.

12. _____ All blood cells are produced in the <u>thymus gland</u>.

13. _____ Red blood cells are very thin to <u>facilitate exchange of oxygen and carbon dioxide</u>.

14. _____ Red blood cells are <u>very long lived</u> cells.

15. _____ <u>Eosinophils</u> are the primary phagocytes on the blood.

16. _____ <u>Monocytes</u> are essential for effective clotting.

Sequence [p.634]

Arrange the following hemostatic events in the correct time sequence. Write the letter of the first event next to 17. The letter of the last event is written next to 22.

17. _____   A. Thrombin converts fibrinogen to fibrin

18. _____   B. Platelets stick together and plug the damaged vessel

19. _____   C. A blood vessel is damaged

20. _____   D. Enzyme cascade activates Factor X

21. _____   E. Fibrin forms a net that collects cells and platelets, forming a clot

22. _____   F. A vascular spasm constricts the vessel

Labeling and Matching [pp.632-633]

Identify the numbered cell types in the accompanying illustration. Complete the exercise by matching and entering the letter of the appropriate function in the parentheses (if any) after the given cell types. A letter may be used more than once.

23. _____ ( )

24. _____ ( )

25. _____ ( )

26. _____ ( )

27. _____ ( )

28. _____ ( )

29. _____ ( )

A. Phagocytosis

B. A role in clotting

C. Immunity

D. O_2, CO_2 transport

E. Immature, unspecialized blood cells

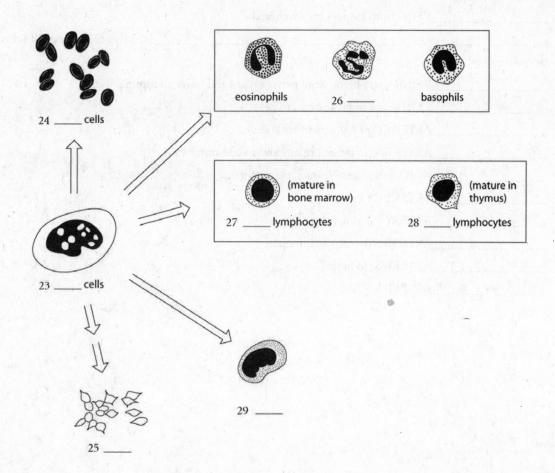

24 _____ cells

eosinophils   26 _____   basophils

(mature in bone marrow)   (mature in thymus)

27 _____ lymphocytes   28 _____ lymphocytes

23 _____ cells

29 _____

25 _____

Short Answer [p.632-633]

30. What is the function of plasma? _____

_____

_____

Fill-in-the-Blanks [p.630]

Fill in the missing words for the labels indicated in the figure. To complete the exercise, color in red on the figure all vessels that contain oxygen-enriched blood, and color in blue all vessels that contain oxygen-poor blood.

31. _____VEINS (from brain, neck, head tissues)

32. _____ _____ _____(from neck, shoulder, arms—

VEINS of upper body)

33. _____VEINS (from lungs to heart)

34. _____VEIN (from kidneys back to heart)

35. _____ _____ _____(receives blood from all veins

below the diaphragm)

36. _____VEINS (carry blood from pelvic organs and lower abdominal wall)

37. _____VEIN (from thigh and inner knee)

38. _____ARTERY (to thigh and inner knees)

39. _____ARTERIES (to pelvic organs, lower abdominal wall)

40. _____AORTA (to digestive tract, pelvic organs)

41. _____ARTERY (to kidney)

42. _____ARTERY (to arm, hand)

43. _____ARTERIES (to cardiac muscle)

44. _____ARTERIES (to lungs)

45. _____ARTERIES

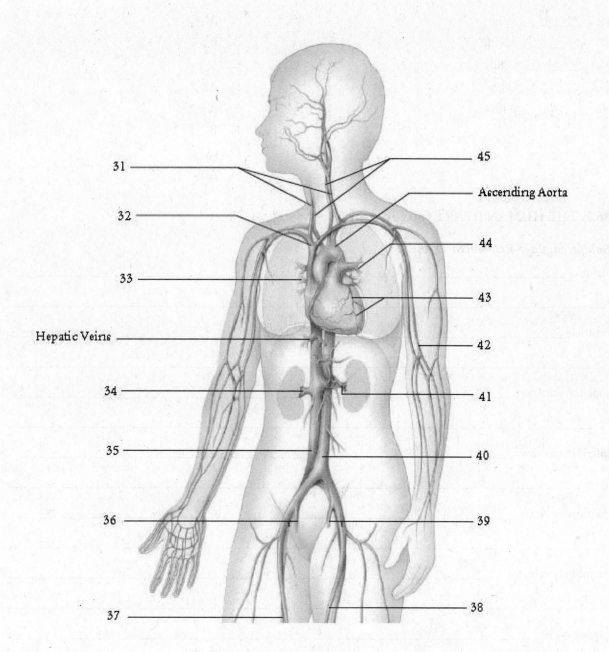

31 ——————

32 ——————

33 ——————

Hepatic Veins ——————

34 ——————

35 ——————

36 ——————

37 ——————

————————— 45

Ascending Aorta

————————— 44

————————— 43

————————— 42

————————— 41

————————— 40

————————— 39

————————— 38

Matching [pp.634-635]

Match each of the following blood types, with the blood type(s) the individual could safely receive in a whole blood (includes plasma and cells) transfusion.

46. _____ A Rh$^+$

47. _____ AB Rh$^+$

48. _____ O Rh$^+$

49. _____ O Rh$^-$

50. _____ B Rh$^-$

A. A$^+$

B. B$^+$

C. O$^+$

D. AB$^+$

E. A$^-$

F. B$^-$

G. O$^-$

H. AB$^-$

36.7. THE HUMAN HEART [pp.636-637]

Boldfaced, Page-Referenced Terms

atrium _____

_____

ventricle _____

_____

superior vena cava _____

_____

inferior vena cava _____

_____

pulmonary artery _____

_____

pulmonary veins _____

_____

cardiac cycle _____

_____

diastole _____

_____

systole _____

_____

sinoatrial (SA) node _____

_____

atrioventricular (AV) node _____

_____

Sequence [p.637]

Arrange the following events in the cardiac cycle in the correct time sequence. Write the letter of the first event next to 1. The letter of the last event is written next to 6.

1. _____   A. AV node receives the signal

2. _____   B. Ventricles contract

3. _____   C. Atria contract

4. _____   D. SA node initiates signal

5. _____   E. Signal spreads across the atria

6. _____   F. Signal sent down bundle fibers to apex of the heart

Labeling [p.636]

Identify each indicated part of the accompanying illustrations. Color in red all vessels and parts of the heart that contain oxygen-rich blood. Color in blue all vessels and parts of the heart that contain oxygen-poor blood.

7. _____

8. _____ _____ _____

9. _____ _____ _____

10. _____ _____

11. _____ _____ _____

12. _____ _____ _____

13. _____ _____ _____

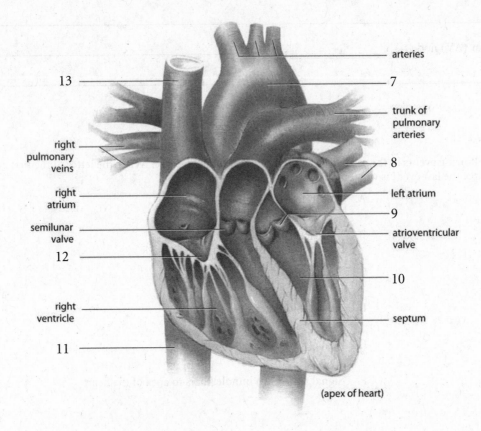

arteries

13

7

trunk of
pulmonary
arteries

right
pulmonary
veins

8

right
atrium

left atrium

9

semilunar
valve

atrioventricular
valve

12

10

right
ventricle

septum

11

(apex of heart)

36.8. BLOOD VESSEL STRUCTURE AND FUNCTION [p.638]

36.9. BLOOD PRESSURE [p.639]

36.10. MECHANISMS OF CAPILLARY EXCHANGE [p.640]

36.11. VENOUS FUNCTION [p.641]

36.2. BLOOD AND CARDIOVASCULAR DISORDERS [pp.642-643]

Boldfaced, Page-Referenced Terms

pulse _____

_____

vasodilation_____

_____

vasoconstriction _____

_____

blood pressure _____

_____

systolic pressure _____

_____

diastolic pressure _____

_____

Fill-in-the-Blanks [p.639]

      Blood pressure is normally high in the (1) _____ immediately after leaving the heart, but then the

pressure drops as the fluid passes along the circuit through different kinds of blood vessels. As blood passes into

smaller-diameter vessels, flow (2) _____ mainly because of increases in (3) _____. (4) _____

guide the flow of blood into various organs. Some signals can make the vessels relax, which causes (5) _____,

and more blood flows into the organ. Other signals cause (6) _____, which decreases blood flow into tissues.

      (7) _____ in the walls of some arteries keep the brain apprised of blood flow. Long-term control is

exerted by hormones that act on the (8) _____.

Labeling [p.638]

Identify each indicated part of the accompanying illustrations.

9. _____

10. _____

11. _____

12. _____

13. _____ _____

14. _____

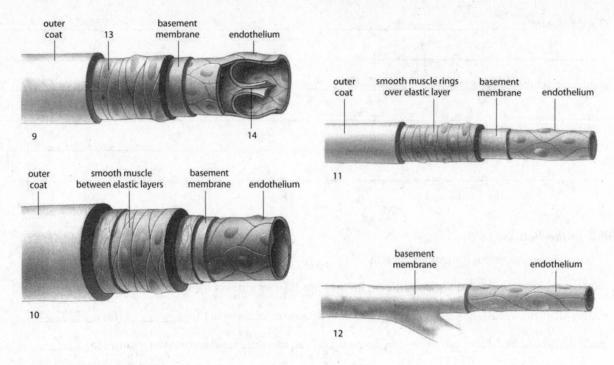

outer coat   13   basement membrane   endothelium

9   14

outer coat   smooth muscle rings over elastic layer   basement membrane   endothelium

11

outer coat   smooth muscle between elastic layers   basement membrane   endothelium

10

basement membrane   endothelium

12

True/False [pp.638, 640-641]

If the statement is true, write a "T" in the blank. If the statement is false, make it correct by changing the underlined word(s) and writing the correct word(s) in the answer blank.

15. _____ Exchange of nutrients and wastes between cells and blood occurs in the <u>venules</u>.

16. _____ Since diffusion works <u>quickly</u>, cells do <u>not need</u> to be close to the blood vessel.

17. _____ Fluids with small solutes and ions leaks out of capillaries between <u>endothelial</u> cells.

18. _____ Veins hold approximately <u>forty</u> percent of blood volume.

19. _____ Because pressure in veins is so low, they have <u>one-way valves</u> to assist with venous return.

20. _____ During exercise there is less blood in the veins because the <u>valves open</u> to allow blood to move.

21. _____ Excess fluid accumulates in the tissues due to <u>vasodilatation</u>.

22. _____ Fluid that accumulates in the tissues is removed by <u>capillary reabsorption</u> and through the <u>lymphatic system</u>.

Short Answer [p.639]

23. Describe the factors that determine blood pressure. _____

_____

_____

_____

24. Your blood pressure is measured at 125/86. Describe what is happening in the heart at each number. _____

_____

_____

_____

Matching [pp.642-643]

Match each term to the most appropriate statement. (One letter is used twice.)

25. _____ atherosclerosis

26. _____ hemolytic anemias

27. _____ hemorrhagic anemias

28. _____ infectious mononucleosis

29. _____ leukemias

30. _____ polycythemias

31. _____ sickle-cell anemia

32. _____ thalassemias

33. _____ hypertension

34. _____ ventricular fibrillation

A. A category of cancers that suppress or impair white blood cell formation in bone marrow

B. When the muscles of the ventricles simply quiver and are unable to move blood out of the heart

C. When blood pressure is chronically high; can lead to an enlarged heart and kidney damage

D. Abnormal hemoglobin formed as a result of a gene mutation

E. Result from a sudden blood loss, as from a severe wound

F. Disorders caused by specific infectious bacteria and parasites as they replicate inside red blood cells and then lyse them

G. An Epstein–Barr virus causes this highly contagious disease, which results from too many monocytes and lymphocytes

H. Disorder involving sluggish blood flow caused by far too many red blood cells; occurs in "blood doping" and some bone marrow cancers

I. Occurs when material collects inside blood vessels and restricts blood flow

J. A change in shape of hemoglobin which leads to multiple health effects, including malformed red blood cells.

36.13. INTERACTIONS WITH THE LYMPHATIC SYSTEM [pp.644-645]

Boldfaced, Page-Referenced Terms

lymph vascular system _____

_____

lymph _____

_____

lymph nodes _____

_____

spleen _____

_____

Matching [pp.644-645]

Choose the correct term for each statement.

1. _____   a huge reservoir of red blood cells and a filter of pathogens and used-up blood cells from the blood

2. _____   delivers water and plasma proteins from capillary beds to the blood vascular system circulation; delivers fats from the small intestine to the blood; delivers pathogens, foreign cells, and material and cellular debris to the organized disposal centers

3. _____   immature T lymphocytes become mature here, and hormones are produced here

4. _____   Contain white blood cells that destroy invading bacteria and viruses gland as they are filtered from the lymph

A. Lymph nodes

B. Lymph

C. Spleen

D. Thymus

Identification/Fill-in-the-Blanks [pp.644-645]

Refer to the illustration on the next page, then supply the missing terms indicated by each answer blank(s).

5. _____

6. _____ gland

7. _____ duct

8. _____

9. _____ _____

10. organized arrays of _____

11. _____ _____

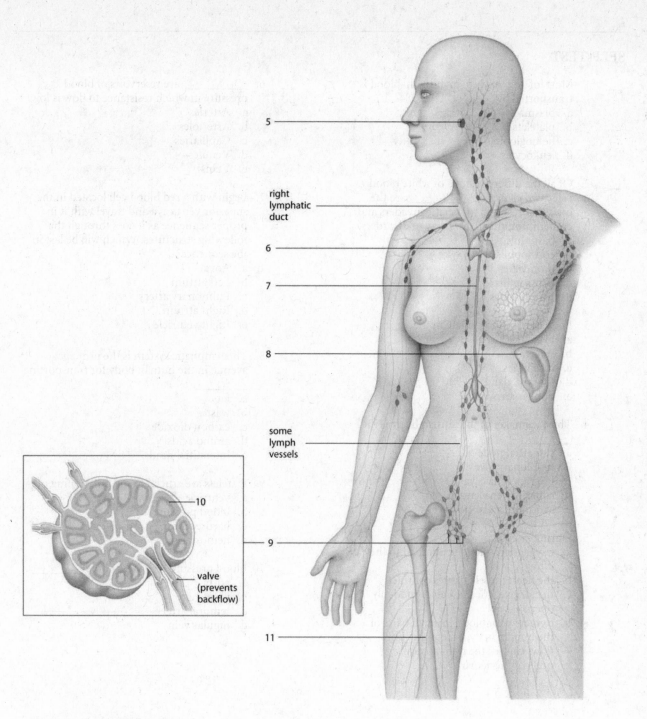

5

right
lymphatic
duct

6

7

8

some
lymph
vessels

9

11

10

valve
(prevents
backflow)

Short Answer [pp.644-645]

12. List the three functions of the lymphatic system. _____

_____

_____

SELF-TEST

1. Most of the oxygen in human blood is transported by _____.
   a. plasma
   b. platelets
   c. hemoglobin
   d. leukocytes

2. Of all the different kinds of white blood cells, two classes of _____ are the ones that respond to specific invaders and confer immunity to a variety of disorders.
   a. basophils
   b. eosinophils
   c. monocytes
   d. neutrophils
   e. lymphocytes

3. Red blood cells originate in the _____.
   a. liver
   b. spleen
   c. capillaries
   d. thymus gland
   e. bone marrow

4. The pacemaker of the human heart is the _____.
   a. sinoatrial node
   b. semilunar valve
   c. inferior vena cava
   d. superior vena cava
   e. atrioventricular node

5. During systole, _____.
   a. oxygen-rich blood is pumped to the lungs
   b. the heart muscle tissues contract
   c. the atrioventricular valves suddenly open
   d. oxygen-poor blood from all parts of the human body, except the lungs, flows toward the right atrium
   e. none of the above

6. _____ are reservoirs of blood pressure in which resistance to flow is low.
   a. Arteries
   b. Arterioles
   c. Capillaries
   d. Venules
   e. Veins

7. Begin with a red blood cell located in the superior vena cava and travel with it in proper sequence as it goes through the following structures. Which will be last in the sequence?
   a. Aorta
   b. Left atrium
   c. Pulmonary artery
   d. Right atrium
   e. Right ventricle

8. The lymphatic system is the principal avenue in the human body for transporting _____.
   a. fats
   b. wastes
   c. carbon dioxide
   d. amino acids
   e. interstitial fluids

9. Platelets are crucial to the functioning of .
   a. ventricles
   b. blood pressure
   c. baroreceptors
   d. hemostasis

10. Blood pressure is highest in the _____ .
    a. aorta
    b. renal vein
    c. femoral artery
    d. jugular vein

Matching

11. _____ agglutination
12. _____ atherosclerosis
13. _____ carotid arteries
14. _____ coronary arteries
15. _____ edema
16. _____ embolism, embolus
17. _____ pulmonary circuit
18. _____ open circulatory system
19. _____ hypertension
16. _____ aorta
20. _____ inferior vena cava
21. _____ jugular veins
22. _____ lymph
23. _____ renal arteries
24. _____ tachycardia
25. _____ thrombosis, thrombus
26. _____ normal blood pressure
27. _____ angioplasty
28. _____ pulse

A. Receive(s) blood from the brain, tissues of the head, and neck

B. The heart's own blood supplier(s)

C. Using a laser or balloon to remove or flatten a plaque in an artery.

D. Much higher than normal rate of heartbeat; occurs during heavy exercising

E. High blood pressure

F. A blood clot that is on the move from one place to another

G. Deliver(s) blood to the head, neck, brain

H. The clumping of red blood cells, or of antibodies with antigens

I. A measurement of 120/80 mm Hg.

J. Delivers blood to kidneys, where its composition and volume are adjusted

K. A blood clot that is lodged in a blood vessel and is blocking it

L. Receive(s) blood from all veins below the diaphragm

M. Progressive thickening of the arterial wall and narrowing of the arterial lumen (space)

N. When blood moves through vessels and mixes with the interstitial fluid

O. Accumulation of excess fluid in interstitial spaces; extreme in elephantiasis

P. The bulging of a an artery with each ventricular contraction.

Q. Fluid that moves through lymph capillaries and vessels.

R. A system of flow that moves blood from the heart to the lungs.

S. The largest artery in the human body.

CHAPTER OBJECTIVES/REVIEW QUESTIONS

1. Distinguish between open and closed circulatory systems. [p.628]
2. Describe the evolutionary changes that occurred in vertebrate circulatory systems as exemplified in fish, amphibians, and birds/mammals. [pp. 628-629]
3. Trace the path of blood in the human body. Begin with the aorta and name all major components of the circulatory system through which the blood passes before it returns to the aorta. [pp.630-631]
4. Describe the composition of human blood, using percentages of volume. [pp.632-633]
5. Distinguish the main types of leukocytes from each other in terms of structure and functions. [pp.632-633]

6. Describe the sequence of events that occurs when a blood vessel is damaged that leads to hemostasis. [p.634]
7. Explain the ABO and Rh blood types and how they are inherited. [p.634-635]
8. Explain what causes a heart to beat. Then describe how the rate of heartbeat can be slowed down or speeded up. [pp.636-637]
9. Describe how the structures of arteries, capillaries, and veins differ. [pp.638, 640-641]
10. Describe what blood pressure is and how it is measured. Correlate the numbers with physical movements in the heart. [p.639]
11. State the significance of high- and low-density lipoproteins to cardiovascular disorders. [p.642]
12. Describe how hypertension develops, how it is detected, and whether it can be corrected. [p.642]
13. Describe the composition and function of the lymphatic system. [pp.644-645]

INTEGRATING AND APPLYING KEY CONCEPTS

1. Suppose humans had two-chambered or three-chambered hearts. How would this affect metabolism and activity levels?
2. A person has elevated blood pressure. She takes two medications—one is a diuretic and the other inhibits calcium release in muscle fibers. Explain each of these can lower the blood pressure.
3. You observe some people who appear as though fluid had accumulated in their lower legs and feet. Their lower extremities resemble those of elephants. You inquire about what is wrong and are told that the condition is caused by the bite of a mosquito that is active at night. Construct a testable hypothesis that would explain (1) why the fluid was not being returned to the torso, as normal, and (2) what the mosquito did to its victims.

37

IMMUNITY

INTRODUCTION

Chapter 37 looks at how the immune system protects us through three layers of defense, all integrated with each other. The first layer is the external barriers that try to prevent entry of antigens. The second layer is an innate response. It has the same basic set of responses to draw from regardless of the antigen. Third, there is the adaptive response that custom builds a response to each individual antigen type. It also looks at problems that arise when the system is not functioning properly, either by not responding to antigens or misidentifying and responding incorrectly.

STUDY STRATEGIES

- Read all of Chapter 37 with the goal of familiarizing yourself with the boldface terms.
- Compare and contrast innate versus adaptive immunity.
- Identify the major modes of defense for an organism against a pathogen.
- Recognize the role of surface barriers and their modes of operation in immunity.
- Discuss the structure and function of the innate immune defenses.
- Explain the role of inflammation and fever in the immune system.
- Describe antigen and antibody interactions.
- Discuss the structure and function of the adaptive immune defenses.
- Explain the antibody-mediated and cell-mediated immune responses.
- Describe some immune system disorders and their possible treatments.
- Explain the relationship between HIV, AIDS, and the immune system.
- Compare and contrast the benefits and risks of vaccines.

FOCAL POINTS

- Table 37.2 [p.653] lists some of the chemical weapons used by the immune system.
- Figure 37.8 [p.654] shows the results of complement action.
- Figure 37.10 [p.656] illustrates the inflammatory response.
- Figure 37.16 [p.662] diagrams the antibody-mediated immune response.
- Figure 37.19 [p.664] illustrates the cell-mediated immune response.

INTERACTIVE EXERCISES

37.1. FRANKIE'S LAST WISH [p.649]

37.2. INTEGRATED RESPONSE TO THREATS [pp.650-651]

37.3. SURFACE BARRIERS [pp.652-653]

37.4. TRIGGERING INNATE DEFENSES [pp.654-655]

37.5. INFLAMMATION AND FEVER [pp.656-657]

Boldfaced, Page-Referenced Terms

immunity _____

_____

antigen _____

_____

complement _____

_____

innate immunity _____

_____

adaptive immunity _____

_____

cytokines _____

_____

neutrophils _____

_____

macrophages _____

_____

dendritic cells _____

_____

eosinophils _____

_____

basophils _____

_____

mast cells _____

_____

B cells _____

_____

T cells _____

_____

cytotixic T cells _____

_____

NK cells _____

_____

normal flora _____

_____

plaque _____

_____

lysozyme _____

_____

chemotaxis _____

_____

inflammation _____

_____

fever _____

_____

Fill-in-the-Blanks [pp.650-653]

Immune responses are provoked by the presence of non-self (1) _____ on the surface of invaders. There are approximately (2) _____ different molecules recognized as non-self. These are called (3) _____. An early chemical response is the release of (4) _____ that binds to invaders, killing them or tagging them for phagocytosis. This fast response is part of (5) _____ immunity. When lymphocytes evolved, they developed a customized response to specific antigens, a process called (6) _____ immunity.

The best way to deal with damaging invaders is to prevent their entry. Several barriers prevent pathogens from crossing the boundaries of your body. Intact (7) _____ and (8) _____ membranes are effective barriers.

(9) _____ is an enzyme that kills many bacteria. (10) _____ fluid destroys many food-borne pathogens in the stomach. Normal (11) _____ residents of the skin, gut, and vagina outcompete pathogens for resources and help keep their numbers under control.

True/False [pp.650-653]

If the statement is true, write a "T" in the blank. If the statement is false, make it correct by changing the underlined word(s) and writing the correct word(s) in the answer blank.

12. _____ A PAMP is a molecular pattern unique to <u>your immune system</u> that(?) enables pathogens to be recognized.

13. _____ Since the body must have openings for nutrients to enter and waste to be removed, these portals are protected by <u>complement molecules</u>.

14. _____ <u>Mucus membranes</u> are sticky and trap many microorganisms which are then swept away.

15. _____ Among the white blood cells, <u>neutrophils</u> make up the largest number.

16. _____ Immune system cells communicate with each other chemically by use of <u>hormones</u>.

Matching [p.651]

Match each of the following cells with its correct function.

17. _____ Eosinophil

18. _____ Mast cell

19. _____ B Lymphocyte

20. _____ Macrophage

21. _____ Basophil

22. _____ Dendritic cell

23. _____ Neutrophil

24. _____ T Lymphocyte

A. Phagocytes that show antigens to naïve T cells

B. Coordinate immune response; recognize infected or abnormal cells and kill those cells by contact

C. Contains enzymes that help remove parasitic worms; circulates

D. Contain histamine granules; contribute inflammation; fixed in tissues

E. Produce antibodies against recognized antigen

F. Most abundant phagocyte; circulates

G. Phagocyte that presents antigens; mature when in tissues

H. Contains histamine granules; contribute inflammation; circulates

Labeling [p.654]

25. Insert the correct immune system process or structure from the figure into the blank next to the appropriate letter.

A. _____

B. _____

C. _____

D. _____

E. _____

a. complement coating

b. pathogen-associated molecular pattern

c. cell lysis

d. antibodies bound to pathogen

e. amplifies response

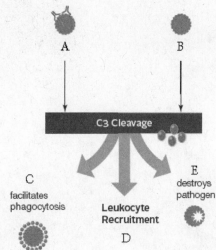

Sequence [pp.656-657]

Arrange the following steps in the inflammatory response in correct time sequence. Write the letter of the first step next to 25. The letter of the last step is written next to 30.

26. _____

27. _____

28. _____

29. _____

30. _____

31. _____

A. Mast cells in tissue release histamine

B. Fluid and proteins in the plasma leak into the tissues causing swelling and pain

C. Neutrophil and macrophages are attracted to area and engulf invaders

D. Capillaries dilate and become more permeable

E. Tissue is damaged and antigens enter the tissue

F. Complement proteins attach to bacteria; clotting factors build a barrier around inflamed area

Labeling [pp.656-657]

Place the letter of the correct structure from the figure of the inflammation response into the blank next each term.

32. _____ mast cell

33. _____ capillary

34. _____ bacterium

35. _____ signaling molecules

36. _____ macrophage

37. _____ compliment

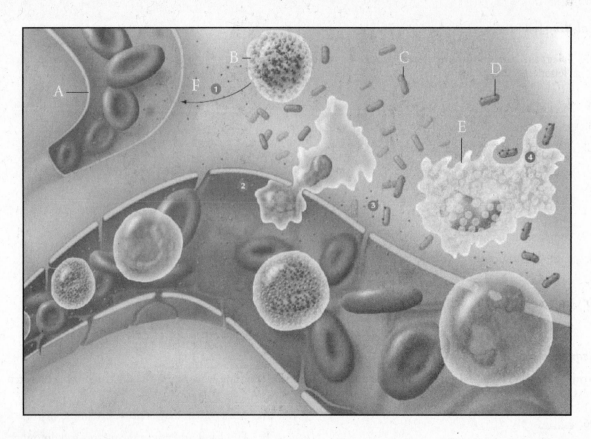

Short Answer [pp.656-657]

38. Explain how a fever helps fight an infection. _____

_____

_____

_____

37.6. ANTIGEN RECEPTORS [pp.650-659]

37.7. OVERVIEW OF ADAPTIVE IMMUNITY [pp.660-661]

37.8. THE ANTIBODY-MEDIATED IMMUNE RESPONSE [pp.662-663]

37.9. THE CELL-MEDIATED IMMUNE RESPONSE [pp.664-665]

Boldfaced, Page-Referenced Terms

T cell receptor (TCR) _____

_____

MHC markers _____

_____

antibodies _____

_____

B cell receptors _____

_____

antibody-mediated immune response _____

_____

cell-mediated immune response _____

_____

effector cells _____

_____

memory cells _____

_____

Fill-in-the-Blanks [pp.658-665]

If the (1) _____ immune response fails to repel the microbial invaders, then the body calls on its (2) _____ immune response , which identifies *specific* targets to kill and *remembers* the identities of its targets. Your own unique(3) _____ patterns identify your cells as "self" cells. Any other surface pattern is, by definition, (4) _____, and doesn't belong in your body.

The principal actors of the adaptive immune system are (5) _____ descended from stem cells in the bone marrow that have two different strategies of action to deal with their different kinds of enemies.(6) _____ mediate the antibody response and act principally against the extracellular enemies that are pathogens in blood or on the cell surfaces of body tissues. (7) _____ defend principally against intracellular pathogens such as (8) _____, and against any cells that are perceived as abnormal or foreign, such as (9) _____ cells and cells of organ transplants.

Each kind of cell, virus, or substance bears particular molecular configurations (patterns) that give it a unique (10) _____. A(n) (11) _____ is any molecular configuration on the pathogen's surface that causes the formation of lymphocyte armies. Any cell that processes and displays a(n) (11) _____together with a suitable MHC molecule is known as a(n) (12) _____ cell and can activate lymphocytes to undergo rapid cell divisions. Lymphocyte subpopulations that fight and destroy enemies are known as (13) _____ cells. Long-

lived (14) _____ cells are also produced and react to future encounters with the same antigen. The antibody-

mediated response is primarily a function of (15) _____ cells, while the cell-mediated response relies on (16)

_____ cells.

True/False [pp.660-661]

If the statement is true, write a "T" in the blank. If the statement is false, make it correct by changing the underlined word(s) and writing the correct word(s) in the answer blank.

17. _____ The first step in adaptive immunity is <u>production of antibodies</u>.

18. _____ Pieces of the antigen are displayed on the cell surface as an <u>antigen-MHC complex</u>.

19. _____ When a T cell binds to an antigen-MHC complex, it secretes <u>antibodies</u> to signal other B and T cells.

20. _____ Antibodies are <u>cells</u> that can inactivate antigens.

21. _____ <u>B cells</u> will attack intracellular pathogens.

22. _____ Lymph nodes swell due to an accumulation of <u>antigens</u> in the node.

Matching [p.658]

Match the antibody type with its function.

23. _____ IgM

24. _____ IgG

25. _____ IgE

26. _____ IgA

27. _____ IgD

A. Produced in exocrine secretions; found on mucus-coated surfaces of the respiratory, digestive, and reproductive tracts where they neutralize pathogens

B. Triggers inflammation when pathogens attack the body; plays a role in allergic responses

C. Activate complement proteins; neutralize many toxins; long-lasting; can cross placenta and protect developing fetus; also present in colostrum (early milk) from mammary glands

D. First to be secreted during immune responses; after binding to antigen, trigger complement cascade; also tag invaders and bind them in clumps for later phagocytosis

E. Acts as the B cell receptor for antigens

Sequence [pp.662-663]

Arrange the following steps in the antibody response in correct time sequence. Write the letter of the first step next to 28. The letter of the last step is written next to 37.

28. _____   A.   Antibody molecules are secreted by effector cells

29. _____   B.   Antigen-presenting cell binds to naïve T cell

30. _____   C.   Naïve B cell covered with identical antibody molecules (receptors)

31. _____   D.   Effector T cell binds to antigen-MHC complex on B cell and secretes cytokines

32. _____   E.   Antigen binds to antibodies (receptors) on naïve B cell and is taken into cell

33. _____   F.   B cell becomes activated B cell with antigen-MHC complex on cell surface

34. _____   G.   Effector and memory B cells mature

35. _____   H.   Cytokine binds to B cells, triggering mitosis of B cell

36. _____   I.   Dendritic cell engulfs same antigen and becomes antigen-presenting cell

37. _____   J.   T cell divides to produce effector and memory T cells

Sequence [pp.664-665]

Arrange the following steps in the cell-mediated response in correct time sequence. Write the letter of the first step next to 38. The letter of the last step is written next to 44.

38. _____   A.   Activated helper T cell secretes cytokines

39. _____   B.   Effector and memory cytotoxic T cells mature

40. _____   C.   Mature cytotoxic T cell binds to antigen-MHC complex of an infected cell

41. _____   D.   Cytotoxic T cells causes infected cell to die

42. _____   E.   Antigen-presenting cell binds to naïve helper T cell and naïve cytotoxic T cell, both become activated

43. _____   F.   Cytokines bind to activated cytotoxic T cell, causing mitosis

44. _____   G.   Dendritic cell engulfs a virus infected cell and becomes antigen-presenting cell

37.10. WHEN IMMUNITY GOES WRONG [pp.666-667]

37.11. HIV AND AIDS [pp.668-669]

37.12. VACCINES [pp.670-671]

Boldfaced, Page-Referenced Terms

allergy _____

_____

allergen _____

_____

auto immune response _____

_____

AIDS _____

_____

immunization _____

_____

vaccine _____

_____

Matching [pp.666-669]

Match each of the following terms with its correct description.

1. _____ Immunization

2. _____ Allergy

3. _____ Anaphylactic shock

4. _____ Vaccine

5. _____ Autoimmune disorder

6. _____ SCID

7. _____ Rheumatoid arthritis

8. _____ Multiple sclerosis

9. _____ AIDS

10. _____ Passive immunity

11. _____ Hashimoto's disease

12. _____ Ulcerative colitis

13. _____ Crohn's disease

14. _____ Lupus erythematosus

15. _____ Graves' disease

A. An autoimmune disorder where the large intestine is attacked by the body.

B. A whole-body reaction where arterioles dilate and fluid leaks out causing a precipitous drop in blood pressure

C. A group of disorders that result when immune system is suppressed by a virus infection

D. Autoimmune disorder who's symptoms include uncontrollable weight loss; rapid, irregular heartbeat; sleeplessness; pronounced mood swings; and bulging eyes.

E. An immune response against healthy body tissues

F. Antibodies attack the DNA and nuclear proteins of the body which leads to the connective tissues being affected.

G. Antibodies produced in another person are administered

H. An autoimmune disorder where the TH synthesis proteins are attacked within the thyroid.

I. A process that induces immunity

J. Occurs when T cells damage myelin sheath of axons

K. An autoimmune disorder where the entire GI tract is attacked by the body.

L. A substance made to contain a weakened antigen that induces a primary immune response to the antigen without causing illness

M. An immune response to a harmless substance

N. An autoimmune disease that causes inflammation of soft tissue in joints

O. A primary immune deficiency

True/False [pp.668-671]

If the statement is true, write a "T" in the blank. If the statement is false, make it correct by changing the underlined word(s) and writing the correct word(s) in the answer blank.

16. _____ A vaccination works because it initiates a <u>secondary</u> immune response.

17. _____ In active immunization, <u>antibodies</u> are administered to produce immunity to an antigen.

18. _____ Failure to distinguish between <u>self and non-self</u> cells is the hallmark of autoimmunity.

19. _____ AIDS is a type of <u>primary</u> immune deficiency.

20. _____ HIV is a retrovirus; that means it requires <u>protease</u> to replicate its genome.

21. _____ Grave's disease has a constellation of symptoms caused by a <u>deficiency</u> of thyroid hormone.

22. _____ Allergens induce a variety of responses mediated by <u>histamine</u> release.

23. _____ The first recorded vaccine in Europe was made by <u>Louis Pasteur</u>.

SELF-TEST

___ 1. Pathogen-associated molecular patterns (PAMPs) include _____ .
   a. bacterial flagellum proteins
   b. prokaryotic cell wall materials
   c. double stranded RNA
   d. bacterial pilus proteins
   e. all of the above

___ 2. The plasma proteins that are activated when they contact a bacterial cell are collectively known as the _____ system.
   a. shield
   b. complement
   c. IgG
   d. MHC
   e. HIV

___ 3. _____ are divided into two groups: T cells and B cells.
   a. Macrophages
   b. Lymphocytes
   c. Platelets
   d. Complement cells
   e. Cancer cells

___ 4. _____ produce and secrete antibodies that set up bacterial invaders for subsequent destruction by macrophages.
   a. B cells
   b. Phagocytes
   c. T cells
   d. Bacteriophages
   e. Thymus cells

___ 5. Antibodies are shaped like the letter _____ .
   a. Y
   b. W
   c. Z
   d. H
   e. E

___ 6. The markers for every cell in the human body are referred to by the letters _____ .
   a. HIV
   b. MBC
   c. RNA
   d. DNA
   e. MHC

7. Effector B cells _____ .
   a. fight against extracellular pathogens and toxins circulating in tissues
   b. develop from antigen-presenting cells
   c. manufacture and secrete antibodies
   d. do not divide and form clones
   e. all of the above

8. The clonal selection hypothesis explains _____ .
   a. how self cells are distinguished from non-self cells
   b. how B cells differ from T cells
   c. how so many different kinds of antigen-specific receptors can be produced by lymphocytes
   d. how memory cells are set aside from effector cells
   e. how antigens differ from antibodies

9. Pathogens include _____.
   a. viruses
   b. bacteria
   c. fungi
   d. worms
   e. all of the above

10. The process of acquired immunity is enhanced via medicine by this action
    a. vaccination
    b. communication
    c. fever
    d. inflammation

11. This disorder occurs when the immune system is not as active as it should be
    a. allergy
    b. autoimmune
    c. immunodeficiency
    d. none of these

12. Surface barriers include
    a. skin
    b. tears
    c. mucus
    d. gastric fluids
    e. all of the above

13. There are ____ different types of complement proteins.
    a. 2
    b. 10
    c. 20
    d. 30

14. Membrane attack complexes
    a. attach to the lipid bilayer of the invading organism
    b. attach to the nuclear membrane of the invading organism
    c. attach to the lipid bilayer of the host
    d. attach to the nuclear membrane of the host

15. A fast, local response that destroys affected tissues and assists the healing process is ____.
    a. fever
    b. inflammation
    c. vaccine
    d. antibiotics

16. _____ work by acting on histamine receptors to reduce their activity.
    a. Fevers
    b. Vaccines
    c. Antibiotics
    d. Antihistamines

17. Currently, approximately 33 million people have been infected by this pathogen which can lead to a total collapse of the immune system.
    a. HIV
    b. Influenza
    c. *Streptococcus sp.*
    d. *Staphylococcus sp.*

18. Which of the following have been virtually eradicated due to vaccination programs?
    a. HIV
    b. Influenza
    c. Smallpox
    d. Hepatitis

Matching

Choose the most appropriate description for each term.

19. _____ allergy

20. _____ antibody

21. _____ antigen

22. _____ macrophage

23. _____ clone

24. _____ complement

25. _____ histamine

26. _____ MHC marker

27. _____ effector B cell

28. _____ T cell

A. Begins its development in bone marrow, but matures in the thymus gland

B. Cells that have directly or indirectly descended from the same parent cell

C. A potent chemical that causes blood vessels to dilate and let protein pass through the vessel walls

D. Y-shaped immunoglobulin

E. A non-self marker

F. A progeny of a turned-on B cell

G. A group of about 30 proteins that participate in the inflammatory response

H. An altered secondary immune response to a substance that is normally harmless to other people

I. The basis for self-recognition at the cell surface

J. Principal perpetrator of phagocytosis

CHAPTER OBJECTIVES/REVIEW QUESTIONS

1. List the general types of cells that form the basis of the vertebrate immune system. [p.651]
2. List and discuss four nonspecific defense responses that serve to exclude microbes from the body. [pp.652-653]
3. Distinguish between the antibody-mediated response pattern and the cell-mediated response pattern. [pp.662-665]
4. Describe the sequence of events that occur during inflammatory responses. [pp.656-657]
5. Explain why the immune system of mammals usually does not attack "self" tissues. Understand how vertebrates (especially mammals) recognize and discriminate between self and non-self tissues. [p.658]
6. Distinguish allergies from autoimmune disorders. [pp.666-667]
7. Describe some examples of immune failures, and identify as specifically as you can which weapons in the immunity arsenal failed in each case. [pp.666-667]
8. Describe how AIDS specifically interferes with the human immune system. [pp.668-669]
9. Describe two ways that people can be immunized against specific diseases. [pp.670-671]

INTEGRATING AND APPLYING KEY CONCEPTS

1. Discuss the advantages of having resident populations of microbes on the skin.
2. What parts of the immune system might slow or stop AIDS if they were enhanced?

38
RESPIRATION

INTRODUCTION

This chapter covers the biological processes involved in the exchange of gases in animals. All animals require an input of oxygen to fuel aerobic respiration, and a release of the waste gas carbon dioxide. This chapter introduces the principles of a respiratory system, as well as the various forms of respiratory systems in animals. It also explores the basis of gas exchange and some of the more common diseases and disorders that may occur when normal gas exchange is interrupted.

STUDY STRATEGIES

- Read all of Chapter 38 with the goal of familiarizing yourself with the boldface terms.
- Identify the important factors affecting diffusion rates in respiration.
- Recognize some of the adaptations of invertebrates for respiration.
- Discuss some of the adaptations of vertebrates for respiration.
- Identify the important structures of the human respiratory system and the role of each.
- Explain the respiratory cycle.
- Describe the exchange and transport of gasses through the respiratory and circulatory systems.
- Discuss how some organisms are adapted to live in hypoxic environments.
- Recognize various respiratory diseases and disorders and their treatments.

FOCAL POINTS

- Figure 38.8[p.680] shows a respiratory system adapted for under water use—gills in fish.
- Figure 38.12 [p.682] illustrates the major structures of the respiratory system in humans. You should note that some of these structures have more than one function in human physiology.
- Figure 38.16 [p.684] diagrams how air moves into and out of the lungs.

INTERACTIVE EXERCISES

38.1. CARBON MONOXIDE – A STEALTHY POISON [p.675]

38.2. THE NATURE OF RESPIRATION [pp.676-677]

38.3. INVERTEBRATE RESPIRATION [pp.678-679]

38.4. VERTEBRATE RESPIRATION [pp.680-681]

Boldfaced, Page-Referenced Terms

respiration _____

_____

respiratory surface _____

_____

respiratory proteins _____

_____

integumentary exchange _____

_____

gills _____

_____

lung _____

_____

tracheal system _____

_____

counter current exchange _____

_____

Labeling [pp.676-677]

For numbers 1-4, identify the structures of the fish gills. For numbers 5-7, identify the fluid and direction of flow. Use Figure 38.8 to guide you.

1. _____

2. _____

3. _____

4. _____

5. _____

6. _____

7. _____

A. respiratory surface

B. gill arch

C. fold with a capillary bed inside

D. oxygenated blood back toward body oxygen-poor blood

E. gill filament

F. oxygen-poor blood from deep in body

G. direction of blood flow

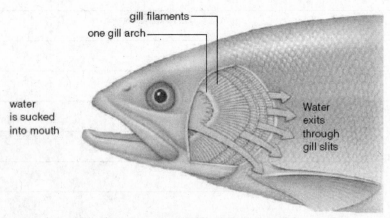

Bony fish with its gill cover removed. Water flows in through the mouth, over the gills, then out through gill slits. Each gill has bony gill arches with many thin gill filaments attached.

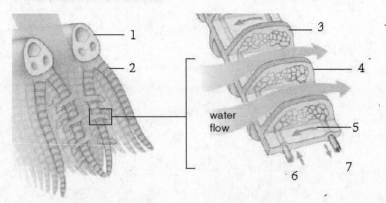

Two gill arches with filaments   Countercurrent flow of water and blood

Matching [pp.676-677]

Choose the most appropriate statement for each of the following terms.

8. _____ hemoglobin

9. _____ respiration

10. _____ surface-to-volume ratio

11. _____ myoglobin

12. _____ concentration gradient

13. _____ ventilation

14. _____ respiratory surface

A. The name given to adaptations that increase gas exchange rates above the level of diffusion

B. The size of animal bodies is governed by this principle

C. The main respiratory pigment of humans

D. The sum of the physiological processes that move oxygen and carbon dioxide

E. Gases enter and leave an organism by first crossing this structure

F. The respiratory pigment found in most muscle cells

G. The variation in concentration of a molecule across a semipermeable membrane.

True/False [p.677]

If the statement is true, write a "T" in the blank. If the statement is false, make it correct by changing the underlined word(s) and writing the correct word(s) in the answer blank.

15. _____ Warm, slow flowing water contains larger amounts of dissolved oxygen.

16. _____ Over-enriched lake waters support rapid growth of algae; as nutrients are used, the algae die and the water is rapidly depleted of oxygen.

17. _____ When oxygen levels in a lake or stream are low, aquatic insect larvae are the last to disappear.

18. _____ Fish will survive in water until the dissolved oxygen level falls below two parts per million.

19. _____ Waters with the lowest oxygen concentrations often contain only sludge worms.

Choice [pp.678-679]

For items 16–20, choose from the following forms of invertebrate respiration.

a. tracheal system   b. integumentary exchange   c. gills   d. book lungs

20. _____ The diffusion of a gas directly across the body covering

21. _____ A form of respiratory system commonly found in aquatic organisms

22. _____ An internal system of tubes that act as a respiratory surface

23. _____ Commonly found in insects, millipedes, and centipedes

24. _____ Thin-walled respiratory surfaces that exchange gases between a body fluid and its surroundings

25. _____ Found in some spiders in place of tracheal tubes

Choice [pp.680-681]

Match each respiratory system characteristic to the group of vertebrates that has it.

    a. fish   b. amphibians   c. birds

26. _____ Utilize a countercurrent flow to effectively exchange gases

27. _____ Utilize a flow-through respiratory system

28. _____ Have gills as larvae, then develop lungs

29. _____ Takes two breaths to move air through the respiratory system

30. _____ These organisms use their skin to supplement gas exchange

38.5. HUMAN RESPIRATORY SYSTEM [pp.682-683]

38.6. CYCLIC REVERSALS IN AIR PRESSURE GRADIENTS [pp.684-685]

Boldfaced, Page-Referenced Terms

pharynx _____

_____

larynx _____

_____

glottis _____

_____

epiglottis _____

_____

trachea _____

_____

bronchus _____

_____

bronchioles _____

_____

alveoli _____

_____

diaphragm _____

_____

intercostal muscles _____

_____

respiratory cycle _____

_____

Heimlich maneuver _____

_____

vital capacity _____

_____

tidal volume _____

_____

Labeling and Matching [pp.682-683]

First, identify each structure indicated in the diagram on the next page. Then match the structure with its correct function and place the corresponding letter in the parentheses. [pp.688-689]

1. _____ ( )

2. _____ ( )

3. _____ ( )

4. _____ ( )

5. _____ ( )

6. _____ ( )

7. _____ ( )

8. _____ ( )

9. _____ ( )

10. _____ ( )

11. _____ ( )

12. _____ ( )

13. _____ ( )

A. A supplemental airway

B. The site of gas exchange

C. Smooth muscle that separates the thoracic and abdominal cavities

D. One of a pair of lobed, elastic organs involved in gas exchange

E. The site of sound production

F. Airway that connects the nasal cavity with the larynx

G. Warms and filters incoming air

H. Skeletal muscles with roles in breathing

I. Airway that connects the larynx to bronchi leading to lungs

J. The fine branches of the "bronchial tree"

K. A double membrane that contains a lubricating fluid

L. Groups of alveoli located at the end of a bronchiole

M. Separates the respiratory system from the digestive system

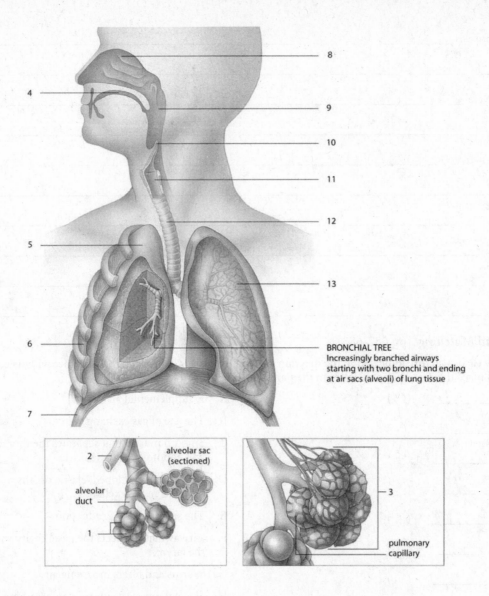

BRONCHIAL TREE
Increasingly branched airways
starting with two bronchi and ending
at air sacs (alveoli) of lung tissue

alveolar sac
(sectioned)

alveolar
duct

pulmonary
capillary

Choice [pp.684-685]

Choose whether each of the following statements is associated with inhalation or exhalation.

a. inhalation   b. exhalation

14. _____ Always requires muscle activity

15. _____ External intercostal muscles contract

16. _____ Occurs passively when one is breathing quietly

17. _____ Diaphragm flattens and moves down

18. _____ Requires muscle activity only during exercise

19. _____ Diaphragm relaxes and returns to dome shape

20. _____ Pressure in alveoli is below atmospheric pressure

21. _____ Pressure in alveoli is greater than atmospheric pressure

Matching [pp.684-685]

Match each of the following terms with its correct description.

22. _____ Tidal volume

23. _____ Medulla oblongata

24. _____ Heimlich maneuver

25. _____ Vital capacity

26. _____ Carbonic acid

27. _____ Carotid arteries

28. _____ Sympathetic nerves

A. A process performed on someone to clear the airway of someone choking

B. Site of chemoreceptors to measure need for changes in breathing rate

C. Volume of air moved in and out lungs in a normal respiratory cycle

D. Produced when CO_2 combines with water in the blood

E. Nerves that increase breathing rate

F. Maximum volume of air that lungs can move in one breath

G. Site of neurons that are control center for breathing

Short Answer [pp.684-685]

29. Briefly explain the action of the Heimlich maneuver.

_____

_____

38.7. GAS EXCHANGE AND TRANSPORT [pp.686-687]

38.8. RESPIRATION IN EXTREME ENVIRONMENTS [pp.688-689]

38.9. RESPIRATORY DISEASES AND DISORDERS [pp.690-691]

Boldfaced, Page-Referenced Terms

respiratory membrane _____

_____

oxyhemoglobin _____

_____

carbonic anhydrase _____

_____

acclimatization _____

_____

erythropoieten _____

_____

Choice [pp.675, 686-687]

For each statement, choose the appropriate gas.

a. oxygen   b. carbon dioxide   c. carbon monoxide

1. _____ When bound to hemoglobin, forms carbaminohemoglobin

2. _____ Heme groups with iron bind with this gas

3. _____ This is released where the blood is warmer and pH is lower

4. _____ The enzyme carbonic anhydrase is involved in the transport of this gas

5. _____ When bound to hemoglobin it forms oxyhemoglobin

6. _____ In water, this gas forms carbonic acid

7. _____ Binds to hemoglobin the most tightly

8. _____ Binds with myoglobin in skeletal and cardiac muscles

Matching [pp.688-691]

Match each of the following disorders or conditions to its correct description.

9. _____ bronchitis

10. _____ lung cancer

11. _____ altitude sickness

12. _____ emphysema

13. _____ nitrogen narcosis

14. _____ bladder cancer

15. _____ hypoxia

16. _____ decompression sickness

17. _____ tuberculosis (TB)

18. _____ asthma

19. _____ pneumonia

20. _____ SIDS

A. The general condition when not enough oxygen reaches the cells

B. General term for lung inflammation caused by infection

C. Ion balances in the cerebrospinal fluid are incorrect due to hyperventilation

D. Known as Sudden Infant Death Syndrome; correlated with sleep apnea in infants

E. Disruption of neural membranes due to incorrect nitrogen levels

F. Nitrogen bubbles form in the blood and tissues

G. Caused by the bacteria *Myobacterium tuberculosis*; many strains are now antibiotic resistant.

H. Cause of death of 4 million people globally each year.

I. Constriction of the airways and inflammation triggered by allergies, irritants, or exercise

J. This risk of acquiring this disease is 7-10 times greater in smokers

K. Inflammation of the epithelium of the bronchioles

L. Destruction of the thin walls of the alveoli

Labeling and Matching [pp.682-683]

Identify each structure indicated in the diagram of the respiratory membrane below.

21. _____
22. _____
23. _____
24. _____
25. _____
26. _____

A. air space inside alveolus

B. alveolar epithelium

C. fused basement membranes of both epithelial tissues

D. red blood cell inside pulmonary capillary

E. capillary endothelium

F. pore for air flow between adjoining alveoli

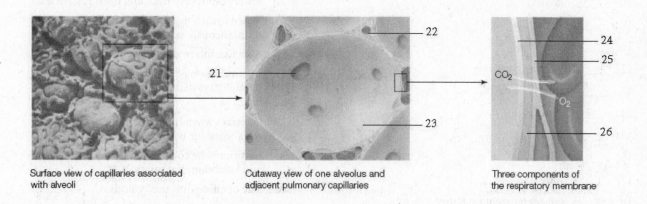

Surface view of capillaries associated with alveoli

Cutaway view of one alveolus and adjacent pulmonary capillaries

Three components of the respiratory membrane

SELF-TEST

Matching

1. _____ bronchioles
2. _____ bronchitis
3. _____ carbonic anhydrase
4. _____ emphysema
5. _____ apnea
6. _____ glottis
7. _____ hypoxia
8. _____ external intercostal muscles
9. _____ larynx
10. _____ oxyhemoglobin
11. _____ pharynx
12. _____ pleurisy
13. _____ tidal volume
14. _____ ventilation
15. _____ vital capacity
16. _____ gills
17. _____ lung
18. _____ countercurrent exchange
19. _____ diaphragm
20. _____ Heimlich maneuver

A. Membrane that encloses human lung becomes inflamed and swollen; painful breathing generally results

B. HbO_2

C. The amount of air inhaled and exhaled during normal breathing of a human at rest; generally about 500 ml

D. manually raising the intra-abdominal pressure of a choking person to dislodge an object stuck in the trachea

E. Throat passageway that connects to *both* the lower respiratory tract *and* the digestive tract

F. When breathing stops and then spontaneously starts repetitively

G. A saclike internal respiratory organ

H. Inflammation of the two principal passageways that carry air into the human lungs

I. Contract when lungs are filling with air, relax when air is leaving the lungs

J. Partitions the coelom into a thoracic cavity and an abdominal cavity

K. The opening into the "voicebox"

L. Finer and finer branchings that lead to alveoli

M. Maximum volume of air that can move out of your lungs after a single, maximal inhalation

N. Filamentous or platelike respiratory organs that increase the surface area available for gas exchange

P. An enzyme that increases the rate of production of H_2CO_3 from CO_2 and H_2O

Q. Lungs have become distended and inelastic so that walking, running, and even exhaling are difficult

R. Where sound is produced by vocal cords

S. Movements that keep air or water moving across a respiratory surface

T. Process of two fluids exchange substances while flowing in opposite directions

U. Too little oxygen is being distributed in the body's tissues

Multiple Choice

___ 21. The respiratory transport molecule that is abundant in muscle and skeletal cells of the human body is _____.
   a. myoglobin
   b. hemoglobin
   c. carbonic acid
   d. carbonic anhydrase

___ 22. _____ is the most abundant gas in Earth's atmosphere.
   a. Water vapor
   b. Oxygen
   c. Carbon dioxide
   d. Hydrogen
   e. Nitrogen

___ 23. With respect to respiratory systems, countercurrent flow is a mechanism that explains how _____.
   a. oxygen uptake by blood capillaries in the lamellae of fish gills occurs
   b. ventilation occurs
   c. intrapleural pressure is established
   d. sounds originating in the vocal cords of the larynx are formed
   e. all of the above

___ 24. A flow-through respiratory system is found in _____.
   a. amphibians
   b. reptiles
   c. birds
   d. mammals
   e. humans

___ 25. During inhalation, _____.
   a. the pressure in the thoracic cavity (intrapleural pressure) is less than the pressure within the lungs (intrapulmonary pressure)
   b. the pressure in the chest cavity (intrapleural pressure) is greater than the pressure within the lungs (intrapulmonary pressure)
   c. the diaphragm moves upward and becomes more curved
   d. the thoracic cavity volume decreases
   e. all of the above

___ 26. Oxygen moves from alveoli to the bloodstream _____.
   a. by diffusion when the concentration of oxygen is greater in alveoli than in the blood
   b. by means of active transport
   c. using the assistance of carbaminohemoglobin
   d. principally due to the activity of carbonic anhydrase in the red blood cells
   e. all of the above

___ 27. Immediately before reaching the alveoli, air passes through the _____.
   a. bronchioles
   b. glottis
   c. larynx
   d. pharynx
   e. trachea

___ 28. Oxyhemoglobin _____.
   a. releases oxygen more readily in metabolically active tissues
   b. tends to release oxygen in places where the temperature is lower
   c. tends to hold onto oxygen when the pH of the blood drops
   d. tends to give up oxygen in regions where partial pressure of oxygen exceeds that in the lungs
   e. all of the above

___ 29. Which of the following is not involved in carbon dioxide transport?
   a. CO_2 is bound to hemoglobin to form carbaminohemoglobin.
   b. CO_2 is dissolved directly in the blood.
   c. CO_2 is transported as bicarbonate ions.
   d. CO_2 binds to myoglobin for transport to alveoli.

___ 30. Which of the following respiratory ailments is common among smokers?
   a. emphysema
   b. bronchitis
   c. cancer
   d. impaired immune system
   e. all of the above

___ 31. Which of the following is always an active endeavor?
   a. inhalation
   b. exhalation
   c. diffusion
   d. closing of the epiglottis

___ 32. Which of the following can use integumentary exchange?
a. amphibians
b. earthworms
c. jellyfish
d. all of the above

___ 33. All of these can affect the rate of diffusion of gasses except
a. surface area
b. gas concentration
c. ventilation
d. willpower

___ 34. Which of the following would promote higher retention of oxygen concentrations?
a. low pH
b. warmer temperatures
c. cooler temperatures
d. slow currents

___ 35. Control over the rate of breathing occurs in the
a. hypothalamus
b. medulla oblongata
c. corpus callosum
d. cerebellum

CHAPTER OBJECTIVES/REVIEW QUESTIONS

1. Understand how the human respiratory system interacts with the digestive, circulatory, and urinary systems. [p.676]
2. Recognize the difference between myoglobin and hemoglobin in human respiration. [pp.676-677]
3. List the types of invertebrate respiratory surfaces that participate in gas exchange and give an example of each. [pp.678-679]
4. Understand the physical properties of gases and the limitations of a respiratory surface. [p.679]
5. Define *countercurrent flow* and explain how it works. [p.680]
6. Recognize the difference in the respiratory system of birds compared to amphibians and mammals. [pp.680-681]
7. List all the principal parts of the human respiratory system, and explain how each structure contributes to transporting oxygen from the external world to the bloodstream. [pp.682-683]
8. Trace oxygen transport from the air to the tissues of the body. [pp.682-683]
9. Describe the respiratory cycle and the processes of inhalation and exhalation. [pp.684-685]
10. Explain the difference between vital capacity and tidal volume. [p.684-685]
11. Explain the factors that influence the release of oxygen to the tissues. [pp.686-687]
12. Trace the transport of carbon dioxide from the tissues of the body to the lungs. [pp.686-687]
13. List some of the challenges of respiration at high altitude and during diving. [pp.688-689]
14. Explain why carbon monoxide is a dangerous gas. [pp.675, 691]
15. Explain the causes of the major breathing disorders. [pp.690-691]
16. Explain how bronchitis, emphysema, and lung cancer are all related to smoking. [pp.690-691]

INTEGRATING AND APPLYING KEY CONCEPTS

1. Consider the amphibians—animals that generally have aquatic larval forms (tadpoles) and terrestrial adults. Outline the respiratory changes that you think might occur as an aquatic tadpole metamorphoses into a land-going juvenile.
2. In the movie *Waterworld,* Kevin Costner's character develops gills to aid his underwater breathing. Explain what other changes would be necessary to his respiratory system in order to make this adaptation effective.
3. Sickle-cell anemia is often considered to be a disease of the circulatory system since it alters the shape of the red blood cells (RBCs) in the body. RBCs are the body's carriers of hemoglobin; thus the disease can also be considered a form of respiratory disease. Explain why.
4. Explain why the circulating fluid in insects does not have respiratory proteins.

©2013 Cengage Learning. All Rights Reserved. May not be scanned, copied or duplicated, or posted to a publicly accessible website, in whole or in part.

39

DIGESTION AND HUMAN NUTRITION

INTRODUCTION

Chapter 39 describes the anatomy and physiology of the digestive system. Although it looks at systems in various animals, it concentrates on humans. It follows food through the system indicating where it is digested and where the resulting products are absorbed. The chapter also discusses what happens to each of the nutrients and human nutrition. Included in nutrition are what nutrients we need and what regulates appetite.

STUDY STRATEGIES

- Read all of Chapter 39 with the goal of familiarizing yourself with the boldface terms.
- Identify the importance of microbes for digestion and those that can cause digestion problems.
- Compare and contrast the various types of digestive systems amongst animals.
- Recognize the structures of the human digestive system and their functions.
- Discuss the relative importance of each type of macromolecule eaten and where it is utilized.
- Explain human nutritional requirements and the importance of vitamins and minerals.
- Describe the importance of maintaining a healthy weight and some healthy ways to do this.

FOCAL POINTS

- Figure 39.2 [p.696] shows various digestive systems in the animal kingdom.
- Figure 39.5[p.698] illustrates the human digestive system and its functions.
- Table 39.1 [p.702] summarizes the function and source of many digestive enzymes.
- Figures 39.8 and 39.9 [p.701] detail the structure of the small intestine.
- Figure 39.10 [p.703] summarizes digestion and absorption in the small intestine.
- Figure 39.12 [p.705] summarizes major pathways in metabolism and liver functions.
- Figure 39.13 [p.706] lists USDA nutritional guidelines.
- Tables 39.3 and 39.4 [pp.708-709] summarize information on major vitamins and minerals needed by humans.
- Section 39.12 [pp.710-711] illustrates how to calculate body mass index (BMI), desired caloric intake, and ideal weight.

INTERACTIVE EXERCISES

39.1. YOUR MICROBIAL "ORGAN" [p.695]

39.2. THE NATURE OF DIGESTIVE SYSTEMS [pp.696-697]

39.3. OVERVIEW OF THE HUMAN DIGESTIVE SYSTEM [pp.698-699]

39.4. DIGESTION IN THE MOUTH [p.699]

Boldfaced, Page-Referenced Terms

probiotic _____

_____

incomplete digestive system _____

_____

complete digestive system _____

_____

esophagus _____

_____

peristalsis _____

_____

sphincter _____

_____

gastrointestinal tract _____

_____

stomach _____

_____

small intestine _____

_____

large intestine _____

_____

anus _____

_____

salivary glands _____

_____

Fill-in-the-Blanks [pp.696-699]

Overall, (1) _____ encompasses the entire range of processes by which food is taken in, digested, absorbed and then used by the body. A digestive system is able to break down food both (2) _____ and (3) _____ to molecules that are small enough for (4) _____. The (5) _____ system can then distribute nutrients to cells throughout the body. A(n) (6) _____ digestive system has only one opening and two-way traffic. Flatworms have a highly branched gut cavity that serves both digestive and (7) _____ functions. A(n) (8) _____ digestive system has a tube or cavity with regional specializations and a(n) (9) _____ at each end. (10) _____ that break up food involve the muscular contraction of the gut wall, and (11) _____ is the release into the lumen of enzyme fluids and other substances required to carry out digestive functions. (12) _____ is the process of uptake of digested nutrients and water into the bloodstream and (13) _____ is the process of ridding the body of undigested materials.

(14) _____ are an important part of the mechanical digestion process. Humans have four types, including (15) _____, canines, premolars, and (16) _____. The pronghorn antelope feeds on plant material that breaks down slowly. Antelopes are (17) _____: hoofed mammals that have multiple stomach chambers; nutrients are released slowly when the animal rests. In comparison with human molars, the antelope molar has a much higher (18) _____; its teeth wear down rapidly because its plant diet is mixed with abrasive bits of dirt. The human digestive system is a tube, 21–30 feet long in an adult, that has regions specialized for different aspects of digestion and absorption; they are, in order, the mouth, pharynx, esophagus, (19) _____, (20) _____, large intestine, rectum, and anus. Various (21) _____ organs secrete enzymes and other substances that are also essential to the breakdown and absorption of nutrients; these include the salivary glands, liver, gallbladder, and (22) _____. Saliva contains water, mucus, and an enzyme called (23) _____ that breaks down starch. A cartilaginous flap called the (24) _____ closes off the trachea when food is swallowed. The (25) _____ is a muscular tube whose contractions propel food to the stomach.

Labeling and Matching [p.698]

Identify each numbered structure in the accompanying illustration. Then match each organ to its function and place it into the parentheses.

26. _____ ( )

27. _____ ( )

28. _____ ( )

29. _____ ( )

30. _____ ( )

31. _____ ( )

32. _____ ( )

33. _____ ( )

34. _____ ( )

35. _____ ( )

36. _____ ( )

A. J-shaped muscular sac that receives food and mixes it with gastric fluid secreted by cells in its lining.

B. Secretes enzymes and bicarbonate (a buffer) into the small intestine.

C. Stores and concentrates bile, then secretes it into the small intestine.

D. Opening through which feces are expelled from the body.

E. It absorbs most remaining water, thus concentrating any undigested waste and forming the feces.

F. Produces bile, which aids digestion and absorption of fats.

G. Muscular tube through which food moves to the stomach

H. Produce and secrete saliva, which moistens food and begins the process of carbohydrate digestion.

I. Entrance to the gut and respiratory system.

J. Most water and products of digestion are absorbed across the highly folded wall of this organ.

K. Mixes food with saliva.

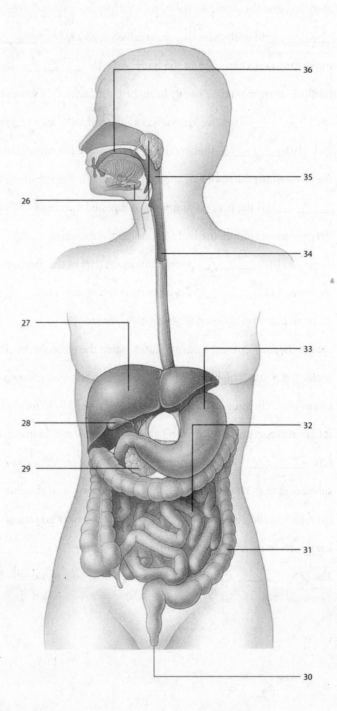

Matching [pp.696-699]

Match each of the following terms with its correct description.

37. _____ digestion

38. _____ secretion

39. _____ mechanical processing and motility

40. _____ elimination

41. _____ absorption

42. _____ enamel

43. _____ dentin

44. _____ molar

45. _____ canine

46. _____ salivary amylase

A. Bone-like substance that makes up teeth

B. An enzyme that breaks down starch

C. Movements that break up, mix and move food through digestive tract

D. Hard material that covers teeth

E. Breakdown of food into smaller particles and nutrient molecules

F. Tooth with flat, grinding surface

G. Release of digestive enzymes into digestive tract

H. Teeth that pierce food

I. Passes digested molecules and water into bloodstream

J. Expulsion of undigested materials

39.5. FOOD STORAGE AND DIGESTION IN THE STOMACH [p.700]

39.6. STRUCTURE OF THE SMALL INTESTINE [p.701]

39.7. DIGESTION AND ABSORPTION IN THE SMALL INTESTINE [pp.702-703]

39.8. ABSORPTION AND STORAGE IN THE COLON [p.704]

39.9. METABOLISM OF ABSORBED ORGANIC COMPOUNDS [p.705]

Boldfaced, Page-Referenced Terms

gastric fluid _____

_____

chyme _____

_____

villi _____

_____

microvilli _____

_____

brush border cells _____

_____

bile _____

_____

gall bladder _____

_____

emulsification _____

_____

feces _____

_____

appendix _____

_____

liver _____

_____

Matching [pp.700-704]

Match each organ with its main function.

1. _____ small intestine
2. _____ colon
3. _____ gallbladder
4. _____ mouth
5. _____ pancreas
6. _____ stomach
7. _____ rectum
8. _____ liver

A. Starts polysaccharide breakdown

B. Stores, mixes, dissolves food; kills many microorganisms; starts protein breakdown; empties in a controlled way

C. Digests and absorbs most nutrients

D. Produces enzymes that break down all major food molecules; produces buffers against hydrochloric acid from stomach

E. Secretes bile for fat emulsification

F. Stores, concentrates bile from liver

G. Stores, concentrates undigested matter by absorbing water and salts

H. Controls elimination of undigested and unabsorbed residues

Complete the Table [pp.700-704]

9. Complete the following table about the human digestive system.

Nutrient	Where Digested	Where Absorbed	Absorbed As
a. carbohydrate			
b. protein			
c. lipid			
d. nucleic acid			

True/False [pp.700-704]

If the statement is true, write a "T" in the blank. If the statement is false, make it correct by changing the underlined word(s) and writing the correct word(s) in the answer blank.

10. _____ Chyme is an alkaline mixture of food and chemicals secreted by the stomach.

11. _____ Gastrin is a hormone secreted by the lining of the small intestine in response to food entering it.

12. _____ Material entering the small intestine through the pyloric sphincter is immediately neutralized by the addition of bicarbonate ions.

13. _____ Cholecystokinin stimulates stomach acid secretion.

14. _____ The folding of the lining of the small intestine increases surface area for absorption.

15. _____ Water is absorbed across the wall of the intestine by active transport.

16. _____ Bile salts assist in the absorption of lipids.

17. _____ Movement of material from the colon to the rectum is stimulated by secretin.

18. _____ The large intestine compacts feces by absorbing sodium ions.

19. _____ The liver is essential in assisting in nucleic acid breakdown.

20. _____ Most of the body's fat soluble vitamins, including A and D, are stored in the liver.

Short Answer [p.705]

21. Why are the liver, pancreas, gall bladder, and salivary glands all considered accessory to the digestive system?

_____

_____

_____

Labeling and Matching [p.700]

Identify each numbered structure in the accompanying illustration of the human stomach.

22. _____ submucosa
23. _____ small intestine
24. _____ pyloric sphincter
25. _____ serosa
26. _____ esophagus
27. _____ mucosa
28. _____ longitudinal muscle
29. _____ circular muscle
30. _____ gastroesophageal sphincter
31. _____ oblique muscle

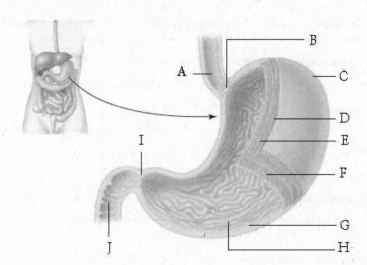

Labeling and Matching [p.704]

Identify each numbered structure in the accompanying illustration of the human colon.

32. _____ descending colon
33. _____ last portion of small intestine
34. _____ transverse colon
35. _____ ascending colon
36. _____ appendix
37. _____ cecum

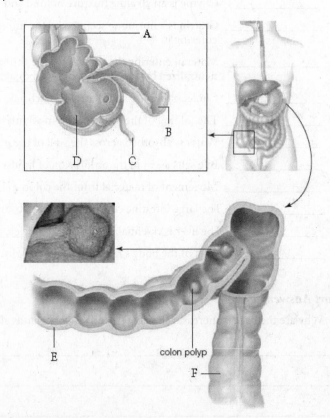

colon polyp

39.10. HUMAN NUTRITIONAL REQUIREMENTS [pp.706-707]

39.11. VITAMINS, MINERALS, AND PHYTOCHEMICALS [pp.708-709]

39.12. MAINTAINING A HEALTHY WEIGHT [pp.710-711]

Boldfaced, Page-Referenced Terms

essential fatty acids _____

_____

essential amino acids _____

_____

vitamins _____

_____

minerals _____

_____

phytochemicals _____

_____

obesity _____

_____

Fill-in-the-Blanks [pp.700, 706-709]

The newest USDA guidelines recommend eating more (1) _____ and _____, fat-free or low-fat

(2) _____ products, and (3) _____. They also recommend lowering intake of (4) _____ grains, (5)

_____ fats, (6) _____ fatty acids, (7) _____ (full of empty calories, and avoiding added (8)

_____. Those who are (9) _____ are advised to consume calcium-rich foods other than dairy products.

It is possible to get all of the necessary (10) _____ from fruits and vegetables. These foods also provide

essential (11) _____ and (12) _____ . Newer products on the market are those labeled (13)

_____, which individuals with celiac disease can eat without fear of pain or discomfort. (14) _____ are

those lipids which the body needs but cannot make. (15) _____ are those subunits of proteins that are

necessary for healthy functioning but the body is unable to produce on its own. (16) _____ are organic

substances needed in small quantities for growth and survival, and (17) _____ are essential inorganic

compounds.

In humans, two hormones help control feeding behavior. (18) _____, secreted by adipose tissue,

suppresses appetite and (19) _____, secreted by the stomach and brain, makes you feel hungry. In the U.S.,

(20) _____, an overabundance of fat in adipose tissue, is a major health problem. Some extremely obese

people opt for (21) _____, in which the digestive tract is surgically shortened.

Matching [pp.708-709]

Match each of the vitamins and minerals with their function.

22. _____ vitamin A

23. _____ biotin

24. _____ folate

25. _____ vitamin C

26. _____ vitamin D

27. _____ vitamin K

28. _____ iodine

29. _____ iron

30. _____ calcium

31. _____ copper

A. Bone and tooth formation; blood clotting; nerve and muscle activity

B. Needed in synthesis of melanin and hemoglobin and in electron transfer chains

C. Thyroid hormone formation

D. Found in hemoglobin and electron transfer proteins

E. Visual pigments; bone and teeth

F. Bone growth and mineralization; calcium absorption

G. Blood clotting; ATP formation

H. Coenzyme in nucleic acid and amino acid metabolism

I. Coenzyme in fat and glycogen metabolism

J. Collagen synthesis; carbohydrate metabolism

Complete the Table [pp.710-711]

32. Complete the following table by determining how many kilocalories the people described should take in daily, given the stated exercise level, in order to maintain their weight.

Height	Age	Sex	Level of Physical Activity	Present Weight (lb)	Number of Kilocalories/Day
5' 6"	25	Female	Moderately active	138	a.
5' 10"	18	Male	Very active	145	b.
5' 8"	53	Female	Not very active	143	c.

33. You are a 19-year-old male, very sedentary (TV, sleep, and computers), 6 feet tall, medium frame, and you weigh 195 pounds. Calculate the number of calories required to sustain your desired weight. Are you underweight, overweight, or just right? _____

_____

_____

_____

34. What may be some drawbacks to using BMI as the sole measure of obesity?

_____

_____

_____

35. What the study of genetics in obesity tell scientists about predisposition?

_____

_____

_____

36. Explain the disorders anorexia nervosa and bulimia.

_____

_____

_____

SELF-TEST

___ 1. The process that moves nutrients into the blood or lymph is _____.
   a. ingestion
   b. absorption
   c. assimilation
   d. digestion
   e. none of the above

___ 2. The enzymatic digestion of proteins begins in the _____.
   a. mouth
   b. stomach
   c. liver
   d. pancreas
   e. small intestine

___ 3. The enzymatic digestion of starches begins in the _____.
   a. mouth
   b. stomach
   c. liver
   d. pancreas
   e. small intestine

___ 4. The greatest amount of absorption of digested nutrients occurs in the _____
   a. stomach
   b. pancreas
   c. liver
   d. colon
   e. small intestine

5. Glucose moves through the membranes of the small intestine mainly by _____. a. peristalsis
   b. osmosis
   c. diffusion
   d. active transport
   e. bulk flow

6. Which of the following is found in bile?
   a. lecithin
   b. salts
   c. digestive enzymes
   d. cholesterol

7. The element needed by humans for blood clotting, nerve impulse transmission, and bone and tooth formation is _____.
   a. magnesium
   b. iron
   c. calcium
   d. iodine
   e. zinc

8. A colonoscopy is used to examine the
   a. stomach
   b. large intestine
   c. small intestine
   d. pancreas

9. A bacteria, _____, is known to contribute to the incidence of ulcers of the stomach.
   a. *E. coli*
   b. *Bifidobacterium*
   c. *H. pylori*
   d. bacteria do not play a role in ulcers

Matching

Match each numbered item with its description in the lettered list.

10. _____ vitamins C and E
11. _____ complex carbohydrates
12. _____ essential amino acids
13. _____ essential fatty acids
14. _____ mineral
15. _____ rickets
16. _____ scurvy
17. _____ vitamin
18. _____ toxic goiter
19. _____ obesity
20. _____ oleic acid
21. _____ brush border cells
22. _____ lipase
23. _____ ghrelin

A. Linoleic acid is one example

B. Phenylalanine, lysine, and methionine are three of eight

C. From olive oil; may produce possible health benefits

D. Combine with free radicals; counteract their destructive effects on DNA and cell membranes

E. Vitamin C deficiency

F. Area of absorption of components of macromolecules in the small intestine

E. Vitamin D deficiency in young children

F. Lack of sufficient iodine in the body

G. Vitamin D deficiency in young children

H. Lack of sufficient iodine in the body

I. Enzyme in the small intestine but secreted by the pancreas that assists in lipid emulsification.

J. Organic substance that helps enzymes to do their jobs; required in small amounts for good health

K. A hormone that stimulates appetite

L. Inorganic substance required for good health

M. Long chains of simple sugars; in grains and white potatoes

N. Characterized by a BMI greater than 30

CHAPTER OBJECTIVES/REVIEW QUESTIONS

1. Distinguish between incomplete and complete digestive systems and tell which is characterized by: (a) specialized regions, (b) two-way traffic, and (c) discontinuous feeding. [pp.696-697]
2. Define and distinguish among *motility, secretion, digestion,* and *absorption.* [pp.696-697]
3. List all parts (in order) of the human digestive system through which food actually passes. Then list the accessory organs that contribute one or more substances to the digestive process. [pp.698-699]
4. Tell which foods undergo digestion in each of the following parts of the human digestive system and state what the food is broken into: mouth, stomach, small intestine, large intestine. [pp.699-704]
5. Describe the cross-sectional structure of the small intestine, and explain how its structure is related to its function. [pp.701-703]
6. List the items that leave the digestive system and enter the circulatory system during the process of absorption. [p.703]
7. State which processes occur in the colon (large intestine). [p.704]
8. Construct an ideal diet for yourself for one 24-hour period. Calculate the number of calories necessary to maintain your weight [pp.710-711] and then use the nutritional guidelines [pp.706-707] to choose exactly what to eat and how much.
9. Name five vitamins and five minerals that are important in human nutrition, and state the specific role of each. [pp.708-709]

INTEGRATING AND APPLYING KEY CONCEPTS

1. Suppose you could not eat food for two weeks and you had only water to drink. List in correct sequential order the measures your body would take to try to preserve your life. Mention the command signals that are given as one after another critical point is reached, and tell which parts of the body are the first and the last to make up for the deficit.
2. What relationship do hormones have to human appetite? How might controlling some of these hormones be used to maintain a proper weight?
3. The bacterium *Vibrio cholorae* is a normal inhabitant of small crustaceans called copepods that live in salt water. It attaches to the intestinal wall and helps the organism regulate the osmotic movement of Na+ and water. In humans, it attaches to the intestinal wall and attempts to do the same thing. Predict the result of this activity in humans.

40

MAINTAINING THE INTERNAL ENVIRONMENT

INTRODUCTION

This chapter looks at kidney function and its relationship to homeostasis. It investigates what the extracellular fluid should contain. The structure of the human urinary system is described and the function of each portion is explained. It also briefly explores the process of temperature regulation.

STUDY STRATEGIES

- Read all of Chapter 40 with the goal of familiarizing yourself with the boldface terms.
- Identify the various substances that can be detected in urine.
- Recognize the role of the urinary system in homeostasis and the various ways that vertebrates regulate their fluids.
- Identify the structures of the human urinary system and their function.
- Explain urine formation including glomerular filtration and tubular reabsorption/secretion.
- Describe how thirst and urination are regulated by the nervous and endocrine systems.
- Discuss the buffer system for acid-base balance.
- Recognize some of the diseases and disorders of the urinary system and ways that these can be prevented and treated.
- Identify mechanisms to regulate internal body temperature as it relates to homeostasis.

FOCAL POINTS

- Figures 40.8 and 40.9 [pp.718-719] diagram the human urinary system.
- Figure 40.10 and 40.11 [pp.720-721] describes the step-by-step process of urine formation.
- Figure 40.12 [p.722] illustrates urinary system regulation.
- Table 40.1 [p.726] list the mammalian responses to heat and cold.

INTERACTIVE EXERCISES

40.1. TRUTH IN A TEST TUBE [p.715]

40.2. REGULATING FLUID VOLUME & COMPOSITION [pp.716-717]

Boldfaced, Page-Referenced Terms

ammonia_____

_____

uric acid _____

_____

kidneys _____

_____

urine_____

_____

urea_____

_____

Fill-in-the-Blanks [pp.715-717]

Urine has been an important diagnostic tool for centuries. (1) _____ was first diagnosed by the large

amounts of sugar in the urine. Alkaline urine is often a sign of (2) _____. One sign of kidney damage is the

excretion of (3) _____ in the urine. A rise in (4) _____ in a woman's urine signals ovulation, and a

woman can also test her urine to see if she is (5) _____ or if she is entering (6) _____. If a person uses

marijuana, testable levels may remain in the urine for up to (7) _____ days.

In most animals, the fluid that fills the spaces between cells and tissues is called (8) _____ fluid. The

fluid called (9) _____ moves substances by way of the circulatory system. Combined, (8) and (9) are called

(10) _____ fluids. In animals a well-developed (11) _____ system helps keep the (12) _____ and

composition of this fluid within tolerable ranges. Other major organ systems interact with the urinary system to

maintain this (13) _____ in the internal environment.

Matching [pp.716-717]

Match each of the following items with its correct description.

14. _____ extracellular fluid

15. _____ solute

16. _____ ammonia

17. _____ intracellular fluid

18. _____ kidneys

19. _____ flame cell

20. _____ coelomic fluid

21. _____ nephridia

22. _____ Malpighian tubule

23. _____ uric acid

A. Cells in the excretory system of flatworms with a tuft of cilia

B. Fluid inside cells

C. Tubular excretory organs of earthworms and other annelids

D. Fluid between cells

E. Fluid inside the body cavity of an earthworm

F. Molecule formed when amino acids and nucleotides are broken down

G. Nitrogen waste product that does not need to be dissolved in large quantity of water to be excreted

H. Long, thin excretory tubes of terrestrial insects

I. Organ that filters blood and produces urine in vertebrates

J. Molecules dissolved in water

True/False [pp.716-717]

If the statement is true, write a "T" in the blank. If the statement is false, make it correct by changing the underlined word(s) and writing the correct word(s) in the answer blank.

24. _____ Marine fish have body fluids that are <u>more</u> salty than seawater.

25. _____ Marine bony fish are constantly <u>gaining</u> water.

26. _____ Life on land requires highly efficient <u>kidneys</u>.

27. _____ Reptiles convert ammonia to <u>urea</u> and mammals convert it to <u>uric acid</u>.

28. _____ Some desert animals are able to survive without taking in extra water; they utilize water produced by their <u>metabolism</u>.

29. _____ Marine mammals have <u>smaller</u> kidneys because they live in sea water.

30. _____ Freshwater bony fish continually <u>lose</u> water because of the osmotic imbalance between their body fluids and the environment.

31. _____ Mammals need <u>more</u> water than the birds or reptiles of a similar size.

Short Answer [pp.716-717]

32. What are some ways that mammals gain water? _____

_____

_____

33. What are some ways that mammals lose water? _____

_____

_____

40.3. THE HUMAN URINARY SYSTEM [pp.718-719]

40.4. HOW URINE FORMS [pp.720-721]

40.5. REGULATING THIRST AND URINE CONCENTRATION [pp.722-723]

40.6. ACID-BASE BALANCE [p.723]

Boldfaced, Page-Referenced Terms

ureter _____

_____

urinary bladder _____

_____

urethra _____

_____

nephrons _____

_____

Bowman's capsule _____

_____

proximal tubule _____

_____

loop of Henle _____

_____

distal tubule _____

_____

collecting tubule _____

_____

glomerulus _____

_____

peritubular capillaries _____

_____

glomerular filtration _____

_____

tubular reabsorption _____

_____

tubular secretion _____

_____

antidiuretic hormone (ADH) _____

_____

aldosterone _____

_____

Labeling [pp.718-719]

Identify each indicated part of the accompanying illustrations.

1. _____

2. _____

3. _____

4. _____

5. _____

6. _____

7. _____

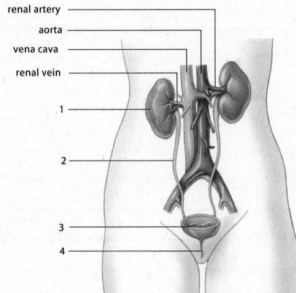

renal artery
aorta
vena cava
renal vein
1
2
3
4

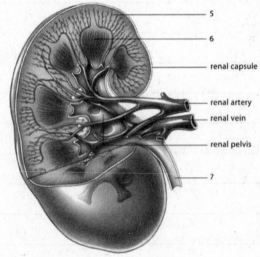

5
6
renal capsule
renal artery
renal vein
renal pelvis
7

Fill-in-the-Blanks [pp.718-721]

Mammalian (8) _____ filter water, mineral ions, organic wastes, and other substances from the blood. Each human kidney contains over a million (9) _____, where water and solutes are filtered from the blood. A nephron starts at the (10) _____, which together with the (11) _____ capillaries is called the (12) _____. Next is the tubular region closest to the capsule, called the (13) _____ tubule. Following this is a hairpin structure called the (14) _____ and a distal tubule. A(n) (15) _____ is a part of the duct system leading to the kidney's central cavity and into the (16) _____.

During this process, most of the filtered material is returned to the blood. The small portion of unclaimed water and solutes is called (17) _____. This fluid flows from each kidney into a(n) (18) _____, which empties into a muscular sac called the (19) _____. Urine is stored here before flowing into the (20)

_____, which opens at the body's surface. Flow from the bladder is a(n) (21) _____ action. A(n) (22)

_____ muscle surrounds the urethra and is under (23) _____ control.

Labeling [pp.720-721]

Identify each indicated part of the illustration. Note that it depicts two nephrons, one of which extends deeply into the medulla, making it possible to greatly concentrate the urine it carries (see Section 40.4).

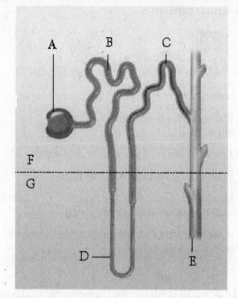

Tubular portion of one nephron, cutaway view. The tubule starts at Bowman's capsule.

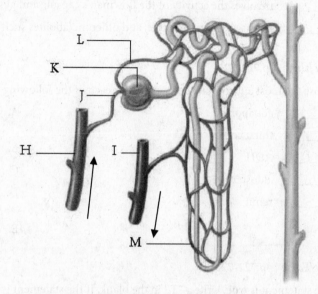

Blood vessels associated with the nephron. The glomerulus is a ball of capillaries that have unusually leaky walls.

24. _____ collecting tubule

25. _____ Loop of Henle

26. _____ glomerulus

27. _____ renal medulla

28. _____ peritubular capillaries

29. _____ Bowman's capsule

30. _____ renal artery

31. _____ proximal tubule

32. _____ renal cortex

33. _____ afferent arteriole

34. _____ distal tubule

35. _____ efferent arteriole

36. _____ renal vein

Choice [pp.720-721]

Choose the process described in each of exercises 31-35.

    a. tubular secretion   b. tubular reabsorption   c. filtration

37. _____ Driven by blood pressure

38. _____ Excess ions are moved into the nephron

39. _____ Water and solutes are reclaimed

40. _____ Involves the activity of the Bowman's capsule and glomerular capillaries

41. _____ Releases drugs, toxicants, and other metabolites such as urea

Matching [pp.722-723]

Choose the most appropriate description for each of the following terms.

42. _____ aquaporins

43. _____ thirst center

44. _____ ADH

45. _____ aldosterone

46. _____ renin

A. Induces the adrenal gland to secrete aldosterone

B. Pressure-induced transporters of water

C. Promotes reabsorption of sodium

D. Hormone released by the pituitary to control water content of the urine

E. A section of the hypothalamus that induces water-seeking behavior

True/False [pp.722-723]

If the statement is true, write a "T" in the blank. If the statement is false, make it correct by changing the underlined word(s) and writing the correct word(s) in the answer blank.

47. _____ A buffer system helps maintain a constant <u>temperature</u>.

48. _____ In the bicarbonate-carbonic acid buffer system, the system will add <u>H^+</u> ions if the solution becomes too acid.

49. _____ Breathing rate affects blood pH by adding or removing <u>CO_2</u> to/from the blood.

50. _____ Acidosis can be caused when the kidney filters too <u>much</u> H^+ from the blood.

51. _____ If the pituitary gland secretes <u>too much</u> ADH, too much fluid will be retained, causing diabetes insipidus.

52. _____ Atrial natriuretic peptide is secreted by heart atrial muscle and when blood volume is <u>too low</u>.

Short Answer [pp. 722-723]

53. What types of neural and endocrine controls are there over water acquisition in humans? _____

_____

_____

40.7. WHEN KIDNEYS FAIL [p.724]

40.8. HEAT GAINS AND LOSSES [p.725]

40.9. RESPONSES TO HEAT AND COLD [pp.726-729]

Boldfaced, Page-Referenced Terms

thermal radiation _____

_____

conduction _____

_____

convection _____

_____

evaporation _____

_____

ectotherm _____

_____

endotherm _____

_____

heterotherms_____

_____

shivering response_____

_____

non-shivering response_____

_____

hibernation _____

_____

estivation_____

_____

Matching [p.724]

Match each of the following terms or conditions with its correct description. One item is used twice.

1. _____ high protein diet
2. _____ diabetes mellitus
3. _____ hemodialysis
4. _____ kidney transplant
5. _____ peritoneal dialysis
6. _____ kidney function measure
7. _____ high blood pressure

A. Patients blood filtered in an artificial kidney
B. Permanent replacement of failed kidney
C. Kidney works extra hard to dispose of nitrogen-rich breakdown products
D. Wastes diffuse across membrane lining body cavity
E. One of the most common causes of kidney failure
F. Rate of glomerular filtration.

Matching [p.725]

Match the most appropriate description to each of the following terms.

8. _____ conduction
9. _____ convection
10. _____ ectotherm
11. _____ endotherm
12. _____ evaporation
13. _____ heterotherm
14. _____ thermal radiation

A. Body temperature determined more by heat exchange with the environment than by metabolic heat.
B. Heat transfer from or toward a body by currents in air or water.
C. Body temperature determined largely by metabolic activity and by precise controls over heat produced and heat lost.
D. Direct transfer of heat energy between two objects in direct contact with each other
E. The emission of energy in the form of infrared or other wavelengths that are converted to heat by the absorbing body
F. Body temperature fluctuating at some times and controlled at other times
G. In a change from the liquid state to the gaseous state, the energy required is supplied by the heat content of the liquid

Choice [pp.726-727]

For statements 15-23, choose from the following terms.

a. fever   b. heat stress   c. cold stress   d. both b and c

15. _____ An inflammatory response to a tissue injury
16. _____ If a response is not successful, hypothermia may occur
17. _____ Response may involve the nonshivering response of brown adipose tissue
18. _____ Response involves peripheral vasodilation
19. _____ May be controlled by the use of anti-inflammatory drugs
20. _____ Response may include rhythmic tremors of skeletal muscles to generate heat
21. _____ Response involves the peripheral vasoconstriction of capillaries
22. _____ Response may involve behavioral responses such as panting
23. _____ If uncontrolled, can result in hyperthermia

24. _____ Response is partially controlled by the hypothalamus

Short Answer [pp.726-727]

25. What is the recommended behavior to avoid heat stress on a hot summer day? _____

_____

_____

26. How do the two processes of dialysis work differently? _____

_____

_____

_____

27. What are some adaptations of birds and mammals to maintain endothermy? _____

_____

_____

_____

_____

SELF-TEST

___ 1. An entire subunit of a kidney that purifies blood and restores solute and water balance is called a _____.
  a. glomerulus
  b. loop of Henle
  c. nephron
  d. ureter
  e. none of the above

___ 2. The last portion of the excretory system that urine passes through before being eliminated from the body is the _____.
  a. renal pelvis
  b. bladder
  c. ureter
  d. collecting ducts
  e. urethra

___ 3. Filtration of the blood in the kidney takes place in the _____.
  a. loop of Henle
  b. proximal tubule
  c. distal tubule
  d. Bowman's capsule
  e. all of the above

___ 4. _____ primarily controls the concentration of water in urine.
  a. Aldosterone
  b. Antidiuretic hormone
  c. Angiotensin II
  d. Glucagon
  e. Renin

___ 5. _____ primarily controls the concentration of sodium in urine.
  a. Insulin
  b. Glucagon
  c. Antidiuretic hormone (ADH)
  d. Aldosterone
  e. Epinephrine

___ 6. Hormonal control over the reabsorption or tubular secretion of sodium primarily affects _____.
  a. Bowman's capsules
  b. distal tubules and collecting ducts
  c. proximal tubules
  d. the urinary bladder
  e. loops of Henle

7. During reabsorption, sodium ions cross the proximal tubule walls into the interstitial fluid principally by means of _____ .
   a. osmosis
   b. countercurrent multiplication
   c. bulk flow
   d. active transport
   e. all of the above

8. In humans, the thirst center is located in the _____.
   a. adrenal cortex
   b. thymus
   c. heart
   d. adrenal medulla
   e. hypothalamus

9. Which of the following represents the process by which there is heat transfer between an animal and another object of differing temperature that is in direct contact with it?
   a. evaporation
   b. conduction
   c. radiation
   d. convection

10. Which of the following terms represents a group of mammals whose high metabolic rates keep them active under a wide variety of temperature ranges?
    a. endotherms
    b. ectotherms
    c. heterotherms
    d. none of the above

11. Which of the following is a response to heat stress?
    a. panting
    b. peripheral vasodilation
    c. evaporative heat loss
    d. all of the above

12. Which of the following is a response to cold stress?
    a. shivering
    b. peripheral vasoconstriction
    c. curling up
    d. decreased muscle action

13. The storage organ of the excretory system that urine is consciously controlled by adults is the _____.
    a. renal pelvis
    b. bladder
    c. ureter
    d. collecting ducts
    e. urethra

14. Which of the following is the transfer of heat between two objects that are touching?
    a. thermal radiation
    b. conduction
    c. convection
    d. evaporation

15. A(n) _____ system maintains the optimal pH system of the body.
    a. acid
    b. base
    c. buffer
    d. biocarbonate

16. Tubular reabsorption occurs between the proximal tubule and the _____.
    a. distal tubule
    b. Bowman's capsule
    c. collecting duct
    d. peritubular capillaries

17. Approximately _____ nephrons are in the kidney.
    a. 10
    b. 1,000
    c. 100,000
    d. 1,000,000

18. Marine fish are _____ to their environment.
    a. hypertonic
    b. isotonic
    c. hypotonic
    d. none of the above

19. Urine tests can detect _____.
    a. drugs
    b. luteinizing hormone
    c. sugar
    d. all of the above

20. The kidneys help to regulate
    a. blood volume.
    b. blood pressure.
    c. blood pH.
    d. all of the above

21. All of the following belong to the urinary systems except the
    a. bladder
    b. urethra
    c. kidney
    d. prostate

CHAPTER OBJECTIVES/REVIEW QUESTIONS

1. List the factors that contribute to the gain and loss of water in mammals. [p.718]
2. List the factors that contribute to the gain and loss of solutes in mammals. [pp.718]
3. Be able to identify the major components of the urinary system from a diagram. [pp.718-719]
4. Trace the flow of fluid through a human nephron, listing the major structures involved. [p.719]
5. Locate the processes of filtration, reabsorption, and tubular secretion along a nephron, and tell what makes each process happen. [pp.720-721]
6. Understand the interaction of the hypothalamus with the urinary system. [pp.722-723]
7. Discuss kidney failure, its causes and treatments. [p.724]
8. Distinguish between evaporation, radiation, conduction, and convection as methods of heat transfer. [p.725]
9. Explain how endotherms, ectotherms, and heterotherms regulate body temperature, and give an example of each. [p.725]
10. Explain the mechanisms by which a mammal may respond to heat stress. [pp.726-727]
11. Explain the mechanisms by which a mammal may respond to cold stress. [pp.726-727]

INTEGRATING AND APPLYING KEY CONCEPTS

1. How might being ectothermic affect human activity? What about being heterothermic?
2. The hemodialysis machine used in hospitals is expensive and time consuming. So far, artificial kidneys capable of allowing people who have nonfunctional kidneys to purify their blood by themselves, without having to go to a hospital or clinic, have not been developed. Which aspects of the hemodialysis procedure do you think have presented the most problems in development of a method of home self-care? If you had an unlimited budget and were appointed head of a team to develop such a procedure and its instrumentation, what strategy would you pursue?
3. Explain why days with high humidity affect our ability to regulate body temperature properly.
4. Explain the symptoms expected in someone with hypothermis.

41

ANIMAL REPRODUCTIVE SYSTEMS

INTRODUCTION

This chapter investigates how animals reproduce. It begins with the distinction between asexual and sexual processes and the advantages of each. The chapter focuses on human reproductive anatomy and the physiological processes that regulate reproductive cycles as well as methods of preventing pregnancy.

STUDY STRATEGIES

- Read all of Chapter 41 with the goal of familiarizing yourself with the boldface terms.
- Compare and contrast the different modes of reproduction and fertilization.
- Identify the major organs, glands, and structures of the male reproductive system.
- Be able to track a sperm from spermatogenesis through ejaculation.
- Explain the process and hormonal control over spermatogenesis.
- Recognize disorders of the male reproductive tract.
- Identify the major organs, glands, and structures of the female reproductive system.
- Be able to track an egg from oogenesis through ovulation and implantation.
- Explain the process and hormonal control over oogenesis, the menstrual cycle and menopause.
- Recognize disorders of the female reproductive tract.
- Explain sexual intercourse and the process of fertilization.
- Compare and contrast the various methods to prevent pregnancy as well as those that prevent STDs.
- Identify the medical options for couples that are dealing with infertility.
- Discuss sexually transmitted diseases, their complications, prevention, and treatments.

FOCAL POINTS

- Figure 41.4[p.735] illustrates the anatomy of the human male reproductive system.
- Figure 41.5 [p.736] diagrams gamete production in males.
- Figure 41.7 [p.737] illustrates hormone regulation of sperm production.
- Figure 41.8 [p.738] illustrates the anatomy of the human female reproductive system.
- Figure 41.9 [p.740] diagrams gamete production in females.
- Figure 41.10 [p.741] illustrates hormone regulation of menstrual cycle.
- Figure 41.12 and Table 41.2[pp.744-745] list and compare the effectiveness of methods of contraception.

INTERACTIVE EXERCISES

41.1. INTERSEX CONDITIONS [p.731]

41.2. MODES OF ANIMAL REPRODUCTION [pp.732-733]

41.3. REPRODUCTIVE SYSTEM OF HUMAN MALES [pp.734-735]

41.4. SPERM FORMATION [pp.736-737]

Boldfaced, Page-Referenced Terms

intersex condition _____

_____

asexual reproduction _____

_____

sexual reproduction _____

_____

hermaphrodites _____

_____

external fertilization _____

_____

internal fertilization _____

_____

placenta _____

_____

gonads _____

_____

testes _____

_____

testosterone _____

_____

epididymis _____

_____

vas deferens _____

_____

penis _____

_____

semen _____

_____

seminiferous tubules _____

_____

gonadotrpoin releasing hormone (GnRH) _____

_____

leutinizing hormone (LH) _____

_____

follicle stimulating hormone (FSH) _____

_____

Matching [pp.731-733]

Match each of the following terms with its correct description.

1. _____ asexual reproduction

2. _____ sexual reproduction

3. _____ fragmentation

4. _____ sequential hermaphrodite

5. _____ simultaneous hermaphrodite

6. _____ external fertilization

7. _____ internal fertilization

8. _____ yolk

9. _____ internal development

10. _____ budding

A. Fluid rich in proteins and lipids

B. A piece of an individual organism forms an entire new organism

C. A single individual forms offspring genetically identical to itself

D. A new organism grows out of the original organism and then breaks free

E. The fertilized egg develops inside the female body, nourished directly from female's circulation

F. Produce offspring containing genetic material from two other organisms

G. Organism that produces both eggs and sperm at the same time

H. Egg fertilized outside the body, usually in water

I. Organisms switch from one sex to another during their life span

J. Egg fertilized inside the female body

Labeling and Matching [pp.734-735]

Identify each numbered part of the following illustration. Then, match the function of each organ and place it into the parentheses.

11. _____ ( )

12. _____ ( )

13. _____ ( )

14. _____ ( )

15. _____ ( )

16. _____ ( )

17. _____ ( )

18. _____ ( )

19. _____ ( )

20. _____ ( )

21. _____ ( )

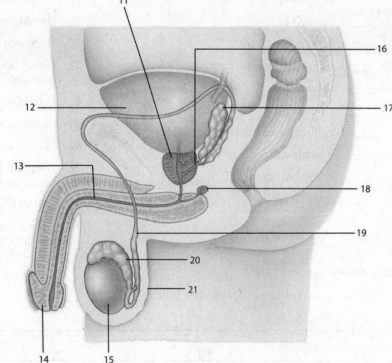

A. Male organ of sexual intercourse

B. A layer of integument and muscle that helps regulate the temperature of the testes

C. One of a pair of ducts that carry sperm to the penis

D. One of a pair of ducts in which sperm mature and are stored

E. One of a pair of exocrine glands that secrete mucus

F. Holds urine from the urinary system

G. An exocrine gland that contributes some fluid to the semen.

H. One of a pair of gonads, packed with small, sperm-producing tubes (seminiferous tubules) and cells that secrete testosterone and other sex hormones

I. Duct with dual functions; channel for ejaculation of sperm during sexual arousal and for excretion of urine at other times

J. One of a pair of exocrine glands that contributes fructose-rich fluid to semen

K. One of a pair of ducts that carry sperm to the penis

Matching [pp.736-737]

Match each of the following terms with its correct description.

22. _____ prostate-specific antigen

23. _____ Leydig's cells

24. _____ Sertoli cells

25. _____ secondary spermatocyte

26. _____ testosterone

27. _____ Luteinizing hormone (LH)

28. _____ Follicle stimulating hormone (FSH)

29. _____ prostate enlargement

30. _____ sperm head

31. _____ sperm midpiece

A. Contains DNA and "digestive" enzymes

B. Developing sperm cell at end of meiosis I

C. Hormone supports sperm development

D. Causes urethra to narrow; causes difficulty urinating

E. Cells in testes that secrete testosterone

F. Contains mitochondria for energy

G. Cells in testes that support developing sperm

H. Hormone that induces development of male secondary sex characteristics

I. Molecule measured to detect prostate cancer

J. Hormone that regulates testosterone production

Short Answer [pp.734-737]

32. What are some risk factors for prostate cancer?

_____

_____

_____

33. What can men do to detect testicular cancer early?

_____

_____

_____

41.5. REPRODUCTIVE SYSTEM OF HUMAN FEMALES [pp.738-739]

41.6. PREPARATIONS FOR PREGNANCY [pp.740-741]

41.7. PMS TO HOT FLASHES [p.742]

Boldfaced, Page-Referenced Terms

ovaries _____

_____

oocyte _____

_____

estrogens _____

_____

progesterone _____

_____

oviduct _____

_____

uterus _____

_____

endometrium _____

_____

cervix _____

_____

vagina _____

_____

menstrual cycle _____

_____

menstruation _____

_____

menopause _____

_____

primary oocytes _____

_____

ovarian follicles _____

_____

secondary oocyte _____

_____

ovulation _____

_____

corpus luteum_____

_____

Labeling and Matching [pp.738-739]

Identify each numbered part of the following illustration. Then, match the function of each organ and place it into the parentheses.

1. _____ ( )

2. _____ ( )

3. _____ ( )

4. _____ ( )

5. _____ ( )

6. _____ ( )

7. _____ ( )

8. _____ ( )

9. _____ ( )

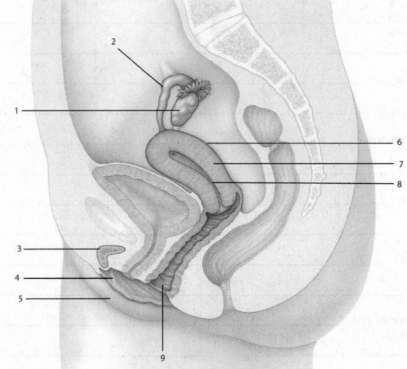

A. smooth muscle layer of the uterus

B. 'womb', chamber in which an embryo develops.

C. one of a pair of fatty outer skin folds

D. one of a pair of ducts through which oocytes are propelled from an ovary to the uterus; usual site of fertilization.

E. one of a pair of inner skin folds

F. female organ of sexual intercourse: birth canal.

G. epithelial lining of the uterus

H. highly sensitive erectile organ; only the tip is externally visible

I. one of two female gonads; makes eggs and secretes female sex hormones (estrogens and progesterone).

Matching [pp.740-741]

Match each of the following terms with its correct description.

10. _____ premenstrual syndrome(PMS)

11. _____ endometriosis

12. _____ fibroids

13. _____ primary oocyte

14. _____ ovarian follicle

15. _____ surge of LH

16. _____ corpus luteum

17. _____ menstrual flow

18. _____ gollicular phase

19. _____ luteal phase

20. _____ menopause

21. _____ uterine cancer

A. A primary oocyte and cells surrounding it

B. Causes ovulation

C. Immature egg that has started meiosis but arrested in prophase I

D. Phase of ovarian cycle where follicle matures

E. Body tissues swell and person may become irritable or depressed

F. Loss of endometrial tissue and blood

G. Benign uterine tumors that may result in pain and long menstrual periods

H. Phase of ovarian cycle where corpus luteum develops

I. Formed from ruptured follicle and produces progesterone

J. Decline in hormones in a woman later in life

K. Growth of endometrial tissues in areas where it should not form

L. Characterized by vaginal bleeding between menstrual periods and after menopause

Sequence [pp.740-741]

Arrange the following levels of organization in nature in the correct time sequence. Write the letter of the first event next to 22. The letter of the last event is written next to 31.

22. _____

23. _____

24. _____

25. _____

26. _____

27. _____

28. _____

29. _____

30. _____

31. _____

A. Follicle secretes estrogen

B. Menstrual flow begins as endometrium tissue dies

C. Ovulation

D. Corpus luteum degenerates

E. Progesterone causes endometrium to complete maturation

F. FSH levels rise

G. LH surge

H. Estrogen stimulates initial growth of endometrium

I. Corpus luteum forms and secretes progesterone

J. Follicle begins to mature

Labeling and Matching [pp.740-741]

32. Identify each numbered part of the following illustration of the ovarian cycle.

A. _____

B. _____

C. _____

D. _____

E. _____

F. _____

follicle cells

Short Answer [pp.739]

33. What is ovarian cancer so deadly for women?

_____

_____

_____

34. What can women do to detect cervical cancer early?

_____

_____

_____

41.8. WHEN GAMETES MEET [pp.742-743]

41.9. PREVENTING OR SEEKING PREGNANCY [pp.744-745]

41.10. SEXUALLY TRANSMITTED DISEASES [pp.746-747]

True/False [pp.742-743]

If the statement is true, write a "T" in the blank. If the statement is false, make it correct by changing the underlined word(s) and writing the correct word(s) in the answer blank.

1. _____ During sexual arousal, <u>sympathetic</u> nervous system signals cause blood vessels to dilate and increase blood flow to penis

2. _____ <u>Seratonin</u> causes rhythmic contractions in male reproductive tract and female vagina

3. _____ <u>Erectile disfunction</u> is a condition where an erection cannot be attained or maintained

4. _____ When sperm reach the egg, <u>several</u> sperm may enter the egg

5. _____ Acrosomal enzymes break down the <u>*zona coporum*</u> to reach egg surface

6. _____ The initial cell after fertilization is a <u>haploid gamete</u>

7. _____ The second meiotic division of the egg occurs just <u>before ovulation</u>

8. _____ Ejaculation can provide 150 to 350 <u>billion</u> sperm

Matching [pp.744-745]
Match each of the following birth control methods with their correct description.

9. _____ rhythm method

10. _____ douching

11. _____ vasectomy

12. _____ condom

13. _____ spermicidal foam

14. _____ diaphragm

15. _____ intrauterine device

16. _____ tubal ligation

17. _____ birth control pill

18. _____ birth control patch

A. An object inserted into the uterus to prevent implantation

B. A mixture of synthetic estrogen and progesterone

C. A material placed in vagina to kill sperm

D. Fallopian tubes are severed

E. Rinsing of the vagina after intercourse

F. Timing intercourse when fertile egg not present

G. An adhesive material applied to skin that delivers estrogen and progesterone

H. Dome-shaped device that blocks the cervix

I. Vas deferens are severed

J. Sheath worn over the penis

Matching [pp.746-747]
Match each of the following sexually transmitted diseases with their correct description.

19. _____ genital herpes

20. _____ trichomoniasis

21. _____ pelvic inflammatory disease

22. _____ Syphillis

23. _____ Gonorrhea

24. _____ Human papilloma virus

25. _____ HIV

26. _____ AIDS

27. _____ Chlamydia

A. A virus that lowers immune system's ability to counteract infections

B. Vaginal discharge; painful urination in both sexes; may be contracted repeatedly

C. Most infected females undiagnosed; many will develop pelvic inflammatory disease

D. A complication of bacterial sexually transmitted diseases that scars the reproductive tract

E. Mucous membranes of genitals have small painful blisters; incurable

F. Compromised immune system leads to opportunistic infections; incurable

G. Causes genital warts and may lead to cervical cancer

H. Flagellated protist infects vagina and urethra

I. Multi stage disease process that starts with chancre and spreads to many other tissues

28. What are the least effective forms of birth control?

_____

_____

29. What are the most effective forms of birth control?

_____

_____

30. What form(s) of birth control also help prevent STDs?

_____

_____

SELF-TEST

___ 1. Reproductive systems are essential for
_____.
   a. maintaining homeostasis
   b. passing genes on to the next
      generation
   c. survival of the individual
   d. immediate responses to
      environmental change

___ 2. Sexual reproduction is advantageous
   because it _____.
   a. produces offspring genetically identical
      to parents
   b. requires finding a mate
   c. produces offspring genetically different
      than parents
   d. is optimal in a stable environment

___ 3. To get from the testes to the outside, sperm
   travel through _____ (in the correct
   order).
   a. vas deferens, ureter, ejaculatory duct
   b. epididymis, urethra, vas deferens
   c. epididymis, vas deferens, ureter
   d. vas deferens, fallopian tubule, urethra
   e. epididymis, vas deferens, ejaculatory
      duct, urethra

___ 4. Sperm finishes maturing in the _____.
   a. epididymis
   b. seminiferous tubule
   c. seminal vesicle
   d. vas deferens

   e. bulbourethral gland

___ 5. Ovulation in humans is triggered by
   _____.
   a. high levels of luteinizing hormone (LH)
   b. high levels of progesterone
   c. the corpus luteum
   d. development of the endometrium
   e. decreasing levels of luteinizing
      hormone (LH)

___ 6. Menstruation is initiated when _____.
   a. luteinizing hormone (LH) levels
      increase
   b. progesterone levels decrease
   c. follicle stimulating hormone (FSH)
      levels increase
   d. the corpus luteum is established
   e. more than two weeks has passed
      since ovulation

___ 7. When a sperm penetrates the egg
   membrane _____.
   a. entry of other sperm is prevented
   b. meiosis II is completed
   c. nuclear membrane of sperm breaks
      down
   d. all of the above occur

8. The most effective options for birth control are _____.
   a. condoms
   b. birth control pills
   c. vasectomy or tubal ligation
   d. rhythm method
   e. IUD with spermicide

9. The estimated number of conceptions that end early due a genetic problem is _____.
   a. less than 5%
   b. 10%
   c. 25%
   d. 50%
   e. 75%

10. The most widespread and fastest growing sexually transmitted disease in the United States is _____.
   a. chlamydia
   b. gonorrhea
   c. genital herpes
   d. human papilloma virus (HPV)

11. The primary organ of intercourse for a male is the _____.
   a. penis
   b. testes
   c. urethra
   d. prostate

12. The primary organ of intercourse for a female is the _____.
   a. vagina
   b. ovary
   c. uterus
   d. clitoris

13. This sexually transmitted disease can cause cervical cancer.
   a. AIDS
   b. gonorrhea
   c. genital herpes
   d. human papilloma virus (HPV)

14. Spermatogenesis takes place in the _____.
   a. penis
   b. testes
   c. urethra
   d. prostate

15. The corpus lutem produces _____.
   a. estrogen
   b. progesterone
   c. FSH
   d. LH

16. Males can produce sperm throughout their lifetime while females are limited to the number of eggs they can produce.
   a. True
   b. False

17. What behaviors can reduce the symptoms of premenstrual syndrome?
   a. balanced diet
   b. regular exercise
   c. birth control pills
   d. all of the above

18. The oxytocin produced during intercourse acts on which region of the brain?
   a. hippocampus
   b. corpus callosum
   c. limbic system
   d. cerebellum

19. Sperm formation takes approximately _____ days.
   a. 10
   b. 36
   c. 65
   d. 120

20. All asexually reproduced individuals are clones of their parent.
   a. True
   b. False

CHAPTER OBJECTIVES/REVIEW QUESTIONS

1. Describe how asexual reproduction differs from sexual reproduction. Know the advantages and problems associated with having separate sexes. [pp. 732-733]
2. Explain why evolutionary trends in many groups of organisms tend toward developing more complex sexual strategies rather than retaining simpler asexual strategies. [pp. 732-733]
3. Describe accessory glands in the male reproductive system and explain the importance of their contributions to semen. [pp.734-735]
4. Describe where immature sperm are found and the function of the cytoplasmic bridges between the cells. [pp.736-737]
5. Explain how sperm production is controlled hormonally. [pp.736-737]
6. Explain the sequence of hormonal changes through the human menstrual cycle and what event each hormone change causes. [pp.740-741]
7. Describe the sequence of events in fertilization from arrival of the sperm at the egg until the first mitotic division begins. [pp.742-743]
8. Describe the various options for birth control and rate each in terms of its effectiveness. [pp.744-745]
9. List the most common agents of sexually transmitted diseases and the symptoms of each. [pp.746-747]

INTEGRATING AND APPLYING KEY CONCEPTS

1. What problems could you foresee for humans if they reproduced asexually instead of sexually?
2. How would evolution be affected if sexual reproduction did not exist?
3. There is increasing evidence that there are chemicals in the environment that are mimicking the affect of human hormones. Predict how estrogen mimics could effect human reproductive processes.
4. Discuss the social consequences of sexually transmitted diseases.

42

ANIMAL DEVELOPMENT

INTRODUCTION

Chapter 42 examines the intricate and fascinating process that leads from single cell zygote to a complex, mature animal composed of multiple tissues, organs, and organ systems. The events of each stage of development are described along with the controlling mechanisms of those events. The chapter then describes human development in more detail.

STUDY STRATEGIES

- Read all of Chapter 42 with the goal of familiarizing yourself with the boldface terms.
- Recognize the various phases of development for animals.
- Discuss the processes of cytoplasmic localization and cleavage.
- Explain cell differentiation and the role of morphogens.
- Describe embryonic induction.
- Discuss the role of master genes in development.
- Identify the major stages of human development.
- Describe early human development through the embryonic stage.
- Describe the development of human organs and systems through week 16.
- Recognize the structures of the placenta.
- Identify the factors that play a role in miscarriages, stillbirths, and birth defects.
- Explain the birth process and lactation.

FOCAL POINTS

- Figure 42.2 [p.752] describes the stages of animal embryonic development.
- Figure 42.3 [pp.752-753] illustrates early embryonic development of vertebrates using frogs as an example.
- Figures 42.5 and 42.6 [p.755] examine cleavage patterns in several different organisms.
- Figures 42.7 and 42.8 [pp.756-757] describe some of the cell movements seen during gastrulation and formation of the neural tube.
- Table 42.2 and Figure 42.11 [p.759] summarizes the stages of human development.
- Figure 42.13 [p.760] describes and illustrates the first two weeks of human development.
- Figure 42.16 [p.762] shows human development for day 14 through day 25.
- Figure 42.17 [p.763] describes interaction between the circulatory systems of mother and fetus through the placenta.
- Figure 42.18 [pp.764-765] illustrates stages of human development through 16 weeks.
- Figure 42.19 [p.766] illustrates times during development when a fetus is sensitive to teratogenic agents.

INTERACTIVE EXERCISES

42.1. MIND BOGGLING BIRTHS [p.751]

42.2. STAGES OF REPRODUCTION AND DEVELOPMENT [pp.752-753]

Boldfaced, Page-Referenced Terms

invitro fertilization _____

_____

cleavage _____

_____

blastula _____

_____

gastrulation _____

_____

gastrula _____

_____

germ layers _____

_____

ectoderm _____

_____

endoderm _____

_____

mesoderm _____

Fill-in-the-Blanks [pp.752-753]

(1) _____ is considered the first stage of animal development. When sperm and egg unite and their DNA mingles and is reorganized, the process is referred to as(2) _____. At the end of (2), a(n) (3) _____ is formed. (4) _____ includes the repeated mitotic divisions of a zygote that segregate the egg cytoplasm into cells known as (5) _____; the entire cluster of cells is known as a blastula. (6) _____ is the process that arranges cells into three primary tissue layers. (7) _____ will eventually give rise to the nervous system and integument; (8) _____ forms the inner lining of the gut and associated digestive glands;(9) _____ forms

many structures including the circulatory system, the muscles, and connective tissues.

Sequence [pp.752-753]

Arrange the following events in correct chronological sequence. Write the letter of the first step next to 10. Write the letter of the last step in the sequence next to 15.

10. _____

11. _____

12. _____

13. _____

14. _____

15. _____

A. Gastrulation

B. Fertilization

C. Cleavage

D. Growth, tissue specialization

E. Organ formation

F. Gamete formation

Matching [pp.752-753]

Match each of the following structures with its correct description.

16. _____ cytoplasmic localization

17. _____ vegetal pole

18. _____ animal pole

19. _____ gray crescent

20. _____ blastomere

21. _____ blastula

22. _____ blastocoel

A. Portion of a yolk-rich egg where there is little yolk

B. Term for cells produced by cleavage process

C. Cavity of a blastula filled with fluid

D. Portion of a yolk-rich egg where most of the yolk is found

E. Collective name of cells produced by cleavage

F. Cytoplasmic components and molecules not evenly distributed in cytoplasm of oocyte

G. In amphibian zygote, region of cell cortex that is lightly pigmented and contributes to organization of development.

H. Cleavage pattern found in most chordates

Short Answer [pp.752-753]

23. Explain how cytoplasmic localization can lead to blastomeres containing different information. _____

_____

_____

_____

42.3. FROM ZYGOTE TO GASTRULA [pp.754-755]

42.4. SPECIALIZED TISSUES AND ORGANS FORM [pp.756-757]

42.5. AN EVOLUTIONARY VIEW OF DEVELOPMENT [p.758]

Boldfaced, Page-Referenced Terms

cytoplasmic localization _____

_____

differentiation _____

_____

morphogens _____

_____

maternal effect gene _____

_____

embryonic induction _____

_____

apoptosis_____

_____

somites _____

_____

Fill-in-the-Blanks [pp.754-758]

By the process of (1) _____ fates of cells change when exposed to signaling molecules from adjacent tissues. Other signaling molecules called (2) _____ affect DNA in cells from a distance. They form a concentration (3) _____, and cells exposed to different levels of (2) develop differently. This pattern formation usually begins with (4) _____ during cleavage. The overall body plan is determined by specialized master genes called (5) _____ genes, which are activated by (6) _____. The actual patterns seen in living organisms are often limited by (7) _____ constraints like the (8) _____ ratio; (9) _____ constraints imposed by the body axes; and (10) _____ constraints imposed on each lineage by the interactions of genes.

True/False [pp.754-758]

If the statement is true, write a "T" in the blank. If the statement is false, make it correct by changing the underlined word(s) and writing the correct word(s) in the answer blank.

11. _____ Transplanting a small segment of the dorsal lip to a new location on the same organism <u>has no effect on development.</u>

12. _____ Gastrulation is characterized by many <u>movements of cells</u> from one place to another in the developing embryo.

13. _____ When one group of cells affects the development of another group of cells we call that <u>transduction</u>.

14. _____ When one group of cells expresses a different select group of genes compared to other cells <u>cell localization</u> is occurring.

15. _____ Master genes produce signaling molecules called <u>homeostats</u>.

16 _____ Morphogens have different effects on cells based on the <u>number of receptors</u>.

17. _____ Somites formed on either side of the neural tube differentiate into <u>spinal nerves</u>.

18. _____ Homeotic genes become active based on their <u>position</u> in the embryo.

19. _____ <u>Architectural</u> constraints on development relate to surface-to-volume ratios.

20. _____ Constraints based on the evolutionary history of an organism are <u>physical</u>.

Short Answer [pp.754-755]

21. Describe what happens in induction. _____

_____

_____

22. Describe the process of apoptosis and its role in development. _____

_____

_____

23. Give the three steps in the general model for development. _____

_____

_____

42.6. OVERVIEW OF HUMAN DEVELOPMENT [p.759]

42.7. EARLY HUMAN DEVELOPMENT [pp.760-761]

42.8. EMERGENCE OF THE VERTEBRATE BODY PLAN [pp.762-763]

42.9. STRUCTURE AND FUNCTION OF THE PLACENTA [p.763]

42.10. EMERGENCE OF DISTINCTLY HUMAN FEATURES [pp.764-765]

Boldfaced, Page-Referenced Terms

amnion _____

_____

chorion _____

_____

placenta _____

_____

allantois _____

_____

human chorionic gonadotropin (HCG) _____

_____

Fill-in-the-Blanks [pp.760-761]

Six or seven days after conception, (1) _____ begins as the blastocyst sinks into the endometrium.

Extensions from the chorion, (2) _____, fuse with the endometrium of the uterus to form a(n) (3)

_____, the organ of interchange between mother and fetus. By the end of the (4) _____ week, the

embryonic stage is completed; the offspring is now referred to as a(n) (5) _____ .

Choice [p.759]

For each of the following stages in development, choose the correct period for that stage.

a. prenatal   b. postnatal

6. _____ Pubescent

7. _____ Fetus

8. _____ Child

9. _____ Newborn

10. _____ Zygote

11. _____ Old age

12. _____ Embryo

13. _____ Adult

14. _____ Infant

15. _____ Adolescent

16. _____ Blastocyst

Sequence [pp.760-762]

Arrange the following events in the correct time sequence. Write the letter of the first event next to 14. The letter of the last event is written next to 25.

17. _____

18. _____

19. _____

20. _____

21. _____

22. _____

23. _____

24. _____

25. _____

26. _____

27. _____

28. _____

A. Fertilization occurs

B. Blastocyst surface cells attach to endometrium

C. Blood-filled spaces form in maternal tissue

D. Somites become visible

E. Amniotic cavity and yolk sac form

F. A compact ball of cells form

G. Blastocoel forms

H. Pharyngeal arches forming

I. Eight loosely arranged cells form

J. Primitive streak forms

K. Somites form

L. Chorionic villi form

Matching [p.763]

Match each of the four extraembryonic membranes with its correct function.

29. _____ Allantois

30. _____ Amnion

31. _____ Chorion

32. _____ Yolk sac

A. Outermost membrane; will become part of the spongy, blood engorged placenta

B. Directly encloses the embryo and holds it a buoyant, protective fluid

C. In humans, some of this membrane becomes a site of blood cell formation; some will become the forerunner of gametes

D. In humans, this membrane serves in development of the umbilical circulation and formation of the urinary bladder

Choice [pp.762, 756-767]

Indicate from which primary tissue each of the following is derived.

a. ectoderm   b. endoderm   c. mesoderm

33. _____ Muscle

34. _____ Nervous tissues

35. _____ Inner lining of the gut

36. _____ Circulatory organs (blood vessels, heart)

37. _____ Outer layer of the integument

38. _____ Reproductive and excretory organs

39. _____ Organs derived from the gut

40. _____ Most of the skeleton

41. _____ Connective tissues of the gut and integument

Short Answer [p.763]

42. Explain the functions of the placenta. _____

_____

_____

43. Explain why the mother's immune system does not make antibodies against the fetal tissues. _____

_____

_____

Labeling and Matching [pp.764-767]

Identify each numbered part of the following illustration of human early development.

44. _____ tail

45. _____ umbilical cord formation

46. _____ future lens

47. _____ lower limb bud

48. _____ developing heart

49. _____ large head with rapid growth

50. _____ developing upper limb

51. _____ forebrain

52. _____ retinal pigment

53. _____ upper limb bud

54. _____ pharyngeal arches

55. _____ foot plate

56. _____ somites

57. _____ rudimentary external ear

58. _____ neural tube formation

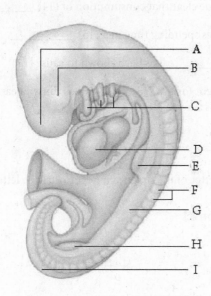
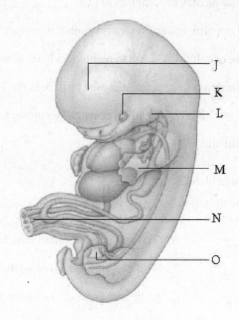

42.11. MISCARRIAGES, STILLBIRTHS AND BIRTH DEFECTS [pp.766-767]

42.12. BIRTH AND LACTATION [pp.768-769]

Boldfaced, Page-Referenced Terms

teratogen _____

_____

Fill-in-the-Blanks [pp.766-767]

Proper nutrition of the mother is vital to the well-being of the fetus as well as the mother. For example, a maternal diet deficient in the B vitamin (1) _____ often leads to neural tube defects in the fetus. A deficiency of (2) _____ can lead to the syndrome of defects in brain function in motor skills called cretinism. Many pregnant women also suffer from episodes of nausea called (3) _____. This condition most often occurs during the time period when (4) _____ are developing in the fetus and is thought to protect the fetus from possible (5) _____ that might be found in certain foods.

Maternal infections can also cause problems in the developmental process of the fetus. (6) _____ are substances that can interfere with development. These include (7) _____ or (8) _____ If a mother becomes infected with a apicomplexan protist could acquire (9) _____ and the fetus could have (10) _____ or be (11) _____ (12) _____ is viral disease that can be dangerous during the early weeks of development. Consumption of (13) _____ can result in a fetus with reduced brain size, facial deformities, mental impairment, and a variety of other problems. It has also become clear that consumption of (14) _____ in coffee or soft drinks leads to higher incidence of miscarriages, as does spending time in a (15) _____. Even some prescription drugs have caused birth defects. The drug (16) _____ has been shown to cause (17) _____ if taken in early pregnancy. Accutane, an effective treatment for (18) _____ has caused heart problems and (19) _____.

Sequence [p.768]

Arrange the following events in the correct time sequence. Write the letter of the first event next to 20. The letter of the last event is written next to 29.

20. _____   A. Oxytocin causes smooth muscles in uterus to contract

21. _____   B. Fetus passes through cervix and vagina

22. _____   C. Additional Oxytocin causes stronger contractions

23. _____   D. Strong muscle contractions cause placenta to detach from uterus

24. _____   E. Receptors in cervix sense pressure

25. _____   F. The umbilical cord is clamped and cut

26. _____   G. Cervix becomes flexible and stretches

27. _____   H. Added pressure from contractions signals more Oxytocin secretion

28. _____   I. "Afterbirth" delivered

29. _____   J. Hypothalamus signals posterior pituitary to release Oxytocin

Matching [pp.768-769]

Match each of the following terms related to milk production with its correct role in nourishing a newborn.

30. _____ prolactin

31. _____ lactation

32. _____ oxytocin

33. _____ progesterone and estrogen

34. _____ suckling

A. Hormone(s) that causes smooth muscle contractions around milk glands

B. Physical stimulus that causes release of Oxytocin

C. Hormone(s) that causes milk synthesis

D. Milk production

E. Hormone(s) that decrease in amount causing milk production to increase

---

SELF-TEST

___ 1. The specialized blastula seen in humans is called a(n) _____.
 a. blastocyst
 b. blastomere
 c. gastrula
 d. cleavage
 e. deuterostome

___ 2. The process of cleavage most commonly produces a(n) _____.
 a. zygote
 b. blastula
 c. gastrula
 d. third germ layer
 e. organ

___ 3. Incomplete cleavage of cells of very yolky eggs is characteristic of the cleavage pattern of _____.
 a. frogs
 b. sea urchins
 c. chickens
 d. humans
 e. none of the preceding

___ 4. The formation of three germ (embryonic) tissue layers occurs during _____.
 a. gastrulation
 b. cleavage
 c. pattern formation
 d. morphogenesis
 e. neural plate formation

___ 5. The differentiation of a body part in response to signals from an adjacent body part is _____.
 a. contact inhibition
 b. ooplasmic localization
 c. embryonic induction
 d. pattern formation
 e. none of the above

___ 6. A homeotic mutation _____.
 a. may cause a leg to develop on the head where an antenna should grow
 b. may affect pattern formation
 c. affects morphogenesis
 d. may alter the path of development
 e. all of the above

___ 7. Shortly after fertilization, the zygote is subdivided into a multicelled embryo during a process known as _____.
 a. meiosis
 b. parthenogenesis
 c. embryonic induction
 d. cleavage
 e. invagination

___ 8. Muscles differentiate from_____ tissue.
 a. ectoderm
 b. mesoderm
 c. endoderm
 d. parthenogenetic
 e. yolky

___ 9. The gray crescent is _____ .
 a. formed where the sperm penetrates the egg
 b. part of only one blastomere after the first cleavage
 c. the yolky region of the egg
 d. where the first mitotic division begins
 e. formed opposite from where the sperm enters the egg

10. The nervous system differentiates from _____ tissue.
    a. ectoderm
    b. mesoderm
    c. endoderm
    d. yolky
    e. homeotic

11. Once implanted, the blastula releases _____ .
    a. prolactin
    b. estrogen
    c. human chorionic gonadotropin
    d. amniotic fluid
    e. oxytocin

12. Which extraembryonic membrane lines the amniotic sac and the yolk sac, becoming part of the placenta?
    a. amnion
    b. chorion
    c. allantois
    d. chorionic villi
    e. urinary bladder

13. In human development, at the end of the eighth week, all organs are formed and we define the individual as a(n) _____.
    a. infant
    b. embryo
    c. zygote
    d. fetus
    e. blastocyst

14. Release of milk from the milk glands requires stimulus of the newborn suckling and _____.
    a. prolactin
    b. progesterone
    c. estrogen
    d. human chorionic gonadotropin
    e. oxytocin

15. The last embryonic tissue layer to develop is the
    a. protoderm
    b. mesoderm
    c. ectoderm
    d. endoderm

16. The embryo stage of development in humans ends after week
    a. 2
    b. 4
    c. 6
    d. 8

17. During labor, which of the following "breaks"?
    a. amnion
    b. chorion
    c. yolk sac
    d. allantois

18. The placenta serves to _____.
    a. provide nutrients
    b. remove wastes
    c. produce hormones
    d. all of the above

19. Oxytocin performs the following function during and after childbirth
    a. signals mammary glands to grow
    b. stimulates muscles in the mammary glands to contract
    c. stimulates contractions in the uterus
    d. both b and c.

20. In animal development, the process of cleavage can best be described as
    a. mitosis
    b. fertilization
    c. differentiation
    d. meiosis

21. The definition of childhood for a human is
    a. the first two weeks after birth.
    b. two weeks to fifteen months.
    c. puberty until about 3 or 4 years later
    d. infancy to about ten or twelve years.

22. Cell differentiation is the process by which cells specialize into their final form and function.
    a. True
    b. False

23. The animal pole is the side of the egg where the yolk is most concentrated.
    a. True
    b. False

24. Diseases (such as rubella), drugs, alcohol, and radiation can disrupt human embryonic development. These exposures have the most severe effects on the fetus if they occur during the first three months of gestation, because at this time _____.
    a. substances can most easily enter the embryo through the placenta
    b. the embryo has not yet been implanted in the wall of the uterus
    c. the mother's hormone levels are highest
    d. the most drastic and rapid changes are occurring in the embryo

CHAPTER OBJECTIVES/REVIEW QUESTIONS

1. Describe the early stages of animal development and the events within each one. [pp.752-753]
2. Explain how cytoplasmic localization leads to blastomeres having different cytoplasmic contents and hence how that can lead to different developmental paths for the cells. [p.754]
3. Explain how the amount of yolk in an ovum can influence an animal's cleavage pattern. [pp.754-755]
4. Define the term *gastrulation* and state what process begins at this stage that did not happen during cleavage. [p.755]
5. Define the term *differentiation* and give two examples of cells in a multicellular organism that have undergone differentiation. [p.756]
6. Discuss the three constraints on development patterns and give one example of each. [pp.756-767]
7. List the stages of human development and briefly describe how each is defined. [p.759]
8. Describe the role of the four extraembryonic membranes found in human development. [p.761]
9. Name each of the three embryonic tissue layers and the organs formed from each. [p.762]
10. Describe the events that occur during the first month of human development. State how much time cleavage and gastrulation require when organ formation begins, and what is involved in implantation and placenta formation. [pp.764-765]
11. Explain why the mother must be particularly careful of her diet, health habits, and lifestyle during the first trimester after fertilization (especially during the first six weeks). [pp.766-767]
12. Describe the series of events during labor that lead to childbirth. [p.768]
13. Describe the events required for successful production and release of milk. [p.769]

INTEGRATING AND APPLYING KEY CONCEPTS

1. If embryonic induction did not occur in a human embryo, how would the eye region appear? What would happen to the forebrain and epidermis? If controlled cell death did not happen in a human embryo, how would its hands appear? How would the face appear?
2. Discuss the implications of the fact that homeotic genes from one organism will also function in a completely different species. For example, human homeotic genes may be expressed in fruit flies.
3. With our increased knowledge of what events and conditions may adversely affect human development, how can a society encourage women to take proper care of themselves while pregnant.

43

ANIMAL BEHAVIOR

INTRODUCTION

This chapter of the text examines the basic forms of behavior, or responses to stimuli. It examines basic behavioral responses, the forms of chemical behavior, signaling mechanisms, and reproductive behavior. It concludes with types of social behavior. its costs and its benefits.

STUDY STRATEGIES

- Read all of chapter 43 with the goal of familiarizing yourself with the boldface terms.
- Explain the ways in which genetics relates to behavior.
- Describe how scientists use intraspecific variation and interspecific variation to understand animal and human behaviors.
- Compare and contrast innate behavior, learned behavior, fixed action pattern, and imprinting.
- Discuss the difference between operant and classical conditioning.
- Recognize the various types of learning including habituation and observational.
- Describe behavioral plasticity and epigenetics.
- Identify the major patterns of motion in animals compare this with the process of migration.
- Discuss the various types of communication between animals.
- Compare and contrast the various types of mating systems and parental care.
- Describe the benefits and drawbacks for animals that live in groups.
- Explain the evolutionary benefits to altruism and eusociality.

FOCAL POINTS

- Section 43.2 [pp.774-775] explores the genetic basis for behavior.
- Section 43.3 [ppp.776-777] compares instinctive and learned behaviors.
- Section 43.5 [p.779] differentiates between taxes and kineses.
- Section 43.6 [pp.780-781] details the various types of signaling.
- Section 43.7 [pp.782-783] explores reproductive behavior.
- Sections 43.8 and 43.9 [pp.784-787] describe various forms of social behavior.

INTERACTIVE EXERCISES

43.1. ALARMING BEE BEHAVIOR [P.773]

43.2. BEHAVIORAL GENETICS [pp.774-775]

43.3. INSTINCT AND LEARNING [pp.776-777]

Boldfaced Terms

pheromone _____

_____

instinctive behavior _____

_____

fixed action pattern _____

_____

imprinting _____

_____

classical conditioning _____

_____

operant conditioning _____

_____

habituation _____

_____

observational learning _____

_____

Matching [pp.773-777]

1. _____ observational learning
2. _____ pheromone
3. _____ instinctive behavior
4. _____ fixed action pattern
5. _____ operant conditioning
6. _____ imprinting
7. _____ habituation
8. _____ classical conditioning

A. Voluntary behavior is modified in response to the consequences of that behavior

B. Chemical signal that influences members of the same species

C. Imitating the behavior of another individual

D. Time-dependent form of learning; triggered by exposure to sign stimuli and usually occurring during sensitive periods of young animals

E. An innate response to a simple stimulus

F. Instinctive movements, triggered by a stimulus, that will run to completion

G. Learning thru experience not to respond to a stimulus

H. Involuntary response to an alternate stimulus

43.4. ENVIRONMENTAL EFFECTS ON BEHAVIORAL TRAITS [p.778]

43.5. MOVEMENTS AND NAVIGATION [p.779]

43.6. COMMUNICATION SIGNALS [pp.780-781]

Boldfaced Terms

communication signals _____

_____

behavioral plasticity _____

_____

taxis _____

_____

kinesis _____

_____

migration _____

_____

Matching [pp.779-781]

1. _____ communication signals
2. _____ behavioral plasticity
3. _____ taxis
4. _____ alarm pheromones
5. _____ sex pheromones
6. _____ threat display
7. _____ courtship display
8. _____ tactile display
9. _____ acoustical signals
10. _____ alarm calls
11. _____ kinesis
12. _____ migration

A. The signaler touches the receiver in a ritualized manner to communicate information

B. Cues for social behavior between members of a species

C. Songs, chirps and other sounds used for attracting a mate or defining a territory

D. Acoustical behavior used to warn of a predator

E. A display of strength and power that avoids conflict

F. Chemical signal that alerts other members of the species to a threatening situation

G. Chemical signal to attract a mate

H. Rituals that must be performed prior to forming a mating pair

I. Behavioral traits are altered by environmental factors

J. Innate directional response

K. Increase or decrease in animal movement regardless of direction

L. Interruption in daily activities to travel to a new habitat

43.7. MATES, OFFSPRING, AND REPRODUCTIVE SUCCESS [pp.782-783]

43.8. LIVING IN GROUPS [pp.784-785]

43.9. WHY SACRIFICE YOURSELF? [pp. 786-787]

Boldfaced Terms

lek _____

_____

territory _____

_____

dominance hierarchy _____

_____

selfish herd _____

_____

theory of inclusive fitness _____

_____

altruistic behavior _____

_____

eusocial animal _____

_____

Complete the Table [pp.782-783]

1. Complete the following table by supplying the common names of the animals discussed in the text that fit the examples of sexual selection.

Animals __   *Description of Sexual Selection*

a.	Males hold strings of eggs around their legs and can mate with, and carry eggs from, more than one female
b.	Females select the males that offer them superior material goods; females permit mating only after they have eaten a "nuptial gift"
c.	Females of a species cluster around a resource; males occupy and defend a territory and will mate with all females in that territory
d.	Males congregate in a lek, or communal display ground; each male stakes out his territory; females are attracted to the lek to observe the male displays and usually select and mate with only one male

Matching [pp.782-787]

Choose the most appropriate statement for each of the following items.

2. _____ disadvantages of sociality

3. _____ dominance hierarchy

4. _____ cooperative predator avoidance

5. _____ the selfish herd

6. _____ cooperative hunting

7. _____ transmission of cultural traits

8. _____ lek

9. _____ territory

10. _____ parental behavior

11. _____ altruism

A. Competition for resources, rapid depletion of food resources, cannibalism, and greater vulnerability to disease

B. A simple society brought together by reproductive self-interest, in bluegill sunfish, the larger, more powerful males tend to claim the central nesting locations

C. Behavior that includes alarm calls of prarie dogs and clustering of sawflies

D. Some individuals of a wolf pack adopt a subordinate status with respect to the other members

E. Social groups of predatory animals

F. Behavior that enhances survival of the young

G. Communal display ground

H. Behavior that enhances another individual's reproductive success at the altruist's expense

I. Young chimpanzees learn to make tools by observing adult behavior

J. A region held by one male

Matching [pp.782]

Choose the most appropriate statement for each of the following terms.

12. _____ promiscuous

13. _____ monogamous

14. _____ polygamous

15 _____ social monogamy

A. One sex has multiple partners

B. A male and female prefer each other's company, but may mate with others

C. Members of both sexes mate with multiple partners

D. One male and one female mate only with each others

Fill-in-the-Blanks [pp.792-793]

A (16) _____ animal lives in a multigenerational family group with a reproductive division of labor. Common examples are (17) _____, _____, and _____. A (18) _____ honeybee is the only fertile female in her hive, Her sterile female daughters are (19) _____ who maintain the hive and collect food. Male honeybees, called (20) _____, die after mating.

(21) _____ workers are indirectly promoting their genes through (22) _____ behavior that benefits their close (23) _____ who share the same genes. This behavior is described in the theory of (24)

_____ _____. All of the individuals in honeybee, termite and ant colonies are members of a great extended (25) _____. Non-breeding family members support siblings, a few of which are the future kings and (26) _____. Although a guard bee dies after she stings, her sacrifice preserves her (27) _____ within her hive mates.

SELF-TEST

___ 1. A behavior is the coordinated response an animal makes to _____.
    a. an instinct
    b. a fixed action pattern
    c. a stimulus
    d. a learned behavior

___ 2. In _____, a particular behavior is performed without having been learned by actual experience in the environment.
    a. natural selection
    b. altruistic behavior
    c. sexual selection
    d. instinctive behavior

___ 3. Newly hatched goslings follow any large moving object to which they are exposed shortly after hatching; this is an example of _____.
    a. homing behavior
    b. imprinting
    c. piloting
    d. migration

___ 4. A young toad flips its sticky-tipped tongue and captures a bumble bee that stings its tongue; in the future, the toad leaves bumblebees alone. This is _____.
    a. instinctive behavior
    b. a fixed action pattern
    c. altruistic behavior
    d. learned behavior

___ 5. Behavior that enhances another's reproductive success at the expense of its own is called _____.
    a. altruism
    b. instinctive
    c. selfish
    d. social

___ 6. Certain individuals receive a larger share of the resources than subordinates in a social system known as _____.]
    a. cooperative predator avoidance
    b. a selfish herd
    c. a dominance hierarchy
    d. an altruistic group

___ 7. An animal interrupts its normal daily pattern of activity to travel toward a new habitat during a _____.
    a. taxis
    b. tropism
    c. migration
    d. kinesis

___ 8. Musk oxen forming a "ring of horns" against predators is an example of _____.
    a. cooperative predator avoidance
    b. a selfish herd
    c. self-sacrificing behavior
    d. dominance hierarchy

___ 9. Competition among members of one sex for access to mates is called _____.
    a. altruism
    b. social behavior
    c. inclusive fitness
    d. sexual selection

___ 10. Alek is a _____.
    a. form of threat display
    b. type of pheromone
    c. communication signal
    d. communal display ground

CHAPTER OBJECTIVES / REVIEW QUESTIONS

1. What explains the fact that costal and inland garter snakes of the same species have different food preferences? [p.774]
2. Describe the intermediate response obtained in Arnold's experiment with costal and inland garter snakes. [p.774]
3. Explain instinctive behavior and give an example. [p.776]
4. Describe and cite an example of a fixed action pattern. [p.776]
5. Distinguish learned behavior from instinctive behavior. [p.776]
6. Explain imprinting. [p.776]
7. Define habituation. [p.777]
8. Explain the differences between classical and operant conditioning. [p.777]
9. What is meant by the term behavioral plasticity? [p.778]
10. Distinguish a taxis from a kinesis. [p.779]
11. Define migration. How and why do animals migrate? [p.779]
12. Understand the various forms of communication signals and displays used by animals. [pp.780-781]
13. What sort of information do bees communicate via the waggle dance? [p.781]
14. Explain the benefits of sexual selection. [p.782]
15. Define *lek* and *territory* [p.783]
16. List the costs and benefits of being a social organism. [pp.784-785]
17. Define *eusocial animal.* [p.786]
18. What is the theory of inclusive fitness? [p.786]

INTEGRATING AND APPLYING KEY CONCEPTS

1. Think about communication signals that humans use, and list them. Do you believe a dominance hierarchy exists in human society? Think of examples of social situations in which dominance hierarchies may exist.
2. Apply the concept of self-sacrifice to parenting. Based on the information in this chapter, what are the benefits of a parent protecting its child?
3. Discuss the advantages and disadvantages of males using a *lek* to attract females. Which position in the *lek* is the most beneficial?
4. How does learned behavior and the transmission of cultural traits help a social species to progress more quickly than through instinctive behavior alone?

44

POPULATION ECOLOGY

INTRODUCTION

This chapter introduces the principles by which biologists study populations of a species. It is important as you progress through the chapter that you distinguish between the many terms associated with describing populations, such as size, density, and structure. It is also essential to understand the effects that other species and the abiotic environment have on populations. The final sections of the chapter apply the knowledge of population structure to humans, and discuss some of the challenges facing human population growth.

STUDY STRATEGIES

- Read all of chapter 44 with the goal of familiarizing yourself with the boldface terms.
- Explain what constitutes population dynamics and how scientists determine these patterns.
- Compare and contrast the various types of population distributions.
- Describe the roles of immigration and emigration in population dynamics.
- Recognize the exponential growth pattern and how it relates to a population's per capita growth rate and biotic potential.
- Identify the various limits to population growth, both density-dependent and density-independent.
- Discuss a population's carrying capacity and how this relates to logistic growth patterns.
- Explain the role of life tables in understanding population dynamics.
- Identify the three major survivorship curves and an organism that typifies each.
- Compare and contrast r-selected and K-selected species.
- Describe the pattern of human population growth, compare this to other patterns, and identify the predictions of human population growth for the future.
- Explain the relationship between demographic transition and resource utilization.

FOCAL POINTS

- Figure 44.5 [p.795] and Figure 44.7 [p.796] together illustrate a key concept, the difference between exponential and logistic growth models.
- Figure 44.9 [p.798] explains the difference between the three forms of survivorship curves.
- Figure 44.13 demonstrates how the growth of the human population has been influenced by historical events.
- Figure 44.15 [p.804] illustrates the different forms of age structure in populations based on their rate of growth.

INTERACTIVE EXERCISES

44.1. A HONKING MESS [p.791]

44.2. POPULATION DEMOGRAPHICS [p.798]

44.3. POPULATION SIZE AND EXPONENTIAL GROWTH [pp.794-795]

Boldfaced Terms

population _____

_____

demographics _____

_____

population size _____

_____

plot sampling _____

_____

mark-recapture _____

_____

population density _____

_____

population distribution _____

_____

age structure _____

_____

reproductive base _____

_____

immigration _____

_____

emigration _____

_____

zero population growth _____

_____

per capita growth rate _____

_____

exponential growth _____

_____

biotic potential _____

_____

Matching [p.792 - 793]

Choose the most appropriate statement for each term.

1. _____ demographics
2. _____ population size
3. _____ population density
4. _____ mark-recapture sampling
5. _____ population distribution
6. _____ age structure
7. _____ reproductive base
8. _____ plot-sampling
9. _____ pre-reproductive, reproductive, and post-reproductive
10. _____ clumped distribution
11. _____ near-uniform distribution
12. _____ random distribution

A. Includes pre-reproductive and reproductive age categories

B. The general pattern in which the individuals of the population are dispersed through a specified area

C. When individuals of a population are more evenly spaced than they would be by chance alone

D. The number of individuals in some specified area or volume of habitat

E. Occurs only when individuals of a population neither attract nor avoid one another, when conditions are fairly uniform through the habitat, and when resources are available all the time

F. The number of individuals in each of several to many age categories

G. Method of estimating population size of organisms that are not very mobile

H. The number of individuals within the population

I. Method of estimating population size of mobile animals

J. Categories of a population's age structure

K. The vital statistics of a population

L. Individuals of a population form aggregations at specific habitat sites; most common dispersion pattern

Short Answer [p.791]

13. Define population ecology. _____

_____

_____

14. Name three characteristics that define a population.

_____

_____

_____

Matching [pp.794-795]

Match each of the following statements to the correct term.

15. _____ growth at a proportional rate/time interval

16. _____ the departure of individuals from a population

17. _____ per capita birth rate

18. _____ a balanced number of births and deaths

19. _____ per capita birth rate minus per capita death rate

20. _____ the arrival of new individuals from other populations

21. _____ the maximum rate of increase per individual under ideal conditions

22. _____ per capita death rate

A. Per capita growth rate, "r"

B. "b"

C. Zero population growth

D. Biotic potential

E. "d"

F. Exponential growth

G. Emigration

H. Immigration

Problems [p.794]

23. Consider the equation $G = rN$, where G = the population growth rate per unit time, r = the net population growth rate per individual per unit time, and N = the number of individuals in the population. Assume that r remains constant at 0.2.
a. As the value of G increases, what happens to the value of N?

_____

b. If the value of G is negative, what happens to the value of N?

_____

c. If the net reproduction per individual stays the same and the population grows faster, then what must happen to the number of individuals in the population?

_____

24. Look at line (a) in the accompanying graph. After seven hours have elapsed, approximately how many individuals are in the population?

_____

25. Look at line (b) in the same graph.

   a. After 24 hours have elapsed, approximately how many individuals are in the population?

   _____

   b. After 28 hours have elapsed, approximately how many individuals are in the population?

   _____

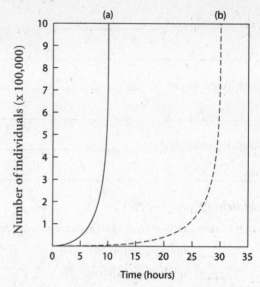

44.4. LIMITS ON POPULATION GROWTH [pp.796-797]

44.5. LIFE HISTORY PATTERNS [pp.789-799]

44.6. EVIDENCE OF EVOLVING LIFE HISTORY PATTERNS [pp.800-801]

Boldfaced Terms

limiting factor _____

_____

carrying capacity _____

_____

logistic growth _____

_____

density-dependent factor _____

_____

density-independent factor _____

_____

life history pattern _____

_____

cohort _____

_____

survivorship curve _____

_____

r-selected species _____

_____

K-selected species _____

_____

Matching [pp.796-799]

Match each of the following terms to the most appropriate statement.

1. _____ limiting factor
2. _____ life history patterns
3. _____ carrying capacity
4. _____ survivorship curve
5. _____ logistic growth
6. _____ density-dependent factor
7. _____ density-independent factor

A. An essential resource that is in short supply

B. The maximum number of individuals of a population that the environment can sustain

C. Inherited traits related to growth, survival and reproduction of a species

D. A small population that initially grows slowly, then rapidly and then the numbers level off

E. Graph line of the age-specific survival of a cohort in a habitat

F. Biotic or abiotic factors that reduce the odds of an individual surviving or reproducing during overcrowding

G. Causes changes in population size regardless of density

Choice [pp.798]

Choose the most appropriate form of survivorship curve for each of the following descriptions.

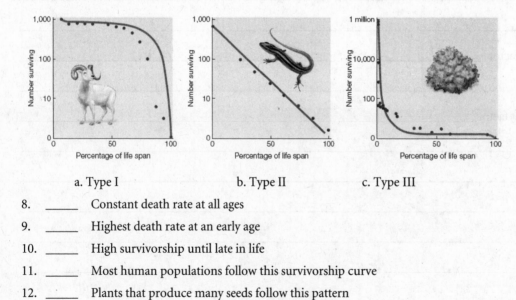

a. Type I   b. Type II   c. Type III

8. _____ Constant death rate at all ages
9. _____ Highest death rate at an early age
10. _____ High survivorship until late in life
11. _____ Most human populations follow this survivorship curve
12. _____ Plants that produce many seeds follow this pattern
13. _____ Lizards and small mammals follow this pattern

44.7. HUMAN POPULATION GROWTH [pp.802-803]

44.8. ANTICIPATED GROWTH AND CONSUMPTION [pp.804-805]

Boldfaced Terms

total fertility rate (TFR) _____

_____

replacement fertility rate _____

_____

demographic transition model _____

_____

ecological footprint_____

_____

Fill-in-the-Blanks [pp.802-805]

Modern humans evolved in Africa about (1) _____ years ago. The size of the human (2) _____ and development of (3) _____ allowed humans to develop skills, communicate those skills to others, and expand into most of the earth's habitats. About 11,000 years ago, the invention of (4) _____ replaced the previous (5) _____-_____ lifestyle, providing humans with a more reliable food supply than before. An agrarian lifestyle allowed humans to replace a wandering lifestyle with a settled one. A pivotal factor was the domestication of wild (6) _____, including the ancestral species to modern (7) _____ and (8) _____. People harvested, stored, and planted the seeds in the same location year after year.

Human population growth surged since the Industrial Revolution due largely to the use of (9) _____ fuels, (10) _____, made from ammonia , and synthetic (11) _____. (12) _____ has historically dampened human population growth. Epidemics swept through overcrowded settlements and outbreaks of (13) _____ and (14) _____ resulted from poor sanitation. In the 1800's, an understanding of microorganisms and illness led to improvements in (15) _____ and (16) _____. With the development of (17) _____ and, more recently, (18) _____, many diseases were no longer as deadly as before. (19) _____ rates began to drop sharply. Birth rates began to exceed death rates and the rapid growth of the human population was underway. World population is currently (20) _____ billion and expected to reach (21) _____ billion by 2050.

The (22) _____ is the average number of children born to a woman during her reproductive years. The (23) _____ is the average number of children a woman must have to replace herself with one daughter who reaches reproductive age. A population (24) _____ as long as the total fertility rate exceeds the replacement rate. Age structure also affects a population's growth rate. The (25) _____ the base of an age structure diagram, the greater the anticipated population growth will be.

Sequence [p.804]

Arrange the following stages of the demographic transition model in correct chronological sequence. Write the letter of the first step next to 26, the letter of the second step next to letter 27, and so on.

26. _____

27. _____

28. _____

29. _____

A. *Industrial stage*: population growth slows and industrialization is in full swing

B. *Preindustrial stage*: harsh living conditions, high birth and death rates, slow population growth

C. *Postindustrial stage*: zero population growth is reached, then birth rate falls below death rate, and population size slowly decreases

D. *Transitional stage*: industrialization begins, food production rises, and health care improves, death rates drop, birth rates remain high, resulting in rapid population growth

SELF-TEST

___ 1. The number of individuals that contribute to a population's gene pool is _____
   a. the population density
   b. the population growth
   c. the population birth rate
   d. the population size

___ 2. The number of individuals in a given area or volume of habitat is _____
   a. the population density
   b. the population growth
   c. the population birth rate
   d. the population size

___ 3. A population that is growing exponentially in the absence of limiting factors can be illustrated accurately by a(n) _____
   a. S-shaped curve
   b. J-shaped curve
   c. curve that terminates in a plateau phase
   d. tolerance curve

___ 4. Assuming the birth rate and death rate remain constant, both can be combined into a single variable, r, or _____.
   a. the per capita rate
   b. the minus migration factor
   c. exponential growth
   d. the net reproduction per individual per unit time

___ 5. Which of the following would you use to determine the population of Sandplain Gerardia in a short grass prairie?
   a. age structure diagram
   b. survivorship curve
   c. plot-sampling
   d. mark-recapture sampling

___ 6. The maximum rate of increase per individual under ideal conditions is called the _____.
   a. biotic potential
   b. carrying capacity
   c. doubling time
   d. population size

7. The maximum number of individuals of a population (or species) that a given environment can sustain indefinitely defines _____.
    a. the carrying capacity of the environment
    b. exponential growth
    c. the doubling time of a population
    d. density-independent factors

8. Which of the following is not characteristic of logistic growth?
    a. S-shaped curve
    b. Leveling off of growth as carrying capacity is reached
    c. Unrestricted growth
    d. Slow growth of a low-density population followed by rapid growth

9. The beginning of industrialization, a rise in food production, improvement of health care, rising birth rates, and declining death rates describes the _____ stage of the demographic transition model.]
    a. preindustrial
    b. transitional
    c. industrial
    d. postindustrial

10. The survivorship curve typical of industrialized human populations is Type _____.
    a. I
    b. II
    c. III
    d. none of the above

11. Which does not fit a r-selection?
    a. Reproduce when young
    b. Lots of offspring
    c. Significant parental care
    d. Maximize number of offspring

12. Which factor had the greatest influence on increasing the human population?
    a. Domestication of wild grasses
    b. Industrialization
    c. Decrease in the death rate
    d. Increase in the birth rates

13. Which country will see the greatest increase in population in the near future?
    a. India
    b. China
    c. Indonesia
    d. Japan

CHAPTER OBJECTIVES / REVIEW QUESTIONS

1. Define the term population. [p.791]
2. Define the following terms: *demographics, population size, population density, population distribution, age structure,* and *reproductive base.* [p.792
3. List and describe the three patterns of distribution illustrated by populations in a habitat. [pp.792-793]
4. Describe plot-sampling and mark-recapture sampling as methods of estimating population size. [p.792]
5. Define *zero population growth* and describe how achieving it would affect population size. [p.794]
6. Distinguish immigration from emigration and and their effects on a population. [p.794]
7. Explain how G= *r*N can be used to predict population growth. [p.794]
8. Describe how the per capita growth rate is determined. [p.794]
9. Explain what is meant by *biotic potential*. [p.795]
10. List several examples of limiting factors, and explain how they influence population curves. [p.796]
11. Explain what is meant by the term *carrying capacity*. [p.796]
12. Explain the meaning of the logistic growth equation. [pp.796-797]
13. Compare logistic and exponential growth. [pp.794-797]
14. Define the term *density-dependent factors* on growth of populations; give an example. [p.797]
15. Define the term *density-independent factors* and list two examples and indicate how the factors would affect a population. [p.797]
16. Explain what is meant by a life history pattern. [p.799]
17. Explain the three survivorship curves. [p.798]
18. Individuals in guppy populations targeted by killifish tend to be larger, less streamlined, and more brightly colored. Individuals in populations targeted by pike-cichlids tend to be smaller, more streamlined, and duller in color patterning. Other life history pattern differences exist between the two groups. After consideration of the research results obtained by Reznick and Endler, provide an explanation for these differences. [pp.800]
19. Define the term, ecological footprint. [p.805]
20. Define the term *total fertility rate*. [p.804]
21. Be able to analyze age structure diagrams to determine patterns of growth. [p.804]

22. List and describe the four stages of the demographic transition model. [p.804]
23. How did the human race redefine the concepts of limiting factors, carrying capacity and biotic potential? [pp.802-803]

INTEGRATING AND APPLYING KEY CONCEPTS

1. The mark-recapture method of estimating population is widely recognized as being inaccurate. If given the resources, how would you go about estimating the size of a population without counting every individual in the population?
2. What measure could have been taken to prevent the reindeer population on St. Matthew's Island from going extinct?
3. If every less-developed country instituted laws / incentives to control population growth (like China), what impact would this have upon the world population?
4. Consider the effects of delayed reproductive age on an age structure diagram of a country. What effect would it have on population growth?

45

COMMUNITY ECOLOGY

INTRODUCTION

This chapter focuses on the variety of mechanisms by which populations of different species interact in a given habitat. You need to understand the principal categories of species interactions presented early in the chapter, which form the basis for the remainder of the chapter's material. The end of the chapter presents some of the consequences that can occur if these interactions are disturbed by human activity or the migration of species.

STUDY STRATEGIES

- Read all of chapter 45 with the goal of familiarizing yourself with the boldface terms.
- Define habitat.
- Recognize the various types of symbiosis.
- Explain mutualism and give an example of this type of symbiosis.
- Define niche.
- Explain the theories of competitive exclusion, resource partitioning, and character displacement as they relate to an organism's niche.
- Discuss predator-prey interactions, functional responses, coevolution, and cycles.
- Identify strategies of predators and prey regarding fitness.
- Compare and contrast predator-prey interactions with herbivore-plant interactions.
- Explain parasitism and give an example of this type of symbiosis.
- Compare and contrast parasitism to predator-prey interactions.
- Explain ecological succession and identify the differences between primary and secondary succession.
- Define a keystone species.
- Recognize the importance of intermediate disturbance events for increased species diversity.
- Identify the problems that exotic species can cause in a new habitat.
- Discuss regional and global patterns of biodiversity.

FOCAL POINTS

- Table 45.1 [p.810] characterizes the different types of direct two-species interactions that are explored in depth in sections 45.3, 45.4, 45.5 and 45.7.
- Section 45.6 [pp.816-817] explores the effects of evolution on co-evolved species.
- Section 45.8 [pp.820-821] describes the process of ecological succession and the role of pioneer species.
- Section 45.9 [pp.822-823] discusses the causes and effects of ecosystem disturbances.

INTERACTIVE EXERCISES

45.1 FIGHTING FOREIGN ANTS [p.809]

45.2. WHICH FACTORS SHAPE COMMUNITY STRUCTURE? [p.810]

45.3. MUTUALISM [p.811]

Boldfaced Terms

community _____

_____

commensalism _____

_____

mutualism _____

_____

habitat_____

_____

symbiosis _____

_____

Fill-in-the-Blanks [p.809-810]

(1) _____ ecology is the study of the interactions of all the species that live in a particular region or (2) _____. Communities may have few or many species. Species diversity has two components: species (3) _____ and species (4) _____. Rainfall, sunlight, temperature and soil quality are (5) _____ factors that can affect a community's structure. This structure is (6) _____, it tends to change slowly over time. Sudden changes in structure may be due to natural or human-induced (7) _____ that may alter the habitat.

Two species may share a long-term close relationship known as (8) _____. Three common types of this close association are (9) _____, (10) _____, and (11) _____.

Matching [p.810-811]

Match each of the following terms to the correct statement.

12. _____ coevolution
13. _____ community
14. _____ species richness
15. _____ symbiosis
16. _____ habitat
17. _____ mutualism
18. _____ commensalism

19. _____ species evenness

A. An interaction between two species that benefits both

B. Close association between two species during part or all of their life cycle

C. Place where a species normally lives

D. The populations of all species in a given habitat

E. Relative abundance of each species in a particular area

F. An interaction that helps one species but does not affect the second species

G. The number of species in a particular area

H. Two species that change over time due to close ecological interactions

45.4. COMPETITIVE INTERACTIONS [pp.812-813]

45.5. PREDATOR-PREY INTERACTIONS [pp.814-815]

45.6. AN EVOLUTIONARY ARMS RACE [pp.816-817]

45.7. PARASITES AND PARASITOIDS [pp.818-819]

Boldfaced Terms

interspecific competition _____

_____

ecological niche _____

_____

competitive exclusion principle _____

_____

resource partitioning _____

_____

character displacement _____

_____

predation _____

_____

prey _____

_____

camouflage _____

_____

mimicry _____

_____

warning coloration _____

_____

herbivory _____

_____

parasitism _____

_____

parasitoids _____

_____

brood parasitism _____

_____

Matching [pp.812-815]

Match each of the following terms to the appropriate statement.

1. _____ resource partitioning
2. _____ interference competition
3. _____ competitive exclusion
4. _____ ecological niche
5. _____ exploitative competition
6. _____ predators

A. Two species reduce the amount of resources available to each other

B. A subdividing of resources that allows two species to coexist

C. All the physical and biological factors that affect a species

D. Consumers that obtain energy and nutrients from living organisms

E. One species prevents or blocks another species' access to the same resources

F. Two species require the same limited resources; the stronger species will outcompete the other for the resources

Choice [pp.816-817]

Choose the most appropriate term for each evolutionary adaptation described in items 7-11.

a. mimicry

b. camouflage

c. chemical defense

d. herbivory

e. warning coloration

7. _____ A blending of body form, color, or behavior with the environment
8. _____ Skunks and darkling beetles exhibit this
9. _____ Protection by pretending to be a dangerous organism
10. _____ Predators learn to avoid organisms that use this defense
11. _____ Caffeine, nicotine and capsaicin evolved as a defense against this

Fill-in-the-Blanks [p.818-819]

In certain symbiotic relationships, one species, the (12) _____ benefits while feeding on another species, the (13) _____, which is harmed. (14) _____ live inside the host, while (15) _____ are attached to the host's exterior surface. A good parasite does not (16) _____ its host until the parasite has produced (17) _____. Weakened hosts are usually more vulnerable to (18) _____ and less attractive to potential (19) _____. Some parasitic infections affect reproduction by causing (20) _____,while others shift the (21) _____ of the host species. Parasites have evolved (22) _____ to locate hosts and feed undetected. Hosts, too, have evolved traits that minimize the (23) _____ effects of a parasite.

A (24) _____ _____ is a species of bird that lays its eggs in the nest of a member of another species. The host then assumes the (25) _____ duties of the parasite bird. The presence of foreign nestlings (26) _____ the reproductive rate of the host species. Insects that lay eggs in other insects are called (27))_____. Their larvae hatch and feed on the host, eventually leading to its death. Parasites and parasitoids are used as a form of (28) _____ pest control. They target specific pest species and have fewer harmful effects than (29) _____ pesticides.

45.8. ECOLOGICAL SUCCESSION [pp.820-821]

45.9. SPECIES INTRODUCTION, LOSS, AND OTHER DISTURBANCES [pp.822-823]

Boldfaced Terms

pioneer species _____

_____

primary succession _____

_____

secondary succession _____

_____

intermediate disturbance hypothesis _____

_____

keystone species _____

_____

indicator species _____

_____

exotic species _____

_____

Choice [p.820-821]

For exercise 1-6 choose from the following terms.

    a. primary succession   b. secondary succession

1. _____ Following a disturbance, a patch of habitat or a community recovers

2. _____ Successional changes begin when a pioneer population colonizes a barren habitat

3. _____ Many plants in this succession arise from seeds or seedlings that are already present when the process began

4. _____ A successional pattern that occurs in abandoned fields, burned forests, and tracts of land destroyed by volcanic eruptions

5. _____ May occur on a new volcanic island or on land exposed by the retreat of a glacier

6. _____ Once established, the pioneers improve conditions for other species, build soils, and often set the stage for their own replacement

Matching [pp.820-823]

Choose the most appropriate statement for each term.

7. _____ pioneer species

8 _____ primary succession

9. _____ secondary succession

10. _____ climax community

11. _____ keystone species

12. _____ exotic species

13. _____ indicator species

A. A species that has been dispersed from its home and become established elsewhere

B. Colonization of barren habitat that does not contain any soil

C. A community of species that will persist over time in a given geographic area

D. Colonizers of newly vacated habitats

E. A successional mechanism in which a disturbed area recovers

F. A species that has a disproportionately large effect on a community relative to its abundance

G. The first to do poorly when environmental conditions change in a community

Short Answer [pp.820-821]

14. Name three factors that affect succession. _____

_____

_____

_____

15. Explain the Intermediate Disturbance Hypothesis.

_____

_____

_____

_____

45.10. BIOGEOGRAPHIC PATTERNS IN COMMUNITY STRUCTURE [pp.824-825]

Boldfaced Terms

area effect _____

_____

equilibrium model of island biogeography _____

_____

distance effect _____

_____

Short Answer [p.824-825]

1. List three factors responsible for creating the higher species-diversity values correlated with distance from the equator, both on land and in the seas.

_____

_____

_____

2. After considering the distance effect, the area effect, and species-diversity patterns related to the equator answer the following question. Two islands (B and C) are the same size, topography and age and are equidistant from the African coast (A), as shown in the diagram. Which island will have the higher species-diversity values?

_____

_____

_____

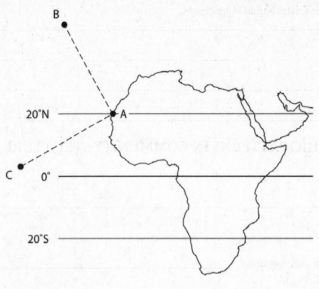

3. How does the distance effect and the area effect help shape species diversity on an island? _____

_____

_____

_____

_____

_____

SELF-TEST

___ 1. All the populations of different species that occupy and are adapted to a given habitat are referred to as a(n) _____.
a. biosphere
b. community
c. ecosystem
d. niche

___ 2. The range of all factors that influence whether a species can obtain resources essential for survival and reproduction is called the _____ of a species.
a. habitat
b. niche
c. carrying capacity
d. ecosystem

___ 3. _____ is one of the one-way relationships in which one species benefits and the other is directly harmed.
a. Commensalism
b. Competitive exclusion
c. Parasitism
d. Mutualism

___ 4. A lopsided interaction that directly benefits one species but does not harm or help the other much, if at all, is _____.
a. commensalism
b. competitive exclusion
c. predation
d. mutualism

___ 5. An interaction in which both species benefit is best described as _____.
a. commensalism
b. mutualism
c. predation
d. parasitism

___ 6. Stinging wasps have yellow and black striped abdomens. Some harmless species of insects also exhibit yellow and black striping on the abdomen to avoid predation. This is an example of_____.
a. warning coloration
b. mimicry
c. camouflage
d. herbivory

7. When an inexperienced predator attacks a yellow-banded wasp, the predator receives the pain of a stinger and will not attack a similar looking insect again. This is an example of _____.
   a. herbivory
   b. camouflage
   c. chemical defense
   d. warning coloration

8. During the process of secondary succession, _____.
   a. pioneer populations grow in habitats that cannot support most species
   b. pioneers set the stage for their own replacement
   c. species will colonize bare soil
   d. a disturbed area recovers

9. Gause used two species of Paramecium in a study that described _____.
   a. interspecific competition and competitive exclusion
   b. resource partitioning
   c. the establishment of territories
   d. coevolved mutualism

10. _____ are insects that lay their eggs in other insects.
    a. Parasites
    b. Social parasites
    c. Vectors
    d. Parasitoids

11. Which organism is the best example of a keystone species?
    a. Sea star
    b. Mussels
    c. Barnacles
    d. Limpets

12. Which of the following is NOT an exotic species?
    a. Kudzu
    b. Gypsy moth
    c. American Elm
    d. Nutria

CHAPTER OBJECTIVES / REVIEW QUESTIONS

1. Define the terms *habitat* and *community*. [pp.809-810]
2. Define the various "two species" interactions. [p.810]
3. Define ecological niche. [p.812]
4. Distinguish between interference competition and exploitative competition. [p.812]
5. Explain how G. Gause's work led to the formation of the competitive exclusion theory. [p.812]
6. Define *resource partitioning*. [p.813]
7. Explain why predator-prey cycles represent coevolution. [pp.814-815]
8. Understand the three models of predator responses to increases in prey density. [p.814
9. Explain how camouflage, mimicry, chemical defenses, and warning coloration all relate to the concept of an evolutionary arms race. Give examples of each. [p.816-817]
10. Explain the influence of a parasite on a population. [p.818
11. Explain the role of a pioneer species in ecological succession. [p.820]
12. Distinguish between primary and secondary succession. [p.820]
13. List the conditions that must be present in order to call a community a climax community. [p.820]
14. Explain the significance of the intermediate disturbance hypothesis. [p.821]
15. List and define the three variables that will influence succession. [p.821]
16. Distinguish between a keystone species and an exotic species. [pp.822-823]
17. How does an indicator species predict environmental health?. [p.823]
18. List three reasons why distance from the equator corresponds to species richness. [p.824]
19. Distinguish between a distance effect and an area effect. [p.824]

INTEGRATING AND APPLYING KEY CONCEPTS

1. Describe the niche that you feel humans belong in. What do you think will happen to community stability as we expand our niche?

2. Give an example of a mutualistic relationship between humans and another species. What would happen to the human population if the other species were to go extinct?

3. What does the section on community instability tell you about the influence of humans on communities in equilibrium? What can be done to lessen this impact?

4. Why do you think exotic species like Kudzu or the Gypsy moth spread uncontrollably throughout a new environment? Consider such concepts as predation, coevolution and competition.

46

ECOSYSTEMS

INTRODUCTION

Ecosystems represent an interaction between the biotic and abiotic factors of a given region. In this chapter you will explore how the abiotic factors, namely energy and nutrients, influence the structure of an ecosystem. The first part of the chapter explains the flow of energy in an ecosystem; the second part takes a global perspective on the flow of nutrients in biogeochemical cycles.

STUDY STRATEGIES

- Read all of chapter 45 with the goal of familiarizing yourself with the boldface terms.
- Define habitat.
- Recognize the various types of symbiosis.
- Explain mutualism and give an example of this type of symbiosis.
- Define niche.
- Explain the theories of competitive exclusion, resource partitioning, and character displacement as they relate to an organism's niche.
- Discuss predator-prey interactions, functional responses, coevolution, and cycles.
- Identify strategies of predators and prey regarding fitness.
- Compare and contrast predator-prey interactions with herbivore-plant interactions.
- Explain parasitism and give an example of this type of symbiosis.
- Compare and contrast parasitism to predator-prey interactions.
- Explain ecological succession and identify the differences between primary and secondary succession.
- Define a keystone species.
- Recognize the importance of intermediate disturbance events for increased species diversity.
- Identify the problems that exotic species can cause in a new habitat.
- Discuss regional and global patterns of biodiversity.

FOCAL POINTS

- Figure 46.3 [p.830] demonstrates how energy flows through an ecosystem while nutrients are cycled within the ecosystem.
- Figure 46.8 [p.836] illustrates the relative population sizes and energy flow in an ecosystem through the use of biomass and energy pyramids.
- Figure 46.9 [p.836] demonstrates energy flow patterns through an aquatic freshwater ecosystem.
- Figure 46.12 [p.837] outlines the water cycle
- Figure46.13 [p.830] is a representation of the carbon cycle. Figure 46.16 [p.841] shows the relationship between rising atmospheric CO_2 levels and increasing global temperatures.
- Figures 46.17 [p.842] and 46.19 [p. 844] illustrate, respectively, the nitrogen and phosphorus cycles.

INTERACTIVE EXERCISES

46.1. TOO MUCH OF A GOOD THING [p.829]

46.2. THE NATURE OF ECOSYSTEMS [pp.830-831]

46.3. THE NATURE OF FOOD WEBS [pp. 832-833]

46.4. ENERGY FLOW [pp. 834-835]

Boldfaced Terms

eutrophication _____

_____

ecosystem _____

_____

primary producers _____

_____

consumers _____

_____

detritivores _____

_____

decomposers _____

_____

trophic levels _____

_____

food chain _____

_____

food web _____

_____

grazing food web _____

_____

detrital food web _____

_____

primary production _____

biomass pyramid _____

_____

energy pyramid _____

_____

Matching [pp.833-835]

Choose the most appropriate statement for each term.

1. _____ primary producers
2. _____ consumers
3. _____ herbivores
4. _____ carnivores
5. _____ decomposers
6. _____ detritivores
7. _____ omnivores
8. _____ grazing food web
9. _____ ecosystem
10. _____ detrital food web
11. _____ primary production
12. _____ biomass pyramid
13. _____ energy pyramid
14. _____ trophic levels
15. _____ food chain
16. _____ food webs

A. Most energy flows from producers to detritivores
B. Consumers that dine on animals and plants
C. Feed on the tissues of other organisms
D. Hierarchy of feeding relationships
E. A system of cross-connecting food chains
F. Feed on g animal tissues
G. Dry weight of all the organisms at each trophic level
H. The autotrophs
I. Energy stored in producer tissues that flows to the herbivores
J. Refers to a straight-line series of steps by which energy stored in autotroph tissues passes on through higher trophic levels
K. Consumers that eat only plants
L. Illustrates how the amount of useable energy decreases as it moves from one trophic level to the next
M. Break down of organic remains and wastes of producers and consumers
N. An array of organisms and their physical environment, interacting by a one-way flow of energy and a cycling of materials
O. Feed upon small particles of organic matter
P. The rate at which producers capture and store energy

Labeling [p.835]

Provide the name of the participant indicated by each number in the following diagram.

17. _____

18. _____

19. _____

20. _____

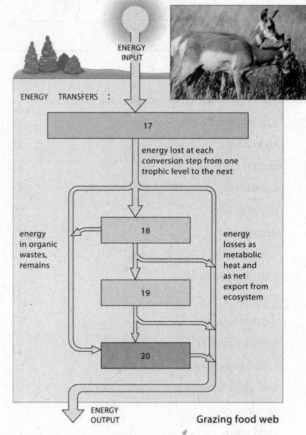

Grazing food web

Short Answer [p.835]

21. Explain why energy transfers in aquatic systems are more efficient than terrestrial ones. _____

_____

_____

_____

22. Define eutrophication. What causes the process to speed up? _____

_____

_____

_____

46.5. BIOGEOCHEMICAL CYCLES [p.836]

47.6. THE WATER CYCLE [pp.836-837]

Boldfaced Terms

biogeochemical cycle _____

_____

water cycle _____

_____

watershed _____

_____

aquifer _____

_____

groundwater _____

_____

runoff _____

_____

soil water _____

_____

Fill-in-the-Blanks [p.836]

In a (1) _____ cycle, an (2) _____ element moves from the environment, through an ecosystem, then back to the environment. (3) _____ and (4) _____ processes move elements to and from environmental (5) _____. In In the (6) _____ cycle, oxygen and hydrogen move globally as molecules of (7) _____ from organisms, land and bodies of water to the atmosphere and back. Gaseous forms of (8) _____ and (9) _____ are available to the ecosystems from the (10) _____. Phosphorus and other solid nutrients that have no gaseous form move in (11) _____ cycles. They accumulate on the (12) _____ and eventually return to land by way of geological (13) _____, which often takes millions of years. The Earth's (14) _____ is the biggest reservoir for nutrients that have sedimentary cycles. Movements of elements among reservoirs is much (15) _____ than movement through the living members of a community.

Matching [pp.836-837]

Match each of the following terms to the appropriate statement.

16. _____ salt water intrusion
17. _____ groundwater
18. _____ precipitation
19. _____ water cycle
20. _____ transpiration
21. _____ watershed
22. _____ aquifer
23. _____ runoff

A. Rain, snow or hail

B. A region where precipitation flows into a specific waterway

C. The loss of water by plants into the atmosphere

D. The movement of water on a global scale among oceans, freshwater reservoirs and the atmosphere

E. Seawater encroaching on groundwater reservoirs

F. The water that is found in soils and aquifers

G. Water that flows over soil without being absorbed

H. Porous rock that contains groundwater

Completion [pp.837]

For each of the following, place the correct numerical value in the blank.

24. _____ of the water Americans use goes directly into agriculture.

25. About _____ of the United States' population relies on groundwater as a source of drinking water.

26. _____ of the Ogallala aquifer has been overdrafted.

46.7. THE CARBON CYCLE [pp.838-839]

46.8. GREENHOUSE GASES AND CLIMATE CHANGE [pp.840-841]

Boldfaced Terms

carbon cycle _____

_____

atmospheric cycle_____

_____

greenhouse gas_____

_____

global climate change _____

_____

Matching [pp.839-841]

Choose the most appropriate statement for each term.

1. _____ greenhouse gases
2. _____ biogeochemical cycle
3. _____ carbon cycle
4. _____ marine sediments and sedimentary rocks
5. _____ greenhouse effect
6. _____ carbon dioxide (CO_2)
7. _____ oceans & the atmosphere
8. _____ global warming

A. The primary reservoir of carbon

B. Largest reservoirs of biologically available carbon

C. The movement of carbon on a global scale

D. One of the main greenhouse gases

E. A nutrient moves among environmental reservoirs and into and out of food webs

F. The warming of the Earth's lower atmosphere as a result of greenhouse gases

G. A long-term increase in temperatures near Earth's surface

H. Prevents the escape of heat energy form the Earth

Short Answer [p.840-841]

9. Describe the Greenhouse Effect. _____

_____

_____

_____

10. Why is the greenhouse effect necessary to life on earth? How can it be harmful? _____

_____

_____

_____

11. What is the primary reason for the accumulation of excess carbon dioxide in the atmosphere? _____

_____

_____

_____

46.9. NITROGEN CYCLE [p.842]

46.10. DISRUPTION OF THE NITROGEN CYCLE [p.843]

46.11. THE PHOSPHORUS CYCLE [p.844]

Boldfaced Terms

nitrogen cycle _____

_____

nitrogen fixation _____

_____

ammonification _____

_____

nitrification _____

_____

denitrification _____

_____

phosphorus cycle _____

_____

sedimentary cycle _____

_____

Matching [pp.842-844]

Choose the most appropriate statement for each term.

1. _____ nitrogen cycle
2. _____ nitrogen fixation
3. _____ ammonification
4. _____ sedimentary cycle
5. _____ nitrification
6. _____ denitrification
7. _____ phosphorus cycle

A. Ammonium is converted to nitrite and nitrate by bacteria

B. Bacteria convert nitrate or nitrite to gaseous forms of nitrogen

C. Microbes break down nitrogen-containing molecules and form ammonium

D. The movement of phosphorus from land to oceans and food webs, then back again

E. The conversion of N_2 to ammonia (NH_3) which is easily taken up by plants

F. The global movement of nitrogen among reservoirs and ecosystems

G. Biogeochemical cycle in which rocks are the major reservoir

Short Answer [p.842]

8. Why are nitrogen-fixing bacteria important to higher plants, and in turn, the consumers that eat them?

_____

_____

_____

_____

9. Describe some of the adverse effects of the accumulation into ecosystems of excess nitrous oxide and nitrate due to human activities.

_____

_____

_____

_____

SELF-TEST

___ 1. An array of organisms and their physical environment, all interacting by a one-way flow of energy and a cycling of materials, is a(n) _____.
   a. population
   b. community
   c. ecosystem
   d. biosphere

___ 2. _____ ingest decomposing particles of organic matter.
   a. Herbivores
   b. Parasites
   c. Detritivores
   d. Carnivores

___ 3. The members of feeding relationships are structured in a hierarchy, the steps of which are called _____.
   a. organism levels
   b. energy source levels
   c. eating levels
   d. trophic levels

___ 4. In a grazing food chain, energy flows from _____.
   a. producers to detritivores and decomposers
   b. primary consumers to detritivores and decomposers
   c. producers to herbivores
   d. primary consumers to herbivores, then through carnivores

___ 5. Which of the following lives in a host and feeds on its tissues?
   a. Parasites
   b. Herbivores
   c. Producers
   d. Carnivores

___ 6. In a natural community, the primary consumers are _____.
   a. herbivores
   b. carnivores
   c. scavengers
   d. decomposers

Ecosystems   **569**

7. A straight-line series of steps of who eats whom in an ecosystem is sometimes called a(n) _____.
   a. trophic level
   b. food chain
   c. ecological pyramid
   d. food web

8. Of the 1,700,000 kilocalories of solar energy that entered an aquatic ecosystem in Silver Springs, Florida, investigators determined that about _____ percent of incoming solar energy was trapped by photosynthetic autotrophs. [
   a. 1
   b. 10
   c. 25
   d. 74

9. A biogeochemical cycle that deals with phosphorus and other nutrients that do not have gaseous forms is the _____ type.
   a. sedimentary
   b. hydrologic
   c. nutrient
   d. atmospheric

10. _____ is the nutrient enrichment of an ecosystem.
    a. Eutrophication
    b. Salinization
    c. Nitrification
    d. Ammonification

11. In the carbon cycle, carbon enters the atmosphere through_____.
    a. carbon dioxide fixation
    b. volcanic eruptions and fossil fuel burning
    c. oceans and accumulation of plant biomass
    d. the process of photosynthesis

12. Human activity has disrupted the nitrogen cycle in all but which of the following ways?
    a. Increase in nitrous oxide
    b. Enhanced Greenhouse Effect
    c. Destruction of the Ozone Layer
    d. Decrease in nitrates

13. Which of the following is NOT true of the Greenhouse Effect?
    a. Atmospheric CO_2 is higher in the summer months in the Northern Hemisphere
    b. Solar energy passing through the atmosphere warms the surface of the earth
    c. Heat energy is re-radiated from the earth into the atmosphere
    d. Greenhouse gases absorb and emit heat back toward the earth

14. Which state in the U.S. has highly overdrawn its groundwater supplies?
    a. Texas
    b. Nevada
    c. Washington
    d. Colorado

CHAPTER OBJECTIVES / REVIEW QUESTIONS

1. Distinguish among herbivores, carnivores, parasites, detritivores, and decomposers. [p.830]
2. Define the term ecosystem. [p.830]
3. Explain the difference between a food chain and a trophic level. [pp.830-831]
4. Understand the flow of energy and the connection between food chains and food webs. [pp.832-833]
5. Define eutrophication [p.829]
6. Discuss the movement of energy and nutrients through the components of an ecosystem. [p.830-831]
7. Distinguish between primary production, gross primary production and net primary production. [p.834]
8. Define biomass. [p.834]
9. Explain the differences between a biomass pyramid and an energy pyramid. [pp.834-835]
10. Using the Silver Springs ecosystem as an example, explain the path of energy flow. [p.835]
11. Explain the role of biogeochemical cycles in nutrient cycling. [p.836]
12. Distinguish between hydrologic, atmospheric, and sedimentary cycles with regards to the major types of reservoirs found in each. [p.836]
13. Explain the water cycle. [p.836-837]
14. Define watershed and aquifer. [p.836]
15. Explain the threats to the global water supply. [pp.836-837]
16. Outline the carbon cycle. [pp.838-839]
17. What is the role of fossil fuels in the carbon cycle? [p.839]
18. Explain how greenhouse gases contribute to the greenhouse effect. [pp.840-841]

19. Outline the evidence for global warming. [p.841]
20. Explain the nitrogen cycle using the terms nitrogen fixation, ammonification, nitrification, and denitrification. [p.842]
21. Explain how the use of fertilizers affects the biosphere. [p.843]
22. Outline the phosphorus cycle and give the biological importance of phosphorus. [p.44]

INTEGRATING AND APPLYING KEY CONCEPTS

1. Pick an ecosystem and indicate 5 organisms that can be found in your ecosystem from each of the following categories: producers, herbivores, carnivores, omnivores, and detritivores. Draw food chains that show the feeding relationship between the various members of the community. Connect the food chains into a larger food web.
2. Investigate the Biosphere project in Arizona. Design your own biosphere. What plant and animal species would you include? How would you control the abiotic factors that affect ecosystems?
3. List everything that you have just eaten for dinner. Indicate the trophic level where each organism is found. How can you decrease your ecological impact by adjusting your meal?

47

THE BIOSPHERE

INTRODUCTION

This chapter takes a big-picture look at the biosphere, or Earth. Starting with an explanation of how climate is influenced by air and ocean currents, and land form formations such as mountains, the chapter progresses to a discussion of the primary surface biomes. The study of water-based biomes follows, from freshwater streams and lakes to the deep ocean. Throughout the chapter you should focus on the key characteristics of each of the biomes. Keep in mind that while biomes might appear discreet and independent of each other, they are interconnected through biogeochemical cycles, atmospheric and hydrologic patterns, and pollutants in the biosphere.

STUDY STRATEGIES

- Read all of chapter 47 with the goal of familiarizing yourself with the boldface terms.
- Explain global atmospheric circulation patterns.
- Describe global oceanic circulation patterns and how these connect to the atmosphere and to land masses.
- Discuss the abiotic and biotic characteristics, as well as locations, of the following terrestrial biomes: deserts, grasslands, dry shrublands/woodlands, broadleaf forests, coniferous forests, tundra.
- Discuss the abiotic and biotic characteristics of the following aquatic biomes: freshwater, coastal, coral reefs, open ocean.
- Identify the adaptations of desert dwellers to arid conditions.
- Explain the unique characteristics of the desert crust communities.
- Compare and contrast temperate grasslands versus savannas.
- Compare and contrast semi-evergreen and deciduous forests to tropical rain forests.
- Compare and contrast arctic and alpine tundras.
- Compare and contrast the freshwater systems of lakes/ponds and streams/rivers.
- Recognize the different zones in the intertidal and open ocean ecosystems.
- Identify the major human impacts on each of the biomes profiled.

FOCAL POINTS

- Sections 47.1, 47.2 and 47.3 [pp. 849-853] explain how climate is influenced by three major factors: air currents, ocean currents, and landforms.
- Sections 47.4 through 47.10 [pp.854-863] outline terrestrial biomes.
- Sections 47.11 through 47.14 [pp.864-871] describe aquatic ecosystems.
- Section 47.15 [pp.872-873] gives one example of the interconnectedness among diverse parts of the biosphere.

INTERACTIVE EXERCISES

47.1. WARMING WATER AND WILD WEATHER [p.849]

47.2. GLOBAL AIR CIRCULATION PATTERNS [pp.850-851]

47.3. THE OCEAN, LANDFORMS AND CLIMATES [pp.852-853]

Boldfaced Terms

climate _____

_____

El Niño _____

_____

La Niña _____

_____

biosphere _____

_____

rain shadow _____

_____

monsoon _____

_____

Fill-in-the-Blanks [p.850]

(1) _____ refers to the average weather conditions over time for a given area. (2) _____ & (3) _____ are two of the most influential factors in determining a region's climate. Due to the rotation of the Earth and its tilt, the intensity of (4) _____ will vary from location to location. As a result of these factors, the Earth will (5) _____ more at the (6) _____ than at the poles.

Regional differences in warming is the start of (7) _____ circulation patterns. In general, warm air holds more (8) _____ than cold air. This helps explain why the tropics are so humid.

Labeling [p.851]

Several terms in the accompanying diagram have been replaced by numbers. Identify each of the missing terms.

9. _____

10. _____

11. _____

12. _____

13. _____

14. _____

15. _____

16. _____

17. _____

18. _____

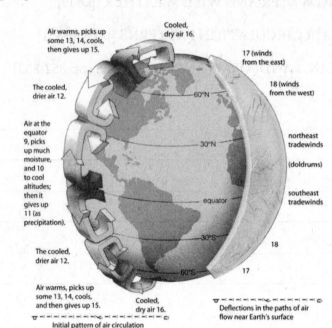

Air warms, picks up some 13, 14, cools, then gives up 15.

Cooled, dry air 16.

17 (winds from the east)

18 (winds from the west)

The cooled, drier air 12.

60°N

Air at the equator 9, picks up much moisture, and 10 to cool altitudes; then it gives up 11 (as precipitation).

30°N

northeast tradewinds

(doldrums)

equator

southeast tradewinds

30°S

18

The cooled, drier air 12.

60°S

17

Air warms, picks up some 13, 14, cools, and then gives up 15.

Cooled, dry air 16.

Deflections in the paths of air flow near Earth's surface

Initial pattern of air circulation

Matching [pp.850-853]

19. _____ Winter solstice

20. _____ Summer solstice

21. _____ Equinox

22. _____ Tradewinds

23. _____ Prevailing winds

A. Longest day of the year in the Northern Hemisphere

B. Winds that blow just north and south of the equator

C. Shortest day of the year in the Northern Hemisphere

D. Equal hours of day and night all over the earth

E. Westerlies and Easterlies

Matching [pp.852-853]

24. _____ monsoons

25. _____ rain shadow

26. _____ leeward

27. _____ doldrums

28. _____ ocean current

A. Air circulation patterns caused by the interaction of the equatorial sun with trade winds causing heavy rain

B. Mountain side facing away from the wind

C. Calm zone around the equator

D. An arid region located leeward of a high mountain

E. Large volumes of water responding to trade winds and prevailing winds

47.4. BIOMES [pp.854-855]

47.5. DESERTS [pp.856-857]

Boldfaced Terms

biome _____

_____

desert _____

_____

Matching [pp854-857]

1. _____ primary production
2. _____ biome
3. _____ soils
4. _____ deserts
5. _____ morphological convergence
6. _____ CAM plants
7. _____ tundra
8. _____ climate
9. _____ grasslands
10. _____ C4 photosynthesis

A. Mixtures of mineral particles and variable amounts of decomposing organic material (humus)

B. Plants that have adapted to dry conditions by opening their stomata at night

C. Most efficient method of production in hot, dry climates

D. Land biome with the thickest layer of top soil

E. Temperature and rainfall patterns for a given area

F. Geographic areas of land that share similar climate and dominant vegetation

G. Areas with the least annual rainfall and the highest temperatures

H. Varies among biomes due to differing climate and soils

I. North American cacti and African euphorbs both have water-storing stems

J. Biome with the lowest temperatures

Matching [pp.856-857]

11. _____ crust community
12. _____ annuals
13. _____ kangaroo rat
14. _____ cactus spines
15. _____ desert soil

A. Cyanobacteria, lichens, mosses, fungi

B. Adaptation to reduce evaporation and water loss

C. Highly efficient kidney designed to conserve water

D. Pebbles, shallow, poor soil

E. Plants that sprout and reproduce over a short growing season

Boldfaced Terms

grassland _____

_____

[dry shrubland _____

_____

dry woodland _____

_____

tropical rainforest _____

_____

temperate deciduous forests _____

_____

boreal forest _____

_____

arctic tundra _____

_____

alpine tundra _____

_____

permafrost _____

_____

Matching [pp.858-863]

1. _____ alpine tundra
2. _____ artic tundra
3. _____ taigas
4. _____ pine barrens
5. _____ tropical rainforests
6. _____ semi-evergreen forests
7. _____ temperate deciduous forests
8. _____ grasslands
9. _____ savannas
10. _____ dry shrublands
11. _____ dry woodlands

A. Sandy acidic soil supports a mixed scrub oak-pitch pine community

B. Characterized by low rainfall (less than 25 centimeters per year), permafrost, and fast plant growth during a brief growing season

C. Also called "swamp forests," these are the boreal forests with long, cold, dry winters and cool summers

D. This develops in the high mountain regions of the world; no permafrost

E. Evergreen broadleaf forests characterized by high rainfall (more than 130 centimeters per year), high humidity and temperature

F. A mix of trees that retain their leaves year-round and those that shed them before the dry season; found in the humid tropics of India and Southeast Asia

G. Characterized by an open canopy, rich soil, long growing season and 50-150 centimeters of rainfall per year

H. Broad belt of grasslands with scattered shrubs and trees

I. Also known as chaparral, fire-adapted vegetation, Short, d 25-60 centimeters of rain per year

J. Forms between deserts and temperate forests, thick, rich soils25-100 centimeters of rain per year

K. Short, drought tolerant trees that don't form a closed canopy, cool, wet winters and dry summers

47.11. FRESHWATER ECOSYSTEMS [pp.864-865]

47-12. COASTAL ECOSYSTEMS [p.866-867]

Boldfaced Terms

lake _____

_____

spring overturn _____

_____

thermocline _____

_____

The Biosphere   **577**

fall overturn _____

_____

estuary _____

_____

Matching [pp.864-867

Match each of the following terms to the appropriate statement.

1. _____ fall overturn
2. _____ eutrophic lakes
3. _____ lake
4. _____ spring overturn
5. _____ oligotrophic lake
6. _____ thermocline
7. _____ estuary
8. _____ mangrove wetlands

A. Newly formed lakes, deep, clear, and nutrient poor

B. Older, shallower, nutrient-rich lakes

C. Winds moving across a lake cause vertical movements of dissolved oxygen and nutrients

D. Cooling of the upper layer of a lake causes vertical movements of dissolved oxygen and nutrients

E. Coastal region where fresh and salt water mix

F. Salt-tolerant woody plants that live in sheltered areas along tropical coasts

G. Thermal stratification of a lake separating the upper, warmer water from the deeper, cooler water

H. Body of standing fresh water

Short Answer [pp.864-865]

9. List the key differences between an oligotrophic and eutrophic lake.

_____

_____

_____

_____

10. Explain what happens during the spring and fall overturns.

_____

_____

_____

_____

47.13. CORAL REEFS [pp.868-869]

47.14. THE OPEN OCEAN [pp.870-871]

47.15. OCEAN-SEA INTERACTIOINS [pp.872-873]

Boldfaced Terms

coral reef _____

_____

Ppelagic province _____

_____

benthic province _____

_____

sea mount _____

_____

hydrothermal vent _____

_____

Matching [pp.872-873]

Match each of the following terms to the appropriate statement.

1. _____ cholera
2. _____ ENSO
3. _____ upwelling
4. _____ Pelagic Province
5. _____ Benthic Province
6. _____ seamounts
7. _____ hydrothermal vents
8. _____ coral reefs
9. _____ symbionts
10. _____ bleaching
11. _____ invasive species

A. Species that lives inside of another

B. Highly diverse marine ecosystem; wave-resistant formations of calcium carbonate built by living organisms of the community

C. Ocean bottom

D. The location of superheated, mineral-rich water

E. The vertical movement of cold, deep, nutrient-rich water to the surface

F. The southern oscillation marked by increases in sea surface temperatures and changes in air circulation patterns

G. A disease caused by the bacteria Vibrio cholerae

H. When coral loses the dinoflagellates that live inside of its tissues

I. Submerged undersea mountains over 1000 meters tall

J. Open waters of the ocean

K. Species that are not native to a region

Labeling [p.884]

Label each numbered item in the accompanying illustration.

12. _____ province

13. _____ province

14. _____ zone

15. _____ zone

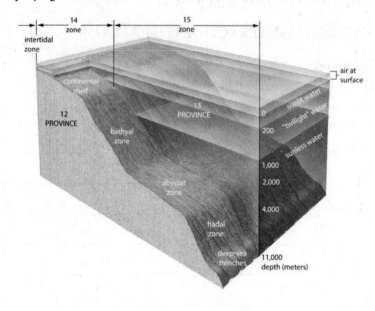

SELF-TEST

___ 1. An oligotrophic lake is _____.
    a. deep and clear
    b. rich in nutrients
    c. rich in phytoplankton
    d. low in oxygen all year

___ 2. In a(n) _____, nutrient-rich fresh water draining from the land mixes with seawater carried in on tides.
    a. pelagic province
    b. rift zone
    c. upwelling
    d. estuary

___ 3. A biome with broad belts of grasslands and scattered tress is known as a _____.
    a. warm desert
    b. savanna
    c. tundra
    d. taiga

___ 4. The _____ biome is located about 30 degrees north and south latitude, has limited vegetation, low rainfall, and rapid surface cooling at night.
    a. shrublands
    b. savanna
    c. taiga
    d. desert

___ 5. In evergreen broadleaf forests, _____.
    a. productivity is high
    b. litter does not accumulate
    c. soils are highly weathered and are poor nutrient reservoirs
    d. decomposition and mineral cycling are extremely rapid
    e. all of the above

___ 6. Permafrost would be found in which of the following?
    a. Alpine tundra
    b. Artic tundra
    c. Savannas
    d. Grasslands
    e. Boreal forests

___ 7. The lake's upper layer cools, the thermocline vanishes, lake water mixes vertically, and once again dissolved oxygen moves down and nutrients move up. This describes the _____.
    a. spring overturn
    b. summer overturn
    c. fall overturn
    d. winter overturn

8. Rain shadows are located on the
_____ side of high mountain ranges.
   a. windward
   b. leeward
   c. both windward and leeward

9. _____ are winds that change direction seasonally; low pressure causes moisture-laden air above an ocean to move inland, resulting in heavy rains.
   a. Geothermal ecosystems
   b. Upwelling
   c. Taigas
   d. Monsoons

10. All of the water above the continental shelves is in the _____.
   a. neritic zone of the benthic province
   b. oceanic zone of the pelagic province
   c. neritic zone of the pelagic province
   d. oceanic zone of the benthic province

11. Forests that consist of a mixture of broadleaf trees that retain leaves year round and deciduous trees are called _____.
   a. tropical deciduous forests
   b. temperate deciduous forests
   c. semi-evergreen forests
   d. evergreen broadleaf forests

12. Chemoautotrophic prokaryotes are the starting point for _____.
   a. hydrothermal vent communities
   b. desert communities
   c. lake communities
   d. coniferous forest communities

13. Which soil type will have an alkaline, deep topsoil layer rich in humus?
   a. grassland
   b. desert
   c. coniferous forest
   d. deciduous forest

14. During which event does the water temperature of the Pacific Ocean rise, evaporation increases, and the air pressure drops?
   a. ENSO
   b. La Nina
   c. Monsoons
   d. fall overturn

CHAPTER OBJECTIVES / REVIEW QUESTIONS

1. Explain the factors that contribute to climate. [p.850]
2. Distinguish between an El Niño and a La Niña event . [p. 849]
3. Explain how ocean currents and mountain ranges influence climate. [pp.852-853]
4. Define a biome. [p.854]
5. Describe how the earth's seasons are a result of the tilt of the earth's axis during its annual path around the sun. [p.850]
6. Give the basic characteristics of a desert, dry shrubland, dry woodland, grassland, and savanna. [pp. 856-859]
7. Give the basic characteristics of the following forest types: evergreen broadleaf, semi-evergreen, temperate deciduous and tropical deciduous. [p.860]
8. Give the general characteristics of a coniferous forest. [p.862]
9. Distinguish between the arctic tundra and the alpine tundra. [p.863]
10. Explain why spring and fall overturns are important for lakes. [p.865]
11. Distinguish between an oligotrophic and eutrophic lake. [p.864]
12. Define eutrophication. [p.864]
13. Describe the structure of a stream. [p.865]
14. Define estuary and explain its importance. [p.866]
15. List the three zones of a rocky shore. [p.867]
16. Describe how coral reefs are formed and their importance. [p.868-869]
17. Report on the status of the coral reefs around the world. [p.869]
18. Explain what is meant by a "southern oscillation." [p.872]
19. Describe how *Vibrio cholerae* causes the disease cholera. [p.872]
20. Rxplain how cholera outbreaks in Louisiana and Bagladesh are connected. [p.873]

INTEGRATING AND APPLYING KEY CONCEPTS

1. Why is deforestation such a threat to tropical rainforests?
2. Consider the course of the Mississippi River System. Explain how farming practices in the Midwest will impact the Gulf of Mexico. What can we do as individuals that will lessen the negative impact we are having on the Gulf of Mexico's estuary?
3. Compare the environmental conditions necessary to form a tropical rainforest versus a deciduous forest.
4. What factors contribute to the immense biodiversity of tropical rainforests?

48

HUMAN IMPACTS ON THE BIOSPHERE

INTRODUCTION

Currently, humans are the dominant species on the planet. There is almost no habitat on the planet that we have not ventured into or impacted in some fashion. Many of our actions have led to damage, destruction or the alteration of many of the Earth's ecosystems. As a result of these changes, many species on Earth are facing extinction in the near future. This chapter focuses on some of the impacts that humans have had on the Earth and the many species that inhabit it. It also tries to address how we can alter our behaviors in order to sustain more of the planet's natural resources.

STUDY STRATEGIES

- Read all of chapter 48 with the goal of familiarizing yourself with the boldface terms.
- Summarize the reasons for the current extinction crisis and causes of species decline.
- Define mass extinction, endemic species, endangered species, and threatened species.
- Discuss the harmful ways that human use land and the types of environmental problems that result from these practices.
- Identify major pollutants, their sources, and the effects on both the environment and humans.
- Explain bioaccumulation and biomagnification.
- Recognize the problems posed by trash and its disposal.
- Discuss the role of the ozone layer.
- Explain the genesis of the ozone hole and near-ground ozone pollution.
- Summarize the global effects of climate change.
- Recognize the role of conservation biology in solving many environmental problems as they relate to biodiversity.
- Identify practices that can reduce the human ecological footprint and those that you are able to participate in.

FOCAL POINTS

- Figure 48.2 [p. 878] is a graphic representation of the major extinctions and adaptive radiations during earth's history. Table 48.1 [p.880] summarizes the Red List of globally threatened species.
- Figure 48.9 [p.884] details the location of acid rain patterns in the U.S.
- Figure 48.12 [p.886] shows the levels of atmospheric CFC pollutants over the last three decades.
- Figure 48.14 depicts the conservation status of ecoregions around the globe.
- Section 48.9 [pp. 890-891] discusses several sustainable solutions to help maintain biodiversity.

INTERACTIVE EXERCISES

48.1 A LONG REACH [p.877]

48.2. THE EXTINCTION CRISIS [pp.878-879]

48.3. CURRENTLY THREATENED SPECIES [pp.880-881]

48.4. HARMFUL LAND USE PRACTICES [pp.882-883]

Boldfaced Terms

mass extinction _____

_____

endangered species _____

_____

threatened species _____

_____

endemic species _____

_____

desertification _____

_____

Fill-in-the-Blanks [pp.878-879]

(1) _____ is a natural process. (2) _____ evolve and become extinct on a regular basis.   It is estimated that (3) _____% of all species ever present on Earth have become extinct.   (4) _____ mass extinctions have occurred and they mark the boundaries for (5) _____ time periods.   The greatest mass extinction occurred at the end of the (6) _____.   It is believed to have been caused by (7) _____. The mass extinction at the end of the Cretaceous may have been caused by a(n) (8) _____.   Scientists believe that we are in the midst of a (9) _____ mass extinction. The current extinctions are due primarily to the activities of (10) _____ and their impact upon the planet.

Matching [pp.894-895]

Match the current threat to species with the appropriate species. Some threats will have more than one species associated with it.

11. _____ habitatfragmentation/loss/destruction

12. _____ habitat degradation

13. _____ overharvesting

14. _____ exotic species

A. Hawaiian Koa bug parasitoid

B. Decrease in the panda population

C. White abalone

D. Rat

E. Decrease in the prairie fringed orchid population

F. Texas blind salamander

G. European brown trout

H. Ivory billed woodpecker

I. Atlantic codfish

J. Passenger pigeon

Short Answer [p.882]

15. Explain how human activities can cause desertification.

_____

_____

_____

_____

16. List several things humans are doing that degrade habitats and threaten species.

_____

_____

_____

_____

17. What are some effects of deforestation?

_____

_____

_____

_____

Interpreting Diagrams [p.878]

Answer questions 18 - 20 based on the following diagram.

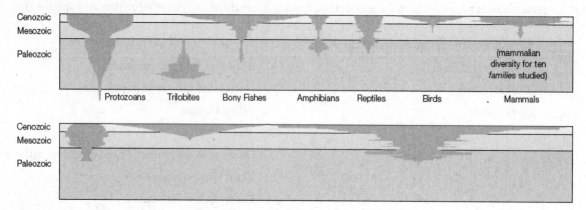

Species diversity over time for a sampling of taxa.

18. During which time periods do you see an adaptive radiation for:

   a. Angiosperms? _____

   b. Mammals? _____

   c. Insects? _____

19. During which time periods do you see a mass extinction for:

   a. Trilobites? _____

   b. Amphibians? _____

   c. Insects?_____

20. Which time period shows the greatest overall biodiversity? _____

48.5. TOXIC POLLUTANTS [pp.884-885]

48.6. OZONE DEPLETION AND POLLUTION [p.886]

48.7. EFFECTS OF GLOBAL CLIMATE CHANGE [p.887]

Boldfaced Terms

pollutants _____

_____

acid rain _____

_____

bioaccumulation _____

_____

biological magnification _____

_____

ozone layer _____

_____

Matching [pp.884-8877]

Match the term with the correct description.

1. _____ chlorofluorocarbons

2. _____ sulfur dioxides and nitrogen oxides combined with water

3. _____ point source

4. _____ bioaccumulation

5. _____ global climate change

6. _____ biomagnifications

A. Rising sea levels, acid seawater, melting glaciers

B. Origin of pollutants

C. Ozone thinning

D. Pollutant concentration increases with trophic level

E. Pollutants stored in tissues of biological organisms

F. Acid rain

Fill-in-the-Blanks [pp.878-879]

Ecosystems are delicately balanced, the health of the community depending on the proper flow of energy and cycling of nutrients. (7) _____ are man-made or natural substances that can enter the air, water or soil in greater than normal amounts and disturb that balance. If the source of the pollutant is easily identifiable, and easy to control, it is known as a (8) _____. Widespread release of pollutants, more difficult to control, come from (9) _____. (10) _____ and (11) _____ are common air pollutants that, when combined with atmospheric water, form (12) _____. Other pollutants find their way into the food chain and (13)

Human Impacts on the Biosphere   **587**

_____ in the tissues of living organisms. When fat-soluble pollutants like DDT are passed up the food chain, they concentrate in increasingly larger amounts, a process called (14) _____.

Natural wastes produced by a community are decomposed and recycled. The mass of garbage produced by humans is dumped into landfills or winds up in (15) _____, such as rivers, streams and the ocean. Items made of plastic do not decompose, and pose a threat to (16) _____, such as waterfowl, seabirds and sea mammals.

The ozone layer of the atmosphere protects the earth from excessive (17)) _____ radiation that can cause (18) _____. The thickness of the ozone layer varies seasonally. Due to the presence of certain pollutants, particularly (19) _____, it has been thinning steadily over the last several decades. The thinnest area over Antarctica is known as a(an) (20) _____. In 1987, a worldwide agreement banned the use of CFCs.

The average temperature of the earth is increasing, largely due to the accumulation of greenhouse gases in the atomosphere. This increase is causing (21) _____. (22) _____ at the poles are melting causing (23) _____ to rise. Due to the presence of fresh meltwater and increased evaporation at the equator, the concentration of (24) _____ in seawater is changing. The (25) _____ of winds is changing as is the pattern of global (26) _____. Marine and terrestrial ecosystems are being stressed. Spring leafing-out and blooming times are changing in (27) _____. (28) _____ patterns and breeding times are shifting in animals. Humans are not exempt from these stresses. Increases in (29) _____ and (30) _____ are expected.

48.8. CONSERVATION BIOLOGY [pp.888-889]

48.9. REDUCING NEGATIVE IMPACTS [pp.890-891]

Boldfaced Terms

biodiversity _____

_____

conservation biology _____

_____

ecological restoration _____

_____

[hot spot _____

_____

Short Answer [p.888]

1. What are the three levels that measure a region's biodiversity?

_____

_____

_____

2. Indicate the 3 ways that Conservation Biology is trying to address the rate of decline in biodiversity.

_____

_____

_____

_____

Fill-in-the-Blanks [pp.888-891]

(3) _____ is considered the biological wealth of a region or nation. A person who monitors biodiversity and seeks to maintain it is a (4) _____. These scientists have identified 867 distinctive land (5) _____, areas characterized by climate, geography and species composition of their communities. By monitoring these areas, biologists can assess regions that are threatened with damage to the environment and the species that inhabit them. A(n) (6) _____ is an area with unique species under great threat of destruction,. Once identified, these areas can become a priority for (7) _____ efforts. These areas may be set aside and protected by governments or individuals in an effort to (8) _____ them before they are destroyed. Sometimes an ecosystem is so damaged or destroyed that conservation efforts cannot maintain its (9) _____. (10) _____ efforts are designed to renew a natural ecosystem that has been destroyed or damaged by restoring integrity of the soil, water, and community. (11) _____ is the concept of meeting the needs of this generation without reducing the ability of the next generation to meet its needs. Our intense use of (12) _____ has a negative impact on biodiversity. Mining minerals removes the surface soil and vegetation from an area creating an ecological (13) _____. (14) _____ is a means of reducing this impact by reducing consumption. Burning fossil fuels contributes to global warming and acid rain. Reducing your (15) _____ is another way to promote sustainability.

SELF-TEST

___ 1. How many major mass extinctions has the Earth experienced in the past?
   a. 5
   b. 7
   c. 2
   d. 8

___ 2. Which species status is of the greatest concern?
   a. Threatened
   b. Vulnerable
   c. Endangered
   d. Stable

___ 3. Which is not a current threat to species?
   a. Habitat loss
   b. Habitat fragmentation
   c. Overharvesting
   d. Ecotourism

___ 4. Which is an example of an extinct species?
   a. Giant Panda
   b. Brown tree snake in Samoa
   c. Golden trout in California
   d. Passenger pigeon

___ 5. An endemic species is a species that _____.
   a. is widespread
   b. lives only in the local region where it evolved
   c. has been around for a long time
   d. is evolving

___ 6. Species may be threatened by _____.
   a. habitat degradation
   b. introduced exotic species
   c. overharvesting
   d. all of the above

___ 7. The average U.S. citizen produces _____ pounds of trash per day.
   a. 2.1
   b. 4.6
   c. 11.6
   d. 8.2

___ 8. _____ is the conversion of grasslands or woodlands to desert-like conditions.
   a. Poaching
   b. Overharvesting
   c. Habitat destruction
   d. Desertification

___ 9. Deforestation results in _____.
   a. flooding
   b. landslides
   c. nutrient reduction
   d. all of the above

___ 10. A hot spot is_____.
   a. an area where the global temperature is greatly increased
   b. an area of increased migratory activity
   c. an area where species are under great threat of destruction
   d. an area marked by all of the above.

CHAPTER OBJECTIVES / REVIEW QUESTIONS

1. Understand how humans are accelerating the rate of extinction. [p.879]
2. Define mass extinction. [p.878]
3. Define adaptive radiation. [p. 878]
4. Describe the criteria for a species to be listed as endangered or threatened. [p.880]
5. Explain how habitat loss, habitat fragmentation, habitat degredation, overharvesting, exotic predators, and exotic species are all contributing to species extinction. [pp.880-881]
6. Define endemic species. [p.880]
7. Describe the process of desertification. [p.882]
8. Name some consequences of deforestation. [pp. 882-883]
9. How is acid rain formed? [p. 884]
10. How does biomagnification affect the presence of certain pollutants in the community? [pp. 884-885]
11. List the measureable levels of biodiversity. [p. 888]
12. List the goals of conservation biologists. [p.888]
13. . Define ecoregion [p.888]
14. Explain the importance of preserving hot spots. [p.888]
15. Identify the critical and endangered ecoregions around the world. [p.888]

16. Discuss the conservation techniques of preservation and restoration. [p. 889]
17. Describe the positive feedback cycle between drought and desertification. [p.882]
18. What is sustainability? [p. 890]
19. Name some activities you can do to promote sustainability. [p.891]

INTEGRATING AND APPLYING KEY CONCEPTS

1. Pick 5 endangered ecoregions and develop a plan to preserve them. Make sure you take into consideration the needs of the local populations.
2. How can we encourage more Americans to reduce, reuse, and recycle their resources? Devise a strategy to cut in half the amount of trash the average American produces each day.
3. Considering the rate of extinction on Earth today, how will humans be affected if this continues?

ANSWER KEY

Chapter 1   Invitation to Biology

1.1. THE SECRET OF LIFE ON EARTH [p.3]
1. Biology; 2. 20; 3. Are

1.2. LIFE IS MORE THAN THE SUM OF ITS PARTS [pp.4-5]
1. E; 2. C; 3. F; 4. H; 5. D; 6. A; 7. I; 8. B; 9. J; 10. G; 11. K; 12. F; 13. H; 14. C; 15. B; 16. D; 17. E; 18. J; 19. I; 20. A; 21. K; 22. G; 23. emergent properties; 24. atom; 25. molecules; 26. cell; 27. organisms; 28. tissues; 29. organs; 30. organ systems; 31. E; 32. C; 33. F; 34. A; 35. D; 36. B

1.3. HOW LIVING THINGS ARE ALIKE [pp.6-7]
1. work; 2. nutrients; 3. producers; 4. photosynthesis; 5. consumers; 6. C; 7. F; 8. D; 9. A; 10. E;11. B

1.4. HOW LIVING THINGS DIFFER [pp.8-9]
1. B; 2. C; 3. C; 4. B; 5. C; 6. A; 7. C; 8. D; 9. B; 10. E; 11. F; 12. A; 13. C

1.5. ORGANIZING INFORMATION ABOUT SPECIES [pp.10-11]
1. Linnaeus; 2. traits; 3. Because it is difficult to know whether or not organisms that are naturally separated could breed. Also, because organisms are constantly evolving, it is difficult to tell when two species may actually diverge from each other.

1.6. THE SCIENCE OF NATURE [pp.12-13]
1.7. EXAMPLES OF EXPERIMENTS IN BIOLOGY [pp.14-15]

1. G; 2. A; 3. D; 4. C; 5. F; 6. E; 7. B.; 8. A; 9. C; 10. D; 11. B; 12. H; 13. G; 14. F; 15. E; 16. critical thinking; 17. science; 18. hypothesis; 19. prediction; 20. experiment; 21. independent; 22. dependent; 23. control; 24. experimental; 25. data; 26. the scientific method

1.8. ANALYZING EXPERIMENTAL RESULTS [pp.16-17]
1. To reduce sampling error; 2. If variables are not isolated, sampling biases, like those described in question 1, might be introduced; 3. It means that the result has a very low probability of having happened due to chance or due to sampling error. 4. T; 5. Probability; 6. large; 7. does not; 8. Subjective

1.9. THE NATURE OF SCIENCE [pp.18-19]
1. Scientific theory refers to a time-tested, well-supported hypothesis in science. The common use of the word "theory" only means a guess; 2. A law of nature is a phenomenon that has been observed to occur in every circumstance without fail but for which we do not have a complete scientific explanation; 3. Limits to science include inability to answer questions with subjective answers or questions that are supernatural.

SELF-TEST
1. a; 2. c; 3. c; 4. a; 5. c; 6. a; 7. b; 8. d; 9. a; 10. a; 11. c; 12. a; 13. b; 14. d; 15. c; 16. b; 17. a; 18. a; 19. c

---

Chapter 2   Life's Chemical Basis

2.1. MERCURY RISING [p.23]
2.2. START WITH ATOMS [p.24-25]
1. K; 2. C; 3. O; 4. D; 5. L; 6. B; 7. F; 8. I; 9. J; 10. E; 11. G; 12. A; 13. M; 14. H; 15. N; 16. electrons; 17.

True; 18. cannot; 19. True; 20. is not; 21. T; 22. carbon

2.3. WHY ELECTRONS MATTER [pp.26-27]
1. B; 2. D; 3. E; 4. A; 5. F; 6. C;

7.

Element	Atomic Number	Atomic Mass	Number of Protons	Number of Neutrons	Number of Electrons
Sodium	11	22	11	11	11
Bromine	35	80	35	45	35
Carbon	6	12	6	6	6
Hydrogen	1	1	1	0	1
Oxygen	8	16	8	8	8
Helium	2	4	2	2	2
Chlorine	17	35	17	18	17

8. atom; 10. ion; 11. electrons; 12. inert; 13. free radicals;
14.

H   C   N

O   P   S

2.4. CHEMICAL BONDS: FROM ATOMS TO MOLECULES [p.28-29]

1. G; 2. D; 3. E; 4. A; 5. C; 6. B; 7. F;
8.

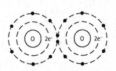

9. In a nonpolar covalent bond, the electrons are shared equally. Examples are hydrogen (H2) and nitrogen (N2). In a polar covalent bond, the atoms do not share the electrons equally, as in a water molecule (H2O).; 10. DNA contains many hydrogen bonds. Although on its own, each hydrogen bond is weak, together they support the structure well.; 11. A molecule is formed when two or more atoms join together chemically. A compound is a molecule that contains at least two different elements.

2.5. HYDROGEN BONDS AND WATER [pp.30-31]

1. polarity; 2. hydrophilic; 3. hydrophobic; 4. Temperature; 5. hydrogen; 6. stabilize; 7. Evaporation; 8. ice; 9. solvent; 10. solute; 11. cohesion; 12. weakest; 13. motion; 14. insulator; 15. intermingling; 16. homogeneous.

2.6. ACIDS AND BASES [pp.32-33]

1. F; 2. C; 3. G; 4. E; 5. B; 6. A; 7. D; 8. H; 9. lemon juice (2), acid rain (4), beer (5), black coffee (3), pure water (7), blood (8), seawater (8), ammonia (11), hair remover (12); drain cleaner (14); 10. If the pH changes from 3 to 5, the hydrogen ion concentration will be less by 100x; 11. A neutralization reaction is occurring; 12. Buffers work to keep an organism's pH in a certain range with little or no change.

SELF-TEST

1. c; 2. c; 3. a; 4. a; 5. d; 6. c; 7. d; 8. c; 9. c; 10. a; 11. b; 12. c; 13. e; 14. b; 15. e; 16. c

---

Chapter 3   Molecules of Life

3.1. FEAR OF FRYING [p.37]
3.2. ORGANIC MOLECULES [pp.38-39]
3.3. FROM STRUCTURE TO FUNCTION [pp.40-41]

1. organic; 2. carbon; 3. hydrogen; 4. organisms; 5. inorganic; 6. laboratories; 7. bonding; 8. four; 9. hydrocarbon; 10. carbon; 11. methane; 12. functional; 13. a carbon atom of an organic molecule; 14. polar; 15. hydrogen; 16. True; 17. The hydroxyl and carbonyl functional groups are polar. 18. True; 19. True; 20. Carbon makes up more than half of the important elements in living organisms4. A; 25. C; 26. D; 27. B; 28. E; 29. F;

H   H   H   H
| | | |
H-C-C-   -C-C-C-H
| | ‖ | |
H   H   O   H   H

31. carbonyl

OH
|
H-C-COOH
|
H-C-COOH
|
OH

32. carboxyl

H   H   H   H
| | | |
H-C-C-C-C=O
| | |
H   H   H

33. carbonyl

H   H   H
| | |
H-C-C-C-C-OH
| | |
H   H   H

34. hydroxyl

H   H   H
| | |
H-C-C-C-H
| | |
H NH₂ H

35. amino

H   O   O
| ‖   ‖
H-C-C-O-P-OH
| |
H   OH

36. phosphate

37. A; 38. G; 39. C; 40. B; 41. H; 42. F; 43. D; 44. I; 45. E; 46. Because carbon is such a versatile element: it can form many different shapes due to its 4 open valence spaces as well as the fact that it is a light element.

3.4. CARBOHYDRATES [pp.40-41]

1.

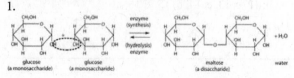

2. b; 3. b; 4. a; 5. c; 6. a; 7. c; 8. b; 9. b; 10. cellulose; 11. glucose; 12. hydrogen; 13. starch; 14. glycogen; 15. C; 16. E; 17. A; 18. G; 19. F; 20. B; 21. D

3.5. GREASY, OILY—MUST BE LIPIDS [pp.42-43]

1. a. unsaturated, b. saturated, c. unsaturated;

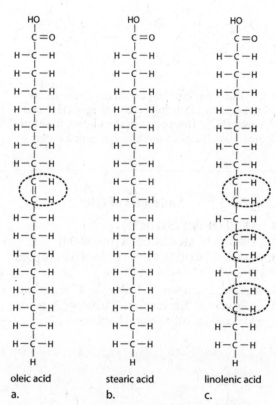

oleic acid a.   stearic acid b.   linolenic acid c.

2. d; 3. a; 4. c; 5. d; 6. a; 7. b; 8. d; 9. b; 10. d; 11. c; 12. d; 13. a; 14. a; 15. a; 16. c; 17. A; 18. Unsaturated; 19. Double; 20. Hydrophobic; 21. Because they can pack so tightly, they are able to more easily form plaques in blood vessels that can constrict blood flow and possibly lead to a heart attack.; 22. They have a hydrophilic head and a hydrophobic tail.;

3.6. PROTEINS—DIVERSITY IN STRUCTURE AND FUNCTION [pp.46-47]
3.7. WHY IS PROTEIN STRUCTURE SO IMPORTANT? [p.48]

1. a. R group, b. amine group, c. carboxyl group; 2. H; 3. D; 4. B; 5. G; 6. A; 7. I; 8. K; 9. E; 10. C; 11. J; 12. F; 13. structure; 14. enzymes; 15. polypeptide; 16. membrane; 17. mutations; 18. characteristics; 19. traits; 20. evolution; 21. a; 22. c; 23. b; 24. b; 25. d; 26. c; 27. d;
28.

amino acid   amino acid   dipeptide

29. Prions are misfolded infectious proteins that may arise spontaneously and can be inherited.

They may cause deterioration of mental and physical function and even death. Bovine spongiform encephalitis and Creutzfeldt-Jakob are two prion diseases.

3.8. NUCLEIC ACIDS [p.49]

1. a. nitrogenous base (adenosine), b. five-carbon sugar, c. phosphate groups; 2. A; 3. C; 4. E; 5. B; 6. D; 7. organic; 8. t; 9. t; 10. different; 11. energy carrier; 12. DNA codes for specific sequencing of amino acids which leads to the formation of a polypeptide.

SELF-TEST

1. c; 2. a; 3. d; 4. b; 5. d; 6. a; 7. b; 8. a; 9. d; 10. a; 11. d; 12. a; 13. c; 14. a; 15. b; 16. d; 17. c; 18. c; 19. b; 20. b; 21. e; 22. c;

Chapter 4   Cell Structure and Function

4.1. FOOD FOR THOUGHT [p.53]
4.2. CELL STRUCTURE [pp.54-55]
4.3. HOW DO WE SEE CELLS? [pp.56-57]

1. C; 2. G; 3. I; 4. F; 5. H; 6. J; 7. D; 8. B; 9. A; 10. E; 11. The four principles of the cell theory are: a) every organism consists of one or more cells, b) the cell is the smallest structural and functional unit of life, c) all cells come from pre-existing cells, d) cells pass hereditary material to their offspring when they reproduce.
12. D; 13. F; 14. B; 15. A; 16. C; 17. G; 18. E; 19. E; 20. D; 21. B; 22. A; 23. C; 24. a. centimeter, b. 100; 25.a. millimeter, b. 1,000; 26.a. micrometer, b. 1,000,000; 27.a. nanometer, b. 1,000,000,000.

4.4. INTRODUCING "PROKARYOTES" [pp.58-59]
4.5. INTRODUCING EUKARYOTIC CELLS [pp.60-61]

1. B; 2. F; 3. D; 4. E; 5. A; 6. H; 7. I; 8. C; 9. G; 10. D; 11. G; 12. A ; 13. H; 14. B; 15. C; 16. E; 17. I; 18. F; 19. cell wall; 20. chloroplast; 21. central water vacuole; 22. nucleus; 23. ribosomes; 24. rough endoplasmic reticulum (rough ER); 25. smooth endoplasmic reticulum (smooth ER); 26. Golgi body; 27. lysosome-like vesicle; 28. plasma membrane; 29. mitochondrion; 30. cytoskeleton; 31. nucleus; 32. ribosomes; 33. rough endoplasmic reticulum (rough ER); 34. smooth endoplasmic reticulum (smooth ER); 35. Golgi body; 36. lysosome; 37. plasma membrane; 38. centrioles; 39. mitochondrion; 40. cytoskeleton; 41. cilia; 42. cell wall; 43. flagellum; 44. capsule; 45. plasma membrane; 46. nucleoid region; 47. cytoplasm; 48. Each organelle has a unique and important

function that interacts intimately with all other organelles.

4.6. THE NUCLEUS [pp.62-63]
4.7. THE ENDOMEMBRANE SYSTEM [pp 64-65]
4.8. LYSOSOME MALFUNCTION [p.66]

1. The DNA of a gene carries the code for correct assembly of a polypeptide like the one used to make the lysosomal enzyme to digest gangliosides. In Tay-Sachs disease, that enzyme misfolds and is destroyed. Gangliosides accumulate in nerve cells instead of being broken down. Within months the build-up begins to damage nerve cells in a baby born with Tay-Sachs disease. Death follows. 2. L; 3. A; 4. H; 5. G; 6. E; 7. M; 8. N; 9. D; 10. B; 11. J; 12. F; 13. K; 14. I; 15. C; 16. 4; 17. 2; 18. 6; 19. 3; 20. 1; 21 7; 22. 5

4.9. OTHER ORGANELLES [pp.66-67]

1. b; 2. a; 3. c; 4. a; 5. e; 6. d; 7. b; 8. c; 9. d; 10. a; 11. c; 12. b; 13. b; 14. a; 15. e; 16. Endosymbiosis is a theory that mitochondria evolved from aerobic bacteria that were eaten (or entered as parasites) but were not digested by a larger heterotrophic eukaryotic cell. Over time, the captive cell evolved inside the host cell into what are now known as mitochondria. The inner mitochondrial membrane is a remnant of the original prokaryotic cell membrane. Mitochondria have their own unique DNA and self-replicate. 17. Because cells are not able to utilize the energy derived from food sources.

4.10. THE DYNAMIC CYTOSKELETON

4.11. CELL SURFACE SPECIALIZATIONS [pp.70-71]

1. C; 2. G. 3. B; 4. I; 5. K; 6. L; 7. F; 8. J; 9. H; 10. A; 11. E; 12. D; 13. E; 14. F; 15. A; 16. K; 17. L; 18. I; 19. G; 20. H; 21. B; 22. C; 23. J; 24. D; 25. tight; 26. tight; 27. adhering; 28; gap

SELF-TEST

1. d; 2. a; 3. d; 4. a; 5. c; 6. a; 7. b; 8. c; 9. b; 10. d; 11. c; 12. c; 13. d; 14. a; 15. b; 16. c; 17. c; 18. b; 19. d; 20. d; 21. b; 22. a;

Chapter 5   GROUND RULES OF METABOLISM

5.1. A TOAST TO ALCOHOL DEHYROGENASE [p.77]
5.2. ENERGY IN THE WORLD OF LIFE [pp.78-79]

1. alcohol; 2. alcohol dehydrogenase; 3. alcoholic hepatitis; 4. digests fats; 5. regulates the body's blood sugar level; 6. breaks down many toxic compounds; 7. binge drinking; 8. I; 9. I; 10. II; 11. I; 12. II; 13. I; 14. II; 15. II; 16. Energy transfers are never completely efficient. Some energy is converted to heat (thermal energy) which is not useful for doing work in biological systems. Therefore, the amount of useful energy is always decreasing.; 17. The measure of how much the energy of a system has become dispersed.; 18. The First Law of Thermodynamics states that energy cannot be created or destroyed. The Second Law of Thermodynamics states that all energy tends to spread out spontaneously, known as entropy; 19. Potential energy is that which is stored. Kinetic energy is the energy in motion; 20. The second law of thermodynamics relates to entropy.; 21. True; 22. True; 23. True; 24. In our world, energy can flow in only one direction, from the sun, to the producers, to the consumers.

5.3. THE ENERGY IN THE MOLECULES OF LIFE [pp.80-81]
5.4. HOW ENZYMES MAKE SUBSTANCES REACT [pp.82-83]

1. free energy; 2. endergonic; 3. exergonic; 4. activation energy; 5. Store; 6. Release; 7. exergonic; 8. endergonic; 9. exergonic; 10. exergonic; 11. exergonic; 12; energy; 13. activation; 14. photosynthesis; 15. reactants; 16. B; 17. E; 18. A; 19. C; 20. D; 21. activation energy without enzyme; 22. activation energy with enzyme; 23. energy released by the reaction; 24. catalyzed; 25. uncatalyzed; 26. substrate; 27. active site; 28. induced fit;

5.5. METABOLISM—ORGANIZED, ENZYME-MEDIATED REACTIONS [pp.84-85]
5.6. COFACTORS IN METABOLIC PATHWAYS [pp.86-87]

1. metabolism; 2. metabolic pathway; 3. linear; 4. cyclic; 5. reverse; 6. reactants; 7. rate; 8. feedback mechanisms; 9. enzymes; 10. inhibit; 11. allosteric site; 12. feedback inhibition; 13. electron; 14. oxidized; 15. reduced; 16. electron transfer chain; 17. most; 18. coenzymes; 19. ATP; 20. B; 21. E; 22. A; 23. C; 24. D; 25. ribose; 26. adenine; 27. triphosphate; 28. ATP; 29. C; 30. B; 31. D; 32. E; 33. A

5.7. A CLOSER LOOK AT CELL MEMBRANES [pp.88-89]

1. B; 2. D; 3. H; 4. I; 5. A; 6. C; 7. J; 8. G; 9. E; 10. L; 11. K; 12. F; 13. The cell membrane is a mosaic of phospholipids, steroids, proteins and other molecules. The behavior of the phospholipids of the plasma membrane contributes to its fluidity; 14. Proteins act as markers, selective transporters, receptors, and enzymes; 15. Membranes assist in homeostasis by maintaining the cellular environment in contrast to the external environment.

5.8. DIFFUSION AND MEMBRANES [pp.90-91]
5.9. MEMBRANE TRANSPORT MEMBRANES [pp.92-93]
5.10. MEMBRANE TRAFFICKING [pp.94-95]

1. concentration; 2. concentration gradient; 3. diffusion; 4. size; 5. temperature; 6. steepness of concentration gradient; 7. charge; 8. pressure; 9. permeability; 10. ions; 11. large, polar molecules; 12. gases; 13. nonpolar molecules; 14. passive; 15. active; 16. against; 17. endocytosis; 18. Exocytosis; 19. F; 20. A; 21. G; 22. E; 23. B; 24. D; 25. H; 26. towards B; 27. From A to B; 28. The sucrose will not move from one side to another due to its size; 29. increase; 30. increase; 31. T; 32. hypotonic; 33. T; 34. hypertonic; 35. D; 36. E; 37. C; 38. A; 39. B

SELF-TEST

1. d; 2. c; 3. e; 4. d; 5. a; 6. c; 7. d; 8. c; 9. d; 10. a; 11. d; 12. a; 13. d; 14. d; 15. d; 16. c; 17. d; 18. b; 19. c; 20. a; 21. d; 22. c; 23. c; 24. a

Chapter 6   Where it Starts – Photosynthesis

Impacts, Issues: Biofuels [p.106]

6.1. BIOFUELS [p.101]
6.2. SUNLIGHT AS AN ENERGY SOURCE [pp.102-103]
6.3. EXPLORING THE RAINBOW [p.104]
1. highest; 2. shorter; 3. longest; 4. G; 5. J; 6. D; 7. E; 8. A; 9; B; 10. I; 11. C; 12. H; 13. F; 14. K; 15. By using algae and bacteria, he was able to see which direction the organisms moved to photosynthesize and it turned out to be the blue and red wavelengths.

6.4. OVERVIEW OF PHOTOSYNTHESIS [p.105]
1. CO_2; 2. O_2; 3. $C_6H_{12}O_6$; 4. 6; 5. carbon dioxide; 6. oxygen; 7. glucose; 8. light-dependent; 9. light-independent; 10. thylakoid; 11. pigments (photopigments); 12. photosystems; 13. ATP; 14. stroma; 15. chloroplast membrane; 16. stroma; 17. grana; 18. Thylakoid; 19. A, C, D, E; 20. B, F, G; 21. E, F, G, I; 22. A, D, H

6.5. LIGHT-DEPENDENT REACTIONS [pp.106-107]
6.6. ENERGY FLOW IN PHOTOSYNTHESIS [p.108]
1. cyclic; 2. noncyclic; 3. ATP; 4. noncyclic; 5. NADPH; 6. oxygen; 7. Photosystem I; 8. Photosystem II; 9. Photosystem I; 10. 700; 11; 680; 12. electrons; 13. electron transfer chain; 14. hydrogen; 15. oxygen; 16. photolysis; 17. gradient; 18. stroma; 19. ATP synthases; 20. ATP; 21. Photosystem I; 22. H^+; 23. NADPH; 24. cyclic; 25. b; 26. c; 27. b; 28. c; 29. b; 30. a; 31. c; 32. b; 33. b; 34. C; 35. 6; 36. 7; 37. 8; 38. 4; 39. 2; 40. 1; 41. 5; 42. 3

6.7. LIGHT-INDEPENDENT REACTIONS: THE SUGAR FACTORY [p.109]
6.8. ADAPTATIONS: DIFFERENT CARBON-FIXING PATHWAYS [pp.110-111]
1. ribulose biphosphate (b); 2. phosphoglyceraldehyde (e); 3. phosphoglycerate (c); 4. plants with 3-carbon PGA as first stable intermediate (f); 5. plants with 4-carbon oxaloacetate as first stable intermediate (a); 6. crassulacean acid metabolism plants (d); 7. Calvin-Benson; 8. stroma; 9. ATP; 10. NADPH; 11. rubisco; 12. RuBP; 13. PGA; 14. ATP; 15. NADPH; 16. PGAL; 17. glucose; 18. RuBP; 19. sucrose; 20. starch; 21. C3 plants; 22. stomata; 23. CO_2; 24. O_2; 25. rubisco; 26. photorespiration; 27. carbon; 28. C4 plants; 29. mesophyll; 30. bundle-sheath cells; 31. Calvin-Benson; 32. CAM; 33. night; 34. C4; 35. CO_2

BIOFUELS [pp.112-113]
1. autotrophs; 2. photoautotroph; 3. photosynthesis; 4. sugar; 5. oxygen; 6. Heterotrophs; 7. chemoautotrophs; 8. methane (hydrogen sulfide); 9. hydrogen sulfide (methane); 10. billion; 11. cyclic photophosphorylation; 12. noncyclic photophosphorylation; 13. oxygen; 14. free radicals; 15. extinct; 16. aerobic respiration; 17. ozone; 18. ultraviolet; 19. ozone layer; 20. species; 21. carbon cycle; 22. industrial revolution; 23. carbon dioxide; 24. fossil fuels; 25. global warming

SELF-TEST
1. c, 2. b, 3. c, 4. a, 5. c, 6. d, 7. c, 8. b, 9. c, 10. d, 11. c, 12. b, 13. d, 14. d; 15. d; 16. a; 17. c; 18. a; 19. b; 20. b; 21. c; 22. c; 23. a

---

Chapter 7   How Cells Release Chemical Energy

7.1. MIGHTY MITOCHONDRIA [p.117]
7.2. OVERVIEW OF CARBOHYDRATE BREAKDOWN PATHWAYS [pp.118-119]
1. Mitochondria are necessary because they produce cellular energy.; 2. Similarities – both aerobic respiration and anaerobic fermentation pathways begin with the same reactions, glycolysis, which breaks down glucose into two 3-carbon molecules called pyruvate. These reactions happen in the cytoplasm. Differences – Aerobic respiration ends in the mitochondria where oxygen serves as the final electron acceptor at the end of the pathway. Aerobic respiration is much more efficient, netting 36 molecules of ATP per molecule of glucose. Fermentation ends in the cytoplasm where a molecule other than oxygen serves as the final electron acceptor at the end of the pathway. Fermentation is much less efficient, netting 2 molecules of ATP per molecule of glucose.; 3. Cells will eventually die without mitochondria that are working efficiently. 3. ATP; 4. metabolic pathways; 5. aerobic; 6. anaerobic; 7. glycolysis; 8. glucose; 9. pyruvate; 10. mitochondria; 11. cytoplasm; 12. 36; 13. 2; 14. Krebs cycle; 15. acetyl-CoA; 16. NAD^+; 17. FAD; 18. electron transport chain; 19. ATP; 20. aerobic; 21. T; 22. 24; 23. T24.

	In	*Out*
Glycolysis	glucose, 2 ATP	4 ATP (2 net), 2 NADH, 2 pyruvate
Krebs Cycle	2 pyruvate	6 CO_2, 2 ATP, 8 NADH, 2 $FADH_2$
Electron Transfer Phosphorylation	2 NADH, 8 NADH, 2 $FADH_2$	32 ATP

7.3. GLYCOLYSIS—GLUCOSE BREAKDOWN STARTS [pp.120-121]

1. autotrophic; 2. glucose; 3. pyruvate; 4. ATP; 5. NADH; 6. glucose; 7. ATP; 8. PGAL; 9. phosphate; 10. hydrogen; 11. ATP; 12. NADH; 13. phosphate; 14. ATP; 15. substrate-level; 16. glucose; 17. pyruvate; 18. three; 19. D; 20. F; 21. B; 22. G; 23. H; 24. A; 25. E; 26. C

7.4. SECOND STAGE OF AEROBIC RESPIRATION [pp.122-123]
7.5. AEROBIC RESPIRATION'S BIG ENERGY PAYOFF [pp.124-125]

1. acetyl-CoA; 2. Krebs; 3. substrate-level; 4. 6 CO_2; 5. 2 ATP; 6. oxaloacetate; 7. electrons; 8. FAD; 9. NAD^+; 10. inner compartment; 11. inner membrane; 12. outer compartment; 13. outer membrane; 14. cytoplasm; 15. ATP; 16. oxygen (O_2); 17. $FADH_2$; 18. NADH; 19. electron transfer chain; 20. electron transfer phosphorylation; 21. active transport (pump); 22. inner; 23. inner; 24. outer; 25. ATP synthase; 26. 32; 27. oxygen; 28. water

7.6. ANAEROBIC ENERGY-RELEASING PATHWAYS [pp.126-127]

1. fermentation; 2. aerobic respiration; 3. glycolysis; 4. pyruvate; 5. CO2; 6. H2O; 7. yield of 2 ATP; 8. alcohol fermentation; 9. *Saccharomyces cerevisiae*; 10. CO2; 11. alcoholic; 12. lactate fermentation; 13. *Lactobacillus acidophilus*; 14. fast-twitch; 15. slow-twitch; 16. many; 17. aerobic respiration; 18. few; 19. lactate fermentation

7.7. ALTERNATIVE ENERGY SOURCES IN FOOD [pp.128-129

1. oxidizing; 2. Carbohydrates; 3. Lipids; 4. Protiens; 5. Coenzymes; 6. fatty acids; 7. glycerol; 8. glycolysis; 9. amino acids; 10. Krebs cycle; 11. acetyl-CoA; 12. pyruvate; 13. e; 14. c; 15. d; 16. a; 17. h; 18. f; 19. j; 20. i; 21. b; 22. g; 23. fats, in the form of fatty acids and glycerol, complex carbohydrates, and proteins.

SELF-TEST

1. c; 2. c; 3. d; 4. b; 5. c; 6. c; 7. d; 8. a; 9. c; 10. a; 11. a; 12. d; 13. c; 14. c; 15. a; 16. d; 17. b; 18. d

---

Chapter 8　　DNA Structure and Function

8.1. A HERO DOG'S GOLDEN CLONES [p.133]
8.2. EUKARYOTIC CHROMOSOMES [pp.134-135]

1. Chromosomes are condensed forms of DNA that, when replicated, form sister chromatids that are joined at the centromere. Histones are the proteins that DNA initially winds around to form structures known as nucleosomes; 2. Diploid refers to having two sets of each different type of chromosome; 3. Autosomes are those chromosomes which are not involved in determining sex and which are the same in males and females. Sex chromosomes differ between males and females.

8.3. THE DISCOVERY OF DNA'S FUNCTION [pp.136-137]
8.5. THE DISCOVERY OF DNA'S STRUCTURE [pp.138-139]

1. deoxyribose sugar, phosphate group, nitrogen base.; 2. guanine (pu); 3. cytosine (py); 4. adenine (pu); 5. thymine (py); 6. deoxyribose (B); 7. phosphate group (G); 8. purine (C); 9. pyrimidine (A); 10. purine (E); 11. pyrimidine (D); 12. mucleotide (F); 13. virus; 14. bacterial; 15. hereditary; 16. genetic; 17. protein; 18. viral DNA; 19. ^{32}P; 20. bacteriophages; 21. ^{35}S; 22. ^{32}P; 23.

DNA; 24. protein; 25. Miescher – found that nuclei contain an acidic substance composed mostly of nitrogen and phosphorus. Later than substance would be called deoxyribonucleic acid (DNA).; 26. Griffith – discovered that there is a substance in cells that encodes the information about traits that parents pass to offspring.; 27. Avery (*et. al*) – discovered that the "transforming principle" is DNA.; 28. Hershey and Chase – determined that DNA, not protein, is the material of heredity common to all life on Earth.; 29. Rosalind Franklin used x-ray crystallography to calculate DNA's diameter, the distance between its chains and between its bases, the pitch of the helix and the number of bases in each coil. This data confirmed the work of Watson and Crick, causing the recognition of the structure of DNA; 30. Because she had died by the time the prize was given and the Nobel is not given posthumously.

8.5. DNA REPLICATION [pp.140-141]

1. DNA unwinds in only one direction. Since the two DNA strands are antiparallel, only one side will add new nucleotides continuously. The other side must add segments of new DNA (Okazaki fragments) a little at a time and join them with DNA ligase.; 2. Most DNA polymerases

immediately reverse synthesis and remove then replace mismatched nucleotides. Errors that are not corrected by the DNA polymerases and checked again by DNA repair mechanisms and most are corrected. Errors that remain even after these measures remain in the DNA as mutations; 3. b; 4. c; 5. d; 6. a; 7. b; 8. e; 9. d; 10. a;

11.   T-A   T-A
    G-C   G-C
    A-T   A-T
    C-G   C-G
    C-G   C-G
    C-G   C-G

12. thymine; 13. semiconservative; 14. T; 15. polymerases; 16. T

8.6. MUTATIONS: CAUSE AND EFFECT[pp.142-143]

Chapter 9   From DNA to Protein

14.1. THE APTLY NAMED RIPS [p.149]
14.2. DNA, RNA, AND GENE EXPRESSION [pp.150-151]

1. sequence (linear order); 2. transcription; 3. translation; 4. transcription; 5. translation; 6. amino acids; 7. gene expression; 8. enzymes; 9. a. rRNA; RNA molecule that associates with certain proteins to form the ribosome, the "workbench" on which polypeptide chains are assembled; b. mRNA; RNA molecule that moves to the cytoplasm, complexes with the tRNA and ribosome, carries the code for the amino acid sequence of the protein; c. tRNA; RNA molecue that moves into the cytoplasm, picks up a specific amino acid, and moves it to the ribosome where tRNA pairs with a specific mRNA code word for that amino acid; 10. RNA differs from DNA in that polymerase where transcriptions uses RNA polymerase, 3) transcription results in a single strand of RNA; DNA replication results in two DNA double helices; 3. pre-mRNAs contain introns that are excised from the final mRNA, a cap is added to the 5' end, and a poly-A tail is added; 4. DNA (E); 5. intron (B); 6. cap (F); 7. exon (A); 8. tail (D); 9. [mature] mRNA (C)

9.4. RNA AND THE GENETIC CODE [pp.154-155]

1. AUG-UUC-UAU-UGU-AAU-AAA-GGA-UGG-CAG-UAG; 2. met-phe-tyr-cys-asn-lys-gly-trp-gln-stop; 3. E; 4. B; 5. F; 6. D; 7. A; 8. C.

9.5. TRANSLATION: RNA TO PROTEIN [pp.156-157]

1. translation; 2. initiation; 3. elongation; 4. termination; 5. messenger; 6. transfer; 7. start

8.7. ANIMAL CLONING[pp.144-145]

1. 4; 2. 3; 3. 1; 4. 5; 5. 2; 6. F; 7. A; 8. C; 9. E; 10. H; 11. D; 12. B; 13. G; 14. Undifferentiated (stem) cells may be used to replace nerve cells in individuals with spinal cord damage. New organs might be "grown" to replace organs for transplant.; 15. Reproductive cloning can be used to produce "twinning" or to achieve a clone of a prize animal; 16. Organisms that are clones tend to age faster than normal and other significant health problems often develop.

SELF-TEST

1. d; 2. d; 3. a; 4. d; 5. c; 6. d; 7. b; 8. a; 9. d; 10. a; 11. a; 12. c; 13. b; 14. d; 15. a; 16. b; 17. c; 18. b; 19. a; 20. c.

it is single-stranded, it utilizes uracil instead of thymine, and it contains ribose as its 5-carbon sugar; 11. The sequence of As, Ts, Cs, and Gs, code for amino acids when are then assembled into polypeptides. These proteins are then used by the cell or delivered for transport outside of the cell.

9.3. TRANSCRIPTION: DNA TO RNA [pp.152-153]

1. Each nucleotide provides energy for its own attachment to the end of a growing strand. Transcription and DNA replication both use one strand of a nucleic acid as a template for synthesis of another.; 2. 1) In DNA replication only part of a DNA strand, not the whole molecule as in RNA, is used as a template, 2) DNA replication uses DNA

(AUG); 8. elongation; 9. polypeptide; 10. termination; 11. releasing; 12. enzyme; 13. polysomes; 14. phosphate-group; 15. GTP; 16. mRNA carries the "message" as it has been transcribed out of the nucleus into the cytoplasm. There, it meets up with a ribosomal subunit (made partially of rRNA) and the tRNAs carry amino acids to the correct codon on the mRNA.; 17. There is one start codon (AUG) for every protein. There are three stop codons that may be used to tell the protein assembly process when to stop.

9.6. MUTATED GENES AND THEIR PROTEIN PRODUCTS [pp.158-159]

1. B; 2. C; 3. B; 4. C; 5. A; 6. A; 7. Deletion is usually more damaging because it causes a frame shift in every codon past the mutation.; 8. Less likely because the third position is many times the "wobble" that still codes for the same amino acid.; 9. At the beginning because it would affect all of the

codons from the mutation to the end of the mRNA strand.; 10. A silent mutation is one that does not affect the protein product or how the protein acts; 11. DNA (H); 12. transcription (J); 13. intron (E); 14. exon (A); 15. mRNA (B); 16. rRNA (F); 17. tRNA (C); 18. ribosomal subunits (G); 19. amino acids (D); 20. anticodon (K); 21. translation (I); 22. polypeptide (L)

SELF-TEST
1. c; 2. b; 3. b; 4. c; 5. a; 6. b. 7. a; 8. c; 9. d; 10. d; 11. d; 12. b; 13. a; 14. a; 15. c; 16. b; 17. a; 18. b

Chapter 10   Gene Control

10.1. BETWEEN YOU AND ETERNITY [p.163]
10.2. SWITCHING GENES ON AND OFF [pp.164-165]
1. B; 2. E; 3. D; 4. A; 5. C; 6. Differentiation is the process by which cells become specialized.; 7. Many factors, including conditions in the cytoplasm and extracellular fluid, the type of cell, stage of development, and number of copies of the gene.; 8. A lot of protein is made quickly.; 9. Enzymes begin disassembling mRNA as soon as it arrives in the cytoplasm. The longer the poly-A tail, the longer the mRNA remains intact to be translated for production of protein; 10. Stability of the mRNA transcript and presence of microRNAs.

10.3. MASTER GENES [pp.166-167]
10.4. EXAMPLES OF GENE CONTROL IN EUKARYOTES [pp.168-169]
1. D; 2. A; 3. F; 4. B; 5. G; 6. E; 7. C; 8. X chromosome inactivation; 9. Barr body; 10. mosaic; 11. incontinentia pigmenti; 12. dosage compensation; 13. XIST; 14. RNA; 15. The presence of the SRY gene on the Y chromosome, which codes for testosterone production; 16. Environmental cues such as daylight; 17. Homeodomains are approximately 60 amino acids that can bind to a promoter or similar area on a chromosome. These are coded for by homeotic genes.; 18. It ensures that only one set of polypeptides is made in any given cell of the woman's body; 19. Knockout experiments help to determine the function of different genes, especially for the homeotic genes.

10.5. EXAMPLES OF GENE CONTROL IN PROKARYOTES [pp.170-171]
1. do not; 2. True; 3. True; 4. True; 5. not absorbed directly by the intestine; 6. True; 7. repressors; 8.declines; 9. transcription; 10. not everyone is lactose intolerant; 11. True; 12. True; 13. regulatory gene (I); 14. operator (B); 15. lactose enzyme genes (F); 16. promoter (H); 17. lactose operon (A); 18. repressor protein (G); 19. repressor-operator complex (E); 20. lactose (C); 21. mRNA (J); 22. RNA polymerase (D);

10.6. EPIGENETICS [pp.172-173]

1. Genes are inactivated by introducing a methyl group into the DNA strand.; 2. Environmental events can shape the protein behavior of individuals conceived during or after a famine. In this case, individuals and grandchildren of those conceived during a harsh winter lived longer than those whose ancestors ate well during the same time; 3. Epigenetics is the study of the heritable changes in gene expression not underlying the DNA sequence. It can tell us a great deal about how offspring can be adapted for environmental stressors more quickly than natural selection.

SELF-TEST
1. a; 2. a; 3. d; 4. b; 5. c; 6. a; 7. a; 8. b; 9. d; 10. b; 11. a; 12. b; 13. b; 14. b; 15. c; 16. b; 17. a; 18. a; 19. d; 20. c.

Chapter 11   How Cells Reproduce

11.1. HENRIETTA'S IMMORTAL CELLS [p.177]
11.2. MULTIPLICATION BY DIVISION [P.178-179]
1. H; 2. I; 3. G; 4. F; 5. A; 6. B; 7. E; 8. C; 9. D; 10. diploid; 11. daughter; 12. chromosomes; 13. one; 14. spindle; 15. microtubules; 16. chromosomes; 17. chromatids; 18. opposite; 19. G1; 20. S phase; 21. G2 phase; 22. prophase; 23. metaphase; 24. anaphase; 25. telophase; 26. mitosis; 27. daughter cells; 28. interphase; 29.

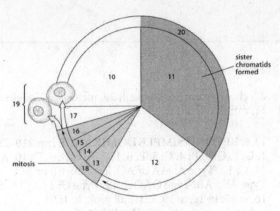

sister
chromatids
formed

mitosis

30. T; 31. G2; 32. G1; 33. T

11.3. A CLOSER LOOK AT MITOSIS [pp.180-181]

1. interphase – daughter cells (F); 2. anaphase (A); 3. late prophase (G); 4. metaphase (D); 5. cell at interphase (E); 6. early prophase (C); 7. transition to metaphase (B); 8. telophase (H); 9. (5) interphase; 10. (6) early prophase; 11. (3) late prophase; 12. (7) transition to metaphase; 13. (4) metaphase; 14. (2) anaphase; 15. (8) telophase; 16. (1) interphase – daughter cells; 17. T; 18. Telophase; 19. two; 20. metaphase

11.4. CYTOPLASMIC DIVISION OF CYTOPLASM [p.182]

1. D; 2. F; 3. E; 4. A; 5. C; 6. B; 7. A; 8. B; 9. A; 10. B; 11. C; 12. A; 13. C; 14. A; 15. D; 16. B; 17. 2; 18. 3; 19. 1; 20. 4

11.5. MARKING TIME WITH TELOMERES [p.182]

1. They act as a biological clock by shortening with each division and limiting the number of divisions of each cell type to avoid cancerous growths; 2. Each division cuts a few bases off of the ends of chromosomes in each M phase that the chromosome moves through. If they get to short, the cell dies; 3. Stem cells continue to make telomeres so they are functionally immortal.

11.6. WHEN MITOSIS BECOMES PATHOLOGICAL [pp.183-184]

1. F; 2. A; 3. E; 4. C; 5. B; 6. D

SELF-TEST

1. a; 2. d; 3. a; 4. b; 5. c; 6. c; 7. e; 8. b; 9. c; 10. b; 11. c; 12. c; 13. a; 14. b; 15. d; 16. c; 17. a; 18.

Chapter 12   Meiosis and Sexual Reproduction

12.1. WHY SEX? [p.189]
12.2. MEIOSIS HALVES THE CHROMOSOME NUMBER [pp.190-191]
INTRODUCING ALLELES [p.156]

1. two; 2. two; 3. maternal; 4. paternal; 5. clones; 6. mutations; 7. combinations; 8. Alleles; 9. H; 10. J; 11. G; 12. F; 13. B; 14. D; 15. E; 16. C; 17. I; 18. A; 19. true; 20. false; 21. false; 22. true; 23. true; 24. false; 25. False; 26. D; 27. C; 28. A; 29. B

12.3. VISUAL TOUR OF MEIOSIS [pp.192-193]

1. a. 3(D), b. 6(D), c. 5(D), d. 4(D), e. 8(H), f. 7(H), g. 1(D), h. 2(D); 2. Anaphase II – H; 3. Metaphase II - A; 4. Metaphase I - F; 5. Prophase II – B; 6. Telophase II - C; 7. Telophase I - G; 8. Prophase I - E; 9. Anaphase I – D.

12.4. HOW MEIOSIS INTRODUCES VARIATIONS IN TRAITS [pp.194-195]

1. crossing over, independent assortment; independent segregation; 2. Prophase I; 3.

Daughter cells produced are identical to the parent cell.; 4. H; 5. E; 6. D; 7. B; 8. G; 9. F; 10. A; 11. C

12.5. FROM GAMETES TO OFFSPRING [pp.196-197]

1. B; 2. C; 3. A; 4. C; 5. B; 6. A; 7. B; 8. A; 9. B; 10. A; 11. B; 12. A; 13. C; 14. A; 15. 1 (2n); 16. 4 (n); 17. 3 (n); 18. 2 (2n); 19. 4 (n); 20. 1 (2n); 21. 3 (n); 22. 2 (2n); 23. Both sperm and egg are haploid (n) and a combination of both maternal and paternal DNA due to crossing over. Four basically equal sperm are formed in spermatogenesis where only one viable ovum is formed with three polar bodies in oogenesis.

12.6. MITOSIS AND MEIOSIS—AN ANCESTRAL CONNECTION? [pp.198-199]

1. C; 2. F; 3. D; 4. A; 5. B; 6. E; 7. 4; 8. 8; 9. 4; 10. 8; 11. 2

SELF TEST

1. a; 2. a; 3. b; 4. c; 5. d; 6. c; 7. a; 8. c; 9. b; 10. d; 11. b; 12. c; 13. b; 14. c; 15. a; 16. d; 17. c; 18. a;

Chapter 13   Observing Patterns in Inherited Traits

13.1. MENACING MUCUS [p.203]
**13.2. MENDEL, PEA PLANTS, AND
INHERITANCE PATTERNS [pp.204-205]**
1. L; 2. A; 3. C; 4. H; 5. G; 6. J; 7. D; 8. E; 9. F; 10. M;
11. B; 12. I; 13. K; 14. T; 15. recessive, dominant; 16.
two; 17. T; 18. The plant is self-fertilizing and a true-
breeder; a true-breeder and has definite dominant
and recessive traits.

**13.3. MENDEL'S LAW OF SEGREGATION [pp.206-
207]**
**13.4. MENDEL'S LAW OF INDEPENDENT
ASSORTMENT [pp.208-209]**
1. Larger sample sizes reduce the effect of chance.; 2.
During meiosis, the genes of each pair on
homologous chromosomes separate, so each gamete
gets one or the other gene; 3. During meiosis, gene
pairs on homologous chromosomes
tend to sort into gametes independently of other
gene pairs; 4. T; 5. T; 6. T; 7. sample size; 8. T; 9.
parents; 10. T; 11. E; 12. B; 13. G; 14. A; 15. F; 16. C;
17. D; 18. I; 19. H; 20. a. 1:2 Tt, 1:2 tt b. 1:2 tall.1:2
dwarf; 21. a. homozygous dominant b. heterozygous;
22. heterozygous normal 1:2, Homozygous recessive
(albino) 1:2; 23. a. F_1: black trotter 1:1, F_2 9 – black
trotter/3 – black pacer/3 – chestnut trotter/1 –
chestnut pacer, b. black trotter, c. BbTt, d. bbtt –
chestnut pacer; 24. a. 1:2:1, b. 75% green, 25% yellow;
25. a. 100% BbSs, b. All black short-haired; 26. genes;

27. chromosome; 28. linked; 29. meiosis, 30. alleles;
31. ratios; 32. Short

13.5. BEYOND SIMPLE DOMINANCE [pp.210-211]
1. D; 2. G; 3. B; 4. C; 5. E; 6. F; 7. A; 8. B; 9. D; 10. A;
11. C; 12. Type A: AA or AO, Type B: BB or BO,
Type AB: AB, Type O: OO; 13. Type O; 14. c; 15. d;
16. a; 17. b; 18. a; 19. a. 1:1 all pink, b. 1:2:1
red:pink:white; 20. a. A, B, AB/b. A, B, AB, O/c. A/d.
O/e. AB, A, B; 21. a. 1:3 colored:white; b. 3:1
colored:white; 22. 3:1:4 black:brown:yellow; 23. 50%
black, 50% chocolate

13.6. NATURE AND NURTURE [p.212-213]
1. Taller plants at low altitude, short plants at middle
altitude, and mid-sized plants at high altitude.; 2.
More melanin (more color) in cooler parts of the
body.; 3. Many predators.; 4. Serotonin

**13.7. COMPLEX VARIATIONS IN TRAITS [pp.214-
215]**
1. b; 2. b; 3. a; 4. b; 5. a

SELF TEST
1. d; 2. b; 3. c; 4. c; 5. b; 6. b; 7. e; 8. a; 9. b; 10. c; 11. c;
12. a; 13. a; 14. b; 15. c; 16. d; 17. a; 18. d; 19. d; 20. c

Chapter 14   Chromosomes and Human Inheritance

14.1. SHADES OF SKIN [p.219]
14.2. HUMAN CHROMOSOMES [p.220-221]
1. pedigree; 2. traits; 3. disorder; 4. dominant; 5. recessive; 6. autosome; 7. sex chromosome; 8. generations; 9. disorders; 10. health; 11. reproductive; 12. mutations; 13. heterozygous;
14. E; 15. B; 16. G; 17. D; 18. A; 19. F; 20. C; 21. Skin pigmentation is a defense again UV light and is a result of over 100 different genes; 22. A genetic abnormality is a rare or uncommon version of a trait that is usually non-life threatening. A disorder is a genetic anomaly that is life threatening; 23; Studying human inheritance is difficult because experiments cannot be conducted ethically but scientists can use pedigrees to study generations of families instead.

14.3. EXAMPLES OF AUTOSOMAL INHERITANCE PATTERNS [pp.222-223]
1. E; 2. A; 3. B; 4. C; 5. D; 6. Aa, Aa, aa; 7. 50%; 8. 0%, it does not run in families but arises through mutations; 9. Skin starts to thin, muscles weaken, and bones soften. Premature baldness occurs at a very young age and individuals usually die in their early teens; 10. lamins; 11. a; 12. b; 13. a; 14. a; 15. b; 16. a; 17. b; 18. a; 19. b; 20. a; 21. b

14.4. EXAMPLES OF X-LINKED INHERITANCE PATTERNS [pp.224-225]
14.5. HERITABLE CHANGES IN CHROMOSOME STRUCTURE [pp.226-227]
1. a; 2. e; 3. b; 4. c; 5. duplication (B); 6. inversion (C); 7. deletion (A); 8. translocation (D); 9. About 350 million years ago, a gene on one of the two homologous chromosomes mutated. The change, which was the beginning of the male sex

determination gene *SRY*, interfered with crossing over during meiosis. A reduced frequency of crossing over allowed the chromosomes to diverge around the changed region. Mutations began to accumulate separately in the two chromosomes. Over evolutionary time, the chromosomes became so different that they no longer crossed over at all in the changed region, so they diverged even more. Today, the Y chromosome is much smaller than the X, and only retains about 5 percent homology with it. The Y crosses over mainly with itself—by translocating duplicated regions of its own DNA; 10. Mutations can alter the phenotypes of organisms significantly enough to prohibit those with and without the mutation to breed, leading to reproductive isolation.

14.6. HERITABLE CHANGES IN THE CHROMOSOME NUMBER [pp.228-229]
1. 100%; 2. 50%; 3. polyploidy: cells have three or more of each type of chromosome, trisomic: cells have three chromosomes where normally there is a pair of chromosomes, monosomic: cells have one chromosome where normally there is a pair of chromosomes; 4. c; 5. b; 6. c; 7. d; 8. a; 9. b; 10. d; 11. c; 12. a; 13. d

14.7. GENETIC ANALYSIS [pp.230-231]
1. A; 2. H; 3. F; 4. G; 5. D; 6. B; 7. C; 8. E; 9. medical; 10. pku; 11. prenatal; 12. amniocentesis; 13. chorionic villi sampling; 14. abortion

SELF-TEST
1. a; 2. d; 3. c; 4. c; 5. a; 6. a; 7. b; 8. b; 9. c; 10. e; 11. c; 12. a; 13. a ; 14. a; 15. c; 16. b; 17. c; 18. a; 19. e; 20. d

---

Chapter 15   Studying and Manipulation Genomes

15.1. PERSONAL DNA TESTING [p.235]
15.2. CLONING DNA [pp.236-237]
1. bacteriophage; 2. restriction enzymes; 3. EcoRI; 4. recombinant; 5. restriction; 6. sticky ends; 7. DNA cloning; 8. plasmids; 9. cloning vectors; 10. clones; 11. mRNA; 12. introns; 13. DNA ligase; 14. reverse transcriptase; 15. cDNA; 16. B; 17. F; 18. E; 19. C; 20. A; 21. D; 22. Because it allowed scientists to cut DNA up in segments at known locations; 23. It allows them to combine genetic information from different organisms.

15.3. ISOLATING GENES [pp.238-239]
1. B; 2. E; 3. D; 4. A; 5. F; 6. C; 7. 6; 8. 4(A); 9. 2(D); 10. 3(E); 11. 5(C); 12. 1(B)

15.4. DNA SEQUENCING [pp.240-241
1. Polymerase randomly adds either a regular nucleotide or a dideoxynuclotide to the end of a growing DNA strand. If a dideoxynuclotide is added, synthesis of that particular strand ends.; 2. An electric field is used to pull a mixture of different length DNA fragments through a semisolid gel matrix. Smaller fragments of DNA work their way through the gel faster than larger fragments. Over time this separates the fragments by size, revealing the DNA sequence; 3. The Human Genome Project is intended to sequence all of the nucleotide of the human genome. It was completed in 2003.

15.5. GENOMICS [pp.242-243]
1. PCR is used to copy a region of a chromosome known to have tandem repeats of 4 to 5 nuclotides. Electrophoresis is used to separate DNA fragments by

size, revealing a unique banding pattern – the individual's DNA fingerpring.; 2. Just a small sample of DNA from blood, semen, hair follicle cells, etc. can be copied by PCR for DNA fingerprinting. Since DNA fingerprints are essentially unique, this evidence is compelling to convict criminals and also to determine if a suspect is innocent; 3. DNA fingerprinting can be used in kinship studies or in identifying a body; 1. F; 2. A; 3. C; 4. G; 5. B; 6. E; 7. D

15.6. GENETIC ENGINEERING [p.244]

1. production of human insulin, cheese production, improving taste and clarity of beer and juices, slows bread staling, modifies fats; 2. This research may help us to discover why some bacteria become resistant to antibiotics.

16.7. DESIGNER PLANTS [pp.244-245]

1. 2; 2. 5; 3. 1; 4. 4; 5. 3; 6. "Frankenfood" is used by groups opposed to genetic engineering.; 7. True; 8. True; 9. False – Ti plasmids; 10. False – Bt plasmids

15.8. BIOTECH BARNYARDS [pp.246-247]
15.9. SAFETY ISSUES [p.247]

1. Integrating the rat growth hormone gene into mice may allow us to incorporate beneficial genes.; 2. Goat milk producing lysozyme may protect infants and children from acute diarrheal disease.; 3. Studying knockout experiments in glucose metabolism of mice may help us to understand how diabetes works in humans.; 4. Xenotransplantation may provide a ready supply of transplantable organs of human use.; 5. Goat milk containing spider silk proteins may be

valuable source for silk to produce fabrics, bulletproof vests, sports equipment and biodegradable medical supplies; 6. Host organisms are designed and used under a narrow range of conditions in the laboratory, DNA from pathogenic or toxic organisms for recombinant experiments are prohibited until proper containment facilities are developed, and release and import of genetically-modified organisms is carefully regulated.

15.10. MODIFIED HUMANS? [p.248]

1. Genes missing or malfunctioning may be introduced into persons suffering from these diseases, resulting in a cure or at least relief of symptoms.; 2. Cystic fibrosis, hemophilia A, several types of cancer, inherited retinal disease, inherited immune disorders.; 3. Unmutated copies of IL2RG gene were inserted into bone marrow for each of the boys. Those transgenic cells reproduced and restored a healthy immune system to 10 of the 11 boys.; 4. Jesse Gelsinger was allergic to the viral vector, an unexpected complication of the therapy. This is a warning because there is always risk to gene therapies and it should not be undertaken lightly.; 5. Benefits: potential cures for severe genetic disorders, Inappropriate uses: for monetary benefit and/or to choose preferences rather than legitimate needs

SELF-TEST
1. a; 2. b; 3. c; 4. b; 5. d; 6. b; 7. a; 8. d; 9. d; 10. a; 11. a; 12. b; 13. b; 14. d; 15. d; 16. d; 17. c; 18. a

---

Chapter 16   Evidence of Evolution

16.1. REFLECTIONS OF A DISTANT PAST [p.253]
16.2. EARLY BELIEFS, CONFOUNDING
        DISCOVERIES [pp.254-255]
16.3. A FLURRY OF NEW THEORIES [pp.256-257]
16.4. DARWIN, WALLACE, AND NATURAL
        SELECTION [pp.258-259]

1. K; 2. M; 3. O; 4. G; 5. D; 6. B; 7. I; 8. N; 9. L; 10. C; 11. E; 12. J; 13. H; 14. A; 15. F; 16. b; 17. c; 18. b; 19. a; 20. c; 21. b; 22. b; 23. a; 24. b; 25. b; 26. a; 27. d; 28. b; 29. c; 30. A; 31. The giraffes that have alleles that produce long necks have more access to food and reproduce more often than those with shorter necks. Their offspring are likely to inherit the long necks and the trait is selected for.

16.5. FOSSILS: EVIDENCE OF ANCIENT LIFE
        [pp.260-261]
16.6. FILLING IN PIECES OF THE PUZZLE
        [pp.262-263]

1. E; 2. D; 3. A; 4. B; 5. C; 6. G; 7. F; 8. 250,000; 9. rare; 10. decay; 11. eaten; 12. oxygen; 13. soft-bodied; 14. 0.5 grams, 16,110 years

16.7. DRIFTING CONTINENTS, CHANGING SEAS
        [pp.264-265]
16.8. PUTTING TIME INTO PERSPECTIVE
        [pp.266-267]

1. D; 2. C; 3. A; 4. H; 5. I; 6. F; 7. B; 8. G; 9. a; fault, b. trench, c. ridge, d. hot spot, e. trench; 10. There were major extinctions at the end of the Ordovician, Devonian, Permian, Triassic, and Cretaceous. Each event led to more biodiversity in the millennia afterward.

SELF-TEST
1. d; 2. d; 3. d; 4. b; 5. a; 6. d; 7. b; 8. c; 9. a; 10. e; 11. b; 12. a; 13. a; 14. b; 15. b; 16. b; 17. b; 18. d; 19. a; 20. c

Chapter 17   Processes of Evolution

17.1. RISE OF THE SUPER RATS [p.271]
17.2. INDIVIDUALS DON'T EVOLVE, POPULATIONS DO [pp.272-273]
17.3. A CLOSER LOOK AT GENETIC EQUILIBRIUM [pp.274-275]

1. F; 2. I; 3. G; 4. K; 5. A; 6. B; 7. H; 8. C; 9. D; 10. L; 11. M; 12. J; 13. E; 14. There is no mutation, the population is very large, the population is isolated from other populations of the same species, all members of the population survive and reproduce, mating is completely random; 15. Evolution acts upon a gene pool, which is a feature of populations, not individuals; 16. a. 0.64 BB, 0.16 Bb, 0.16 Bb, 0.04 bb; b. genotypes: 0.64 BB, 0.32 Bb, and 0.04 bb; phenotypes: 965 black, 4% gray

c.

Parents (F_1)	B sperm	b sperm
0.64 BB	0.64	0
0.32 Bb	0.16	0.16
0.04 bb	0	0.04
Totals =	0.80	0.20

17.a. 18% [$2pq = 2$ x (0.9) x (0.1) = 2 x (0.09) = 0.18 = 18% heterozygotes], b. 0.9 [p^2 = 0.81, p = 0.9 = the frequency of the dominant allele], c. 0.1 [p + q = 1, q=1.00 − 0.9 = 0.1 = the frequency of the recessive allele]; 18.a. 128 [homozygous dominant = p^2 x 200 = $(0.8)^2$ x 200 = 0.64 x 200 = 128 individuals], b. 8 [q = (1.00 −p) = 0.20 = homozygous recessive, q^2 x 200 = $(0.2)^2$ x 200 = (0.04) x (200) = 8 individuals], c. 64 [heterozygotes = 2pq x 200 = 2 x 0.8 x 0.2 x 200 = 0.32 x 200 = 64 individuals] Check: 128 + 8 + 64 = 200; 20. 30% [If p=0.70, and p+q = 1, 0.70 + q = 1; then q = 0.30, or 30%]; 21. 48% [If p = 0.60, and p+q = 1, 0.60 +q=1; then q= 0.40; thus, 2pq = 0.48 or 48%]; 19. 30%; 20. 0.24; 21; 0.09

17.4. PATTERNS OF SELECTION REVISITED [p.275]
17.5. DIRECTIONAL SELECTION [pp.276-277]
17.6. STABILIZING & DISRUPTIVE SELECTION [pp.278-279]

1. a. directional selection - the allele frequencies shift in a consistent direction, b. disruptive selection - the extreme ends of the variations are selected for, c. stabilizing selection - the average form of a trait is selected for 2. a; 3. b; 4. b; 5. c; 6. b; 7. a; 8. c; 9. b; 10. This is creating antibiotic resistant bacteria within a short period of time. We are selecting out the weak bacteria and leaving the "strong" ones. The "strong" ones have little competition for resources so they are able to reproduce easily creating new "antibiotic resistant"

strains; 11. The melanistic phenotype of the peppered moth does best under high pollution conditions such as those during the early days of coal burning in the industrial revolution. As pollution controls have improved, so has the frequency of the peppered phenotype of the moth.

17.7. FOSTERING DIVERSITY [pp.280-281]
17.8. GENETIC DRIFT AND GENE FLOW [pp.282-283]
17.9. REPRODUCTIVE ISOLATION [pp.284-285]

1. b; 2. d; 3. c; 4. d; 5. d; 6. b; 7. a; 8. c; 9. b; 10. e; 11. c; 12. e; 13. frequency; 14. homozygous; 15. loss; 16. population; 17. small; 18. fixed; 19. frequencies; 20. gene; 21. a; 22. b; 23. a; 24. a; 25. b; 26. A; 27. Nonrandom mating fosters diversity by allowing for different genetic combinations with each offspring; 28. multiple alleles are maintained in a population at relatively high frequency; 29. As a population is reduced, the individual remaining are the foundation for the population in the future, possibly reducing genetic variability; 30. Because of heterozygote advantage; 31. D; 32. E; 33. A; 34. C; 35. F; 36. B; 37. G

17.10. ALLOPATRIC SPECIATION [pp.286-287]
17.11. OTHER SPECIATION MODELS [pp.288-289]
17.12. MACROEVOLUTION [pp.290-291]

1. C; 2. I; 3. B; 4. G; 5. F; 6. A; 7. D; 8. J; 9. e; 10. h; 11. a; 12. c; 13. b; 14. a; 15. b; 16. b; 17. A; 18. Stasis; 19. Preadaptation; 20. adaptive radiation; 21. key innovation; 22. extinct; 23.In allopatric speciation, a physical barrier arises and separates two populations, ending gene flow between them. In sympatric speciation, populations inhabiting the same geographic region speciate in the absence of a physical barrier between them. In parapatric speciation, adjacent populations speciate despite being in contact across a common border; 24. Mass extinction is when a multitude of species is lost from the planet in a very short period. There have been 5 previous extinctions; 25. Coevolution is the evolution of two species in concert with one another. On example is the large blue butterfly and ants.

SELF-TEST

1. a; 2. d; 3. a; 4. d; 5. e; 6. b; 7. c; 8. a; 9. a; 10. d; 11. a; 12. d; 13. c; 14. b; 15. c; 16. e; 17. a; 18. c; 19. d; 20. a

Chapter 18   Organizing Information About Species

18.1. BYE BYE BIRDIE [p.295]
18.2. PHYLOGENY [pp.296-297]
18.3. COMPARING BODY FORM AND

FUNCTION [pp.298-299]

1. B; 2. H; 3. J; 4. E; 5. C; 6. F; 7. I; 8. A; 9. L; 10. D; 11. K; 12. G; 13. M; 14. differences; 15. cladistics; 16. fewest; 17. parisomony; 18. matrix; 19. tree of life; 20. A, B, D; 21. C

18.4. COMPARING BIOCHEMISTRY [pp.300-

301]

18.5. COMPARING PATTERNS OF DEVELOPMENT [pp.302-303]
18.6. APPLICATIONS OF PHYLOGENY RESEARCH [pp.304-305]

1. embryo; 2. master gene expression; 3. master gene expression; 4. conserved; 5. lineages; 6. limb buds; 7. tail; 8. Hox; 9. divide; 10. differentiate; 11. Dlx; 12. Hox; 13. Dlx; 14. A & C; 15. A & D; 16. A(oldest) evolved into C, C evolved into B, B evolved into D (youngest); 17. b; 18. c; 19. a; 20. a; 21. b; 22. c; 23. C; 24. finding ancestral connections can help species that are still living by helping prioritize funding and conservation efforts; 25. An increased understanding of how this virus evolves may help us find a way to prevent it from spreading.

SELF-TEST

1. c; 2. a; 3. a; 4. b; 5. d; 6. c; 7. a; 8. d; 9. c; 10. a; 11. c; 12. a; 13. a; 14. d; 15. b; 16. a

---

Chapter 19   Life's Origin and Early Evolution

19.1. LOOKING FOR LIFE [p.309]
19.2. THE EARLY EARTH [p.310]
19.3. FORMATION OF ORGANIC MOLECULES [p.311]
19.4. FROM POLYMERS TO PROTOCELLS [pp.312-313]

1. e; 2. f; 3. a; 4. b; 5. b; 6. d; 7. c; 8. c; 9. a; 10. a; 11. e; 12. f; 13. The origins of prokaryotes are hypothesized to have begun approximately 3.8 billion years ago. Early organisms transformed the oceans and atmosphere during the first 2 billion years of the Earth's history, including producing oxygen; 14. The iron-sulfur world hypothesis holds that the first metabolic reactions began on the surface of rocks around hydrothermal vents. The genome may have arisen via RNA as it can act as an enzyme and an information carrier. The plasma membrane would have formed from protocells that were membrane-bound collections of interacting molecules; 15. in one of three ways: at hydrothermal vents, through lightening-fueled atmospheric reactions, or via meteors from space.

19.5. LIFE'S EARLY EVOLUTION [pp.314-315]

19.6. HOW DID EUKARYOTIC TRAITS EVOLVE? [pp.316-317]
19.7. TIME LINE FOR LIFE'S ORIGIN AND EVOLUTION [pp.318-319]
20.6. ABOUT ASTROBIOLOGY [p.328]

1. b; 2. b; 3. a; 4. a; 5. b; 6. B; 7. A; 8. A; 9. 3.8 billion years ago; 10. heterotrophic; 11. oxygen; 12. oxygen; 13. anaerobic; 14. aerobic; 15. releasing; 16. Ozone layer;17. UV radiation; 18. land; 19. D; 20. F; 21. E; 22. B; 23. C; 24. A; 25. I; 26. G; 27. J; 28. c; 29. a; 30. a; 31. b; 32. b; 33. b; 34. See Figure 19.13; 35. Endosymbiosis states that mitochondria and chloroplasts were originally independent organisms that were engulfed by a heterotroph in evolutionary history and now they are held as obligate mutualists; 36. They may have developed from the infolding of the plasma membrane.

SELF-TEST

1. b; 2. b; 3. c; 4. d; 5. a; 6. d; 7. b; 8. a; 9. a; 10. c; 11. d; 12. d; 13. d; 14. c; 15. d; 16. b; 17. d; 18. d

---

Chapter 20   Viruses, Bacteria, and Archaea

20.1. EVOLUTION OF A DISEASE [p.323]
20.2. VIRUSES AND VIROIDS [pp.324-325]
20.3. VIRAL REPLICATION [pp.326-327]
20.4. VIRUSES AS HUMAN PATHOGENS [pp.328-329]

1. C; 2. D; 3. E; 4. B; 5. F; 6. A; 7. The 3 hypotheses of viral origin are: a. viruses are descendants of cells that were parasites inside other cells, b. viruses are genetic elements that have escaped from cells, c. viruses represent a separate evolutionary branch and arose independently; 8. Steps of viral replication include: a. attachment, b. penetration, c.

replication and synthesis, d. assembly, e. release; 9. The lytic cycle is when a virus attaches to a host cell and injects its DNA into the host. Viral genes direct the host cell to make viral DNA and proteins which assemble as viral particles. Eventually the host cell ruptures and the new viruses are released. The lysogenic cycle is when the virus attaches to the host cell but the virus enters a latent state that extends the multiplication cycle. Viral genes are integrated into the host DNA and are copied along with the host DNA. The viral DNA is passed to all of the descendent cells.; 10. The principal features of a virus are a. noncellular, no cytoplasm, ribosomes, or other typical cell components, b. genetic material may be DNA or RNA, c. can only replicate inside of a host cell, d. small (about 25 to 300 nanometers); 11. c; 12. e; 13. h; 14. b; 15. a; 16. f; 17. g; 18. d; 19. Washing hands with regular soap and water on a regular basis is the best way to combat disease; 20. Symptoms of viral diseases include irritation to mucous membranes, leading to a runny nose and cough, and sometimes irritation of the gut and stomach, leading to vomiting and diarrhea; 21. a. 5, b. 9, c. 8, d. 7, e. 1, f. 6, g. 2, h. 3, i. 4.

20.5. "PROKARYOTES" – ENDURING, ABUNDANT, & DIVERSE [pp.330-331]
20.6. STRUCTURE AND FUNCTION OF PROKARYOTIC [pp.332-333]

1. d; 2. c; 3. a; 4. b; 5. The four main features of Prokaryotic cells are: (1) no nucleus; the chromosome is found in the nucleoid region, (2) generally they contain a single chromosome as well as circular loops of DNA called plasmids, (3) cell wall is present in most species, (4) ribosomes are distributed within the cytoplasm.; 6. Transduction is when a virus picks up DNA from one Prokaryotic cell that it is infecting and then transfers that DNA to another Prokaryotic cell. Transformation is when a Prokaryotic cell picks up DNA from the environment.; 7. Bacteria occasionally exchange DNA through a pilus in a process known as conjugation; 7. c; 8. b; 9. e; 10. f; 11. a; 12. g; 13. d; 14. twenty; 15. chromosome; 16. fission; 17. horizontal gene transfer; 18. conjugation;

20.7. BACTERIAL DIVERSITY [pp.334-335]
20.8. ARCHAEA [pp.336-337]

1. b; 2. c; 3. b; 4. a; 5. c; 6. b; 7. d; 8. c; 9. d; 10. a; 11. Bacteria release oxygen, help cycle nutrients, and fix nitrogen; 12. Diseases caused by bacteria include tuberculosis, whooping cough, syphilis, and tetanus; 13. Extreme salt, temperature, pressure, and hypoxic environments.

SELF-TEST

1. a; 2. b; 3. c; 4. d; 5. c; 6. a; 7. d; 8. a; 9. c; 10. b; 11. b; 12. c; 13. c; 14. b; 15. b; 16. c; 17. c; 18. a; 19. c; 20. d

---

Chapter 21   "Protists" – The Simplest Eukaryotes

21.1. MALARIA – FROM TUTANKHAMUN TO TODAY [p.351]
21.2. THE MANY PROTIST LINEAGES [p.342-343]

1. E; 2. P; 3. B; 4. E; 5. P; 6. E; 7. single; 8. heterotrophic; 9. water; 10. soil; 11. Autotrophic; 12. photosynthetic; 13. heterotrohic; 14. environment; 15. asexually; 16. sexually; 17. haploid; 18. zygote

21.3. FLAGELLATED PROTOZOANS [pp.344-345]
21.4. FORAMINIFERANS & RADIOLARIANS [p.346]
21.5. THE CILIATES [p.347]
21.6. DINOFLAGELLATES [p.348]
21.7. CELL-DWELLING APICOMPLEXANS [p.349]
21.8. THE STRAMENOPILES [pp.350-351]

1. Diplomonads; 2. parabasalids; 3. mitochondria; 4. Trypanosomes; 5. vector; 6. eyespot; 7. calcium carbonate; 8. plankton; 9. alveolates; 10. Paramecium; 11. dinoflagellates; 12. apicomplexans; 13. anopheles; 14. plasmodium; 15. I (I, II, IV, V); 16. F (IV); 17. J (III, IV); 18. E, C (VI); 19. D (II, V); 20. H (V); 21. G, B (II, V); 22. A (I, II, IV, V, VI); 23. C; 24. I; 25. H; 26. D; 27. B; 28. E; 29. G; 30. A; 31. F; 32. C; 33. I; 34. F; 35. J; 36. H; 37. G; 38. B; 39. A; 40. D; 41. E;

21.9. RED ALGAE [p.352]
21.10. GREEN ALGAE [pp.352-353]
21.11. AMOEBOZOANS AND CHOANOFLAGELLATES [p.354-355]

1. G; 2. F; 3. C, H; 4. B; 5. E; 6. D; 7. A; 8. a; 9. b; 10. c; 11. C; 12. a; 13. c; 14. b; 15. b; 16. a; 17. b; 18. c; 19. a; 20. a; 21. b

SELF-TEST

1. b; 2. b; 3. b; 4. a; 5. c; 6. a; 7. d; 8. a; 9. d; 10. d; 11. a; 12. b; 13. a; 14. b; 15. a; 16. c; 17. d; 18. d

Chapter 22   The Land Plants

22.1. SPEAKING FOR THE TREES [p.359]
22.2. PLANT ANCESTRY & DIVERSITY [pp.360-361]
22.3. EVOLUTIONARY TRENDS AMONG PLANTS [pp.362-363]
1. D; 2. F; 3. B; 4.G; 5. E; 6. A; 7. C; 8. B; 9. C; 10. D; 11. E; 12. A; 13. N; 14. A; 15. O; 16. K; 17. G; 18. D; 19. I; 20. C; 21. M; 22. J; 23. B; 24. H; 25. E; 26. L; 27. F.

22.4. THE BRYOPHYTES [pp.364-365]
22.5. SEEDLESS VASCULAR PLANTS [pp.366-367]
22.6. HISTORY OF THE VASCULAR PLANT [pp.368-369]
1. mosses; 2. liverworts; 3. nonvascular; 4. lignin; 5. rhizoids; 6. water; 7. gametophytes; 8. sperm; 9. sporophytes; 10. gametophytes; 11. fertilization; 12. zygote; 13. D; 14. H; 15. H; 16. H; 17. D; 18. H; 19. c; 20. b; 21. e; 22. a; 23. d; 24. c; 25. e; 26. e; 27. b; 28. d; 29. e; 30. sporophyte (2n); 31. rhizome (2n); 32. sporangia (2n); 33. gametophyte (n); 34. male gametangium (n); 35. female gametangium (n); 36.

4; 37. 3; 38. 6; 39. 7; 40. 5; 41. 1; 42. 2; 43. F; 44. D; 45. A; 46. B; 47. E; 48. C.

22.7. GYMNOSPERMS [pp.370-371]
1. b; 2. d; 3. a; 4. b; 5. c; 6. b; 7. d; 8. d; 9. a; 10. c; 1. C; 11. D; 12. H; 13. E; 14. C; 15. A; 16. G; 17. F; 18. B; 19. I

22.8. ANGIOSPERMS—THE FLOWERING PLANTS [pp.372-373]
22.9. ANGIOSPERM DIVERSITY AND IMPORTANCE [p.374-375]
1. F; 2. B; 3. C; 4. E; 5. D; 6. A; 7. C; 8. H; 9. D; 10. B; 11. J; 12. E; 13. A; 14. K; 15. I; 16. G; 17. F; 18. mature female gametophyte; 19. cell with 2 nuclei; 20. egg; 21. mature male gametophyte; 22. pollen tube; 23. sperm; 24. seed coat; 25. embryo; 26. endosperm; 27. seed; 28. eudicot; 29. seeds; 30. 16%; 31. amino acids; 32. drought; 33. frost; 34. salty soil

SELF-TEST
1. b; 2. e; 3. a; 4. d; 5. d; 6. e; 7. a; 8. d; 9. e; 10. d; 11. b; 12. c; 13. b; 14. d; 15. d; 16. c; 17. b; 18. b; 19. a; 20. c

---

Chapter 23   Fungi

23.1. HIGH-FLYING FUNGI [p.379]
23.2. FUNGAL TRAITS AND CLASSIFICATION [pp.380-381]
1. F; 2. G; 3. C; 4. E; 5. B; 6. D; 7. A; 8. heterotroph; 9. chitin; 10. yeast; 11. mycelium; 12. hyphae; 13. digestive enzymes; 14. saprobe; 15. Cycled; 16. The major characteristics of fungi include multicellularity, decomposing heterotrophs, walled cells, sessile, and produce via spores. May be saprobes, parasites, or mutualists; 17. The major groups of fungi include cyrtrids, zygomycetes, glomeromycetes, ascomycetes, and basidomycetes.

23.3. THE FLAGELLATED FUNGI [p.381]
23.4. ZYGOSPORE FUNGI AND RELATED GROUPS [pp.382-383]
1. nuclear fusion; 2. zygospore (2n); 3. meiosis; 4. spores (n); 5. spore sac (n); 6. rhizoids (n); 7. asexual reproduction; 8. zygospore (n); 9. a; 10. c; 11. b; 12. c; 13. a; 14. b; 15. c; 16. a; 17. b; 18. d

23.5. SAC FUNGI-ASCOMYCETES [pp.384-385]
23.6. CLUB FUNGI-BASIDIOMYCETES [pp.386-387]

23.7. FUNGAL PARTNERS AND PATHOGENS [pp.388-389]
1. C; 2. E; 3. D; 4. A; 5. B; 6. F; 7. G; 8. H; 9. nuclear fusion; 10. meiosis; 11. spore (n); 12. cytoplasmic fusion; 13. hyphae (n); 14. gill (n); 15. H; 16. F; 17. D; 18. A; 19. G; 20. B; 21. C; 22. E; 23. D; 24. B; 25. E; 26. C; 28. F; 29. Examples may include fungi that form partnerships with algae to form lichen, chytrids that live in the guts of grazers to help them break down cellulose, and types of glomeromycetes that form mycorrhiza with plant roots to assist in water and nutrient absorption; 30. Powdery mildew (affects leaves), shelf fungus (will often kill its host), honey mushroom (will attack and kill living trees), apple scab/peach leaf curl (affect leaves of living plants), some sac fungi (parasitize grains and can cause ergotism); 31. Some microsporidium can infect those with already weakened immune systems, some fungus can cause yeast infections in the warm, moist places of the human body, and athlete's foot will destroy keratin in the outer layers of the skin.

SELF-TEST

1. b; 2. c; 3. a; 4. b; 5. a; 6. d; 7. a; 8. c; 9. c; 10.
d; 11. a; 12. b; 13. b; 14. c; 15. c; 16. c; 17. d; 18. c;

19. d; 20. d;

Chapter 24   Animal Evolution – The Invertebrates

24.1. OLD GENES, NEW DRUGS [p.393]
24.2. ANIMAL TRAITS & BODY PLANS [pp.394-395]
24.3. ANIMAL ORIGINS & ADAPTIVE RADIATION [pp.396-397]

1. venom; 2. pain; 3. harpoon; 4. zinconotide; 5.
fruit flies; 6. 500 million; 7. M; 8. B; 9. F; 10. D; 11.
E; 12. I; 13. G; 14. C; 15. K; 16. J; 17. A; 18. H; 19. L;
20. N; 21. F; 22. A; 23. H; 24. C; 25. E; 26. G; 27. D;
28. B;

24.4. THE SPONGES [p.397]
24.5. CNIDARIANS—PREDATORS WITH STINGING CELLS [pp.398-399]

1. D; 2. G; 3. B; 4. F; 5. C; 6. E; 7. A; 8. cnidarians; 9.
Nematocysts; 10. medusa; 11. polyp; 12.
gastrovascular cavity; 13. gastrodermis; 14. nerve
net; 15. mesoglea; 16. dinoflagellates; 17. coral
bleaching; 18. feeding polyp; 19. reproductive
polyp; 20. female medusa; 21. ciliated bilateral
larva; 22. F; 23. H; 24. C; 25. A; 26. I; 27, G; 28. E;
29. B; 30. D

24.6. FLATWORMS--SIMPLE ORGAN SYSTEMS [pp.400-401]

1. b, c; 2. c; 3. c; 4. a; 5. a; 6. b, c ; 7. b; 8. a; 9. c; 10.
a; 11. c; 12. c; 13. branching gut; 14. pharynx; 15.
brain; 16. nerve cord; 17. ovary; 18. testis; 19.
planaria; 20. no; 21. yes; 22. pseudocoelomate; 23.
bilateral; 24. Larvae; 25. Intermediate; 26. Humans;
27. Intestine; 28. Proglottids; 29. Organs; 30.
Proglottids; 31. Feces; 32. Larval; 33. Intermediate

24.7. ANNELIDS—SEGMENTED WORMS [pp.402-403]

1. f; 2. j; 3. b; 4. d; 5. g; 6. k; 7. c; 8. I; 9. a; 10. h; 11.
e; 12. The earthworm uses its hydrostatic skeleton
to extend longitudinal muscles in the anterior end
while segments in the posterior end have their setae
fixed into the wall of the burrow. This pushes the
anterior end of the worm forward.   Next the worm
fixes the anterior setae and releases the posterior
setae while shortening its segments. This pulls the
posterior end forward. The worm then fixes the
posterior setae and repeats the motion.;13. *Hirudo
medicinalis* is used post surgery to help remove
blood clots from area that have tiny capillary beds
that need to heal.; 14. Clitellum; 15. Mouth; 16.
Brain; 17. Crop; 18. Intestine; 19. Ventral nerve
cord; 20. Coelom; 21. Gut; 22. Pharynx; 23. Anus;

24. Longitudinal muscle; 25. earthworm; 26;
Annelida; 27. segmentation and closed circulatory
system; 28. yes; 29. bilateral; 30. yes

24.8. MOLLUSKS—ANIMALS WITH A MANTLE [pp.404-405]

1. mollusk; 2. shell; 3. mantle; 4. gills; 5. foot; 6.
radula; 7. eyes; 8. tentacles; 9. nudibranchs; 10.
nematocysts; 11. jet propulsion; 12. III; 13. I; 14.
III; 15. Mouth; 16. Anus; 17. Gill; 18. Heart; 19.
Radula; 20. Foot; 21. Shell; 22. Stomach; 23.
Mouth; 24. Gill; 25. Mantle; 26. Retractor muscle;
27. Foot; 28. eye; 29. mantle; 30. sensory tentacle;
31. opening to lung; 32. foot; 33. d; 34. b; 35. a; 36.
d; 37. d; 38. b; 39. c; 40. b; 41. d; 42. d; 43. b; 44. a

24.9. ROTIFERS & TARDIGRADES—TINY & TOUGH [p.406]
24.10. ROUNDWORMS—UNSEGMENTED WORMS THAT MOLT [p.407]
24.11. ARTHROPODS—ANIMALS WITH JOINTED LEGS [p.408]

1. b; 2. e; 3. f; 4. c; 5. g; 6. d; 7. a; 8. roundworm; 9.
no; 10. pseudocoelom (false coelom); 11. d; 12. a;
13. d; 14. a; 15. d; 16. b; 17. a; 18. b; 19. c; 20. e; 21.
e

24.12. CHELICERATES AND THEIR RELATIVES [p.409]
24.13. MYRIAPODS [p.410]
24.14. CRUSTACEANS [p.410-411]

1. an exoskeleton; 2. spiders; 3. ticks; 4. lobsters
and crabs; 5. copepods; 6. barnacles; 7. barnacles; 8.
molts; 9. millipedes; 10. centipedes; 11. centipedes;
12. Malphigian tubules; 13. horseshoe crabs
14. abdomen; 15. cephalothorax; 16. pedipalps; 17.
walking legs; 18. arachnids; 19. cephalothorax; 20.
abdomen segments; 21. swimmerets; 22. walking
legs; 23. 1st legs; 24. antennae

24.15. INSECT - DIVERSE AND ABUNDANT [pp.426-427]

1. F; 2. C; 3. B; 4. E; 5. D; 6. A; 7. E; 8. C; 9. B; 10. D;
11. A; 12. antenna; 13. head; 14. thorax; 15.
abdomen; 16. compound eye; 17. brain; 18. heart;
19. ovary; 20. malpighian tubules; 21. digestive
system; 22. mouth

24.16. THE SPINY-SKINNED ECHINODERMS

1. deuterostomes; 2. calcium carbonate; 3. radial; 4. bilateral; 5. brain; 6. nervous; 7. stomach; 8. enzymes; 9. tube; 10. water-vascular; 11. muscle; 12. ampulla; 13. anus; 14. sea urchin; 15. sea cucumber; 16. internal organs; 17. Lower stomach; 18. Upper stomach; 19. Anus; 20. Gonad; 21. Coelom; 22. Digestive gland; 23. Eyespot; 24. starfish; 25. deuterostome

SELF-TEST

1. a; 2. d; 3. b; 4. a; 5. a; 6. c; 7. c; 8. a; 9. a; 10. a; 11. d; 12. c; 13. a; 14. b; 15. d; 16. d; 17. b; 18. c; 19. b; 20. d; 21. g, G; 22. b, DHJ; 23. f, BE; 24. h, F; 25. d, K; 26. I, M; 27. e, CL; 28. a, I; 29. c, A; 30. Placozoans have four different cell types that form two layers within the body. Their cells have specialized functions (secrete digestive enzymes, movement, etc.). They also have genes that are similar to the ones that encode for signaling molecules in human nerves. These features suggest that placozoans represent an early branch on the animal family tree.

Chapter 25   Animal Evolution – The Chordates

25.1. TRANSITION WRITTEN IN STONE [p.419]
25.2. CHORDATE TRAITS AND EVOLUTIONARY TRENDS [pp.420-421]
25.3. JAWLESS LAMPREYS [p.422]
25.4. THE EVOLUTION OF JAWED FISHES [p.423]

25.5. MODERN JAWED FISH [pp.424-425]
1. transitional forms; 2. Archaeopteryx; 3. Radiometric; 4. biological evolution; 5. nerve cord; 5. gill slits; 7. notochord; 8. lancelets; 9. filter feeding; 10. gut; 11. Tunicates; 12. tunic; 13. metamorphosis; 14. vertebral column; 15. fish; 16. jaws; 17. fins; 18. gills; 19. lungs; 20. circulatory; 21. lancelet; 22. invertebrate; 23. notochord; 24. dorsal nerve cord; 25. pharynx with gill slits; 26. tail; 27. anus; 28; eyespot; 29. backbones; 30. jaws; 31. swim bladder or lung(s); 32. bony appendages; 33. four limbs; 34. amniote egg; 35. hagfish; 36. lamprey; 37. sharks; 38. gill slits; 39. ray-finned; 40. lungfish; 41. swim bladder; 42. coelacanths; 43. amphibians; 44. ovary; 45. swim bladder; 46. kidney; 47. nerve cord; 48. brain; 49. gills; 50. heart; 51. liver; 52. stomach; 53. intestine; 54. anus

25.6. AMPHIBIANS—FIRST TETRAPODS ON LAND [pp.426-427]
25.7. THE AMNIOTES [p.428]

25.8. NON-BIRD REPTILES [p.429]
1. land; 2. three chambered heart; 3. lungs; 4. inner ear; 5. eyelids; 6. insects; 7. salamanders; 8. carnivores; 9. water; 10. reproduce; 11.

respiratory surface; 12. human; 13. Amniotes; 14. eggs; 15. water loss; 16. synapsids; 17. archosaurs; 18. birds; 19. cretaceous; 20. dinosaurs; 21. d; 22. b; 23. c; 24. a; 25. kidney; 26. gonads; 27. vertebral column; 28. spinal cord; 29. main parts of brain; 30. olfactory lobe; 31. peglike teeth; 32. esophagus; 33. lung; 34. heart; 35. liver; 36. stomach; 37. intestine; 38. cloaca

25.9. BIRDS—THE FEATHERED ONES [pp.430-431]
25.10. THE RISE OF MAMMALS [pp.432-433]
25.11. MODERN MAMMALIAN DIVERSITY [pp.434-435]
1. yolk sack; 2; embryo; 3. amnion; 4. chorion; 5. allantois; 6. albumin; 7. shell; 8. pectoral girdle; 9. pelvic girdle; 10. sternum; 11. humerus; 12. radius; 13. ulna; 14. dinosaurs; 15. Jurassic; 16. scales; 17. feathers; 18. sternum; 19. air cavities; 20. metabolic; 21. oxygen / blood; 22. four; 23. 1.6; 24. ostrich; 25. migration; 26. Antarctic; 27. hair; 28. mammary glands; 29. teeth; 30. synapsids; 31. dinosaurs; 32. monotremes; 33. marsupials; 34. placental; 35. Morphological convergence; 36. c; 37. c; 38. b; 39. b; 40. a; 41. a; 42. b; 43. c

SELF-TEST

1. a; 2. A; 3. D; 4. B; 5. C; 6. B; 7. C; 8. D; 9. A; 10. A; 11. D; 12. C; 13. A; 14. C; 15. B; 16. d, H; 17. b, B; 18. i, F; 19. g, D; 20. h, A; 21. a, I; 22. e, G; 23. f, C

Chapter 26   Human Evolution

26.1. A BIT OF A NEANDERTHAL [p.439]
26.2. PRIMATES: OUR ORDER [pp.440-441]
26.3. THE APES [p.442-443]

26.4. RISE OF THE HOMININS [p.444-445]
26.5. EARLY HUMANS [pp.446-447]
26.6. RECENT HUMAN LINEAGES [pp.448-449]

1. primates; 2. prosimians; 3. Anthropoids; 4.
monkeys; 5. humans; 6. daytime; 7. forward
facing; 8. smell; 9. prehensile; 10. opposable; 11.
grip; 12. tools; 13. upright walking; 14. teeth; 15.
brain; 16. behaviors; 17. culture; 18. Africa; 19.
Australopiths; 20. Homo; 21. 2.5; 22. brain; 23.
toolmaker; 24. Europe (Asia); 25. Asia (Europe);
26. 195,000; 27. Homo sapiens; 28. 100,000; 29.
Neanderthals; 30. replacement; 31. Homo erectus;
32. Homo sapiens; 33. multiregional; 34. Homo
sapiens; 35. c; 36. a; 37. b; 38. b; 39. b; 40. a; 41.
B; 42. I; 43. G; 44. B; 45. A; 46. D; 47. F. 48. H; 49.
C; 50. E; 51. G; 52. G; 53. H; 54. G; 55. H; 56. H. 57.
G; 58. H; 59. *H. erectus* in Africa and other regions
evolved into populations of *H. sapiens* gradually,

over more than a million years. Gene flow among
populations maintained the species through the
transition to fully modern humans; 60. *H. sapiens*
arose from a single *H. erectus* population in sub-
Saharan Africa within the past 200,000 years. Later,
bands of *H. sapiens* entered regions occupied by *H.
erectus* populations, and replaced them

SELF-TEST

1. a; 2. d; 3. c; 4. a; 5. a; 6. a; 7. b; 8. b; 9. a; 10. a; 11.
e; 12. c; 13. e; 14. b; 15. e; 16. B; 17. H; 18. C; 19. D;
20. A; 21. E; 22. G; 23. F

Chapter 27   Plant Tissues

27.1 SEQUESTERING CARBON [p.453] 27.2. THE PLANT BODY [pp.454-455]

1. ground tissues; 2. vascular tissues; 3. dermal
tissues; 4. shoots; 5. roots; 6. E; 7. H; 8. B; 9. F; 10.
A.; 11. D; 12. C; 13. G; 14. J; 15. I; 16. K; 17. radial;
18. tangential; 19. transverse

27.3. PLANT TISSUES [pp.456-457]

1. simple; 2. complex; 3. complex; 4. a; 5. c; 6. a; 7.
a; 8. b; 9. b; 10. c; 11. a; 12. a; 13. c; 14. (a) tracheid,
(b) dead, (c) xylem; 15. (a) sieve tube member, (b)
alive, (c) phloem; 16. (a) companion cells, (b) alive,
(c) phloem; 17. (a) vessel member, (b) dead, (c)
xylem

27.4. PRIMARY SHOOT STRUCTURE [pp.458-459]

1. epidermis; 2. vascular bundle; 3. pith; 4. cortex;
5. parenchyma; 6. sclerenchyma; 7. sieve tube ; 8.
vessel; 9. monocot; 10. eudicot; 11. apical
meristem; 12. immature leaf; 13. protoderm; 14.
procambium; 15. ground meristem

27.5. A CLOSER LOOK AT LEAVES [pp.460-461]

1. blade; 2. petiole; 3. lateral bud; 4. node; 5. stem;
6. sheath; 7. eudicot; 8. monocot; 9. palisade
mesophyll; 10. spongy mesophyll; 11. lower
epidermis; 12. stomata; 13. leaf vein

27.6. PRIMARY ROOT STRUCTURE [pp.462-463]

1. epidermis; 2. cortex; 3. pith; 4. vascular cylinder;
5. xylem; 6. phloem; 7. endodermis; 8 pericycle; 9.
primary root; 10. root cap; 11. epidermis; 12. root
hairs; 13. surface area; 14. vascular cylinder (stele);
15. endodermis; 16. pericycle; 17. lateral roots; 18.
taproot; 19. fibrous; 20. adventitious

27.7. SECONDARY GROWTH [pp.464-465]
27.8. TREE RINGS AND OLD SECRETS [p.466]

1. C; 2. J; 3. K; 4. E; 5. L; 6. F; 7. B; 8. A; 9. H; 10. G;
11. I; 12. D; 13. Hardwoods are eudicot trees from
temperate and tropical zones; they have vessels,
tracheids and fibers in their xylem. Conifers are
softwoods with less dense wood than hardwoods;
their xylem contains tracheids and parenchyma
rays, but no vessels or fibers. .; 14. The bark consists
of all the living and dead tissues outside the
vascular cambium. It is composed of secondary
phloem and periderm (parenchyma, cork, and cork
cambium). 15. Vascular cambium produces
secondary xylem, secondary phloem and
parenchyma rays. Cork cambium produces
periderm (cork, parenchyma and more cork
cambium).

27.9. VARIATIONS ON A STEM [p.467]
1. e; 2. f; 3. b; 4. a; 5. d; 6. c

SELF-TEST

1. b; 2. a ; 3. b; 4. c; 5. a; 6. d; 7. d; 8. c; 9. c; 10. b;
11. c; 12. b; 13. d; 14. b

Chapter 28   Plant Nutrition and Transport

28.1. LEAFY CLEANUP CREWS [p.471]
28.2. PLANT NUTRIENTS AND AVAILABILITY IN SOIL [pp.472-473]

1. f; 2. c; 3. i; 4. b; 5. h; 6. a; 7. j; 8. e; 9. g; 10. d; 11.
The thin layers of water and negatively charged
crystals that make up clay attract and bind the

positively charged mineral ions (for example, iron,
calcium, potassium, copper, and magnesium)
found in soil water. The larger particles of sand
and silt allow space for air and keep the tiny clay
particles from packing too tightly together. 12. Soil
erosion is the removal of soil due to wind, water,

and glacial action. In areas of deforestation or poor plant cover, much of the organic topsoil can be lost impoverishing the soil for future plant growth. Leaching is the process by which water percolating through the soil removes nutrients to lower soil horizons where plants cannot access them. More leaching occurs in sandy, well-drained soils than in humus or clay soils that are less porous.; 13. c; 14. a; 15. d; 16. b

28.3. HOW DO ROOTS ABSORB WATER AND NUTRIENTS? [pp.474-475]

1.b; 2. a; 3. c; 4. c; 5. a; 6. d; 7. b; 8. c; 9. 2; 10. 5; 11. 1; 12. 4; 13. 6; 14. 3; 15. primary xylem; 16. Mineral ions are actively transported across root hair plasma membranes. They diffuse from cell to cell via plasmodesmata until they reach the vascular cylinder and enter the xylem.

28.4. WATER MOVEMENT INSIDE PLANTS

[pp.476-477]
28.5. WATER-CONSERVING ADAPTATIONS OF STEMS AND LEAVES [pp.478-479]

1. vessels; 2. pits; 3. perforation plates; 4. members; 5. cohesion-tension; 6. xylem; 7. tension; 8. roots; 9. transpiration; 10. stomata; 11. negative pressure; 12. hydrogen bonds; 13. cohesion; 14. tension; 15. leaves; 16. stems; 17. roots; 18. soil; 19. d; 20. f; 21. e; 22. a; 23. b; 24. g; 25. c

28.5. MOVEMENT OF ORGANIC COMPOUNDS IN PLANTS? [pp.480-481]

1. d; 2. f; 3. b; 4. a; 5. c; 6. e; 7. g; 8. fibers, sieve tubes and companion cells; 9. nucleus; 10. companion cell; 11. translocation; 12. source; 13. sink; 14. Pressure Flow; 15. loading; 16. hypertonic; 17. osmosis; 18. turgor; 19. sink; 20. unloaded

SELF-TEST

1. c; 2. c; 3. d; 4. b; 5. c; 6. d; 7. b; 8. b; 9. a; 10. b; 11. c; 12. d; 13. c

---

Chapter 29   Life Cycles of Flowering Plants

29.1. PLIGHT OF THE HONEYBEE [p.485]29.2. REPRODUCTIVE STRUCTURES [pp.486-487]

1.sporophyte; 2. meiosis; 3. male gametophyte; 4. female gametophyte; 5. sperm; 6. eggs; 7. fertilization; 8. zygote in seed; 9. Regular flowers are symmetrical around their central axis (daisy). Irregular flowers are not radially symmetrical around their central axis (orchid).; 10. A complete flower has all four sets of modified leaves: sepals, petals, stamens & carpals (pistils). An incomplete flower is missing one or more of the previously mentioned sets of modified leaves.; 11. A perfect flower has both male (stamens) and female (carpel or pistil) reproductive organs.; 12. filament; 13. anther; 14. stigma; 15. style; 16. ovary; 17. ovule; 18. receptacle; 19. sepal; 20. petal; 21. stamens; 22. carpel, pistil; 23. calyx; 24. corolla

29.3. FLOWERS AND THEIR POLLINATORS [p.488-489]

1. pollination vector; 2. stigma; 3. wind; 4. pollinators; 5. pollen; 6. Birds (insects, bats); 7. insects (birds, bats); 8. bats (birds, insects); 9. coevolved; 10. bees; 11. UV (ultraviolet); 12. bats (moths); 13. moths (bats); 14. beetles (flies); 15. flies (beetles); 16. nectar; 17. honey; 18. floral tube; 19. birds (butterflies, moths); 20. butterflies (birds, moths); 21. moths (butterflies, birds)

29.4. A NEW GENERATION BEGINS [pp.490-491]
29.5. FLOWER SEX [p.492]

1. 4; 2. 2; 3. 6; 4. 5; 5. 1; 6. 3; 7. haploid; 8. diploid; 9. haploid; 10. haploid; 11. haploid; 12. diploid; 13. 5; 14. 2; 15. 6; 16. 1; 17. 3; 18. 4; 19. sporopollenin; 20. recognition proteins; 21. adhesion proteins; 22. dormant; 23. pollen tube; 24. furrows (pores); 25. pores (furrows); 26. style; 27. chemical signals; 28. egg; 29. species-specific; 30. fertilize

29.6. SEED FORMATION [p.493]
29.7. FRUITS [pp.494-495]
29.8 EARLY DEVELOPMENT [pp.496-497]

1. shoot tip (shoot apical meristem); 2. cotyledons; 3. endosperm; 4. root tip (root apical meristem); 5. seed coat (integuments); 6. b; 7. a; 8. b; 9. a; 10. c; 11. A simple fruit develops from one ovary in one flower (pea pod). An aggregate fruit develops from several ovaries in one flower (strawberry). A multiple fruit forms from the fusion of ovaries from separate flowers (pineapple).; 12. A true fruit forms only from the ovary wall and its contents (cherry). Accessory fruits are composed of other flowers parts in addition to the ovary. An apple core is the ovary, the fleshy edible part is derived from the receptacle.; 13. c (peanut, peas); 14. e (acorn, corn); 15. a (cherry, peach); 16. d (grape, tomato); 17. b (apple, pear); 18. d; 19. c; 20. b; 21. a; 22. c; 23. b; 24. a; 25. b; 26. 4; 27. 2; 28. 5; 29. 1; 30. 3

29.9. ASEXUAL REPRODUCTION OF FLOWERING PLANTS [pp.498-499]

1. e; 2. d; 3. h; 4. g; 5. b; 6. i; 7. a; 8. f; 9. j; 10. c; 11. d; 12. b; 13. e; 14. c; 15. a

1. b; 2. c; 3. d; 4. c; 5. c; 6. d; 7. b; 8. c; 9. c; 10. d; 11. d; 12. d; 13. c; 14. d

Chapter 30   Communication Strategies in Plants

30.1. PRESCRIPTION: CHOCOLATE [p.505]30.2. INTRODUCTION TO PLANT HORMONES [pp.506-507]

1. environment; 2. cocoa; 3. chocolate; 4. flavonoids; 5. epicatechin; 6. immunity; 7. heart attack; 8. stroke; 9. cancer; 10. cocoa; 11. c; 12. d; 13. a; 14. e

30.3. AUXIN: THE MASTER GROWTH HORMONE [pp.508-509]
30.4. CYTOKININ [p.510]
30.5. GIBBERELLIN [p.511]
30.6. ABSCISIC ACID [p.512]
30.7. ETHYLENE [p.513]

1. c, d; 2. a; 3. e.; 4. b; 5. a; 6. d; 7. b, e; 8. a; 9. e; 10. b; 11. a; 12. d; 13. e; 14. a; 15. a; 16. c; 17. c; 18. a; 19. b; 20. d; 21. auxin; 22. cytokinin; 23. gibberellin; 24. ethylene; 25. abscisic acid (ABA); 26. gibberellin; 27. amylase; 28. endosperm; 29. indoleacetic acid (IAA); 30. transport proteins; 31. Turgor; 32. shoot tips; 33. young leaves; 34. phloem; 35. efflux carriers; 36. Strigolactone; 37. apical dominance; 38. cytokinin; 39. abscisic acid (ABA); 40. ethylene

30.8. TROPISMS [pp.514-515]

1. b; 2. e; 3. a; 4. a; 5. b; 6. c; 7. d; 8. c; 9. tropism; 10. stimulus; 11. hormones; 12. down; 13. up; 14. gravitropism; 15. statoliths; 16. auxin; 17. phototropism; 18. auxin; 19. phototropins; 20. blue light; 21. shaded; 22. elongate; 23. tendrils; 24. thigmotropism; 25. elongation; 26. Mechanical stress

30.9. SENSING RECURRING ENVIRONMENTAL CHANGES [pp.516-517]
30.10. RESPONSES TO STRESS [pp.518-519] []

1. g; 2. b; 3. i; 4. d; 5. k; 6. c; 7. l; 8. f; 9. e; 10. j; 11, a; 12. h; 13. abiotic; 14. biotic; 15. abscisic acid; 16. nitric oxide ; 17. ethylene; 18. stomata; 19. hypersensive ; 20. systemic acquired resistance; 21. salicylic acid; 22. epicatechin; 23. jasmonic acid; 24. air; 25. egg

SELF-TEST

1. b; 2. d; 3. a; 4. c; 5. c; 6. a; 7. d; 8. c; 9. d; 10. d; 11. a; 12. c; 13. b; 14. d

Chapter 31   Animal Tissues and Organ Systems

31.1. STEM CELLS – IT'S ALL ABOUT POTENTIAL [p.523]
31.2. ORGANIZATION OF ANIMAL BODIES [pp.524-525]

1. stem cells; 2. embryonic; 3. adult; 4. Anatomy; 5. physiology; 6. Animal ; 7. Epithelial; 8. Connective; 9. Muscle; 10. Nervous; 11. extracellular; 12. Wastes; 13. Interstitial; 14. Blood;

31.3. EPITHELIAL TISSUE [pp.526-527]

1. Epithelial; 2. Basement membrane; 3. One; 4. Many; 5. simple epithelial; 6. Stratified epithelium; 7. Exocrine; 8. epithelial; 9. Endocrine; 10. hormones; 11. circulatory system; 12. Tight; 13. Adhering; 14. cell layers; 15. T; 16. protection; 17. T; 18. blood

31.4. CONNECTIVE TISSUES [pp.528-529]
31.5. MUSCLE TISSUES [pp.530-531]
31.6. NERVOUS TISSUE [p.531]

1. b; 2. c; 3. a; 4. c; 5. b; 6. a; 7. c; 8. c; 9. A; 10, C; 11. D; 12. B; 13. smooth; 14. smooth; 15. skeletal; 16. Skeletal; 17. skeletal; 18. striped; 19. cause

movement in internal organs; 20. cardiac; 21. neurons; 22. Neuroglia; 23. neurons; 24. connective, D, h. j; 25. epithelial, G, a, e, k; 26. muscle, I, g, (j); 27. muscle, J, d, (j); 28. connective, E, f, j, (m); 29. connective, B, I; 30. epithelial, H, a, e, k; 31. connective, K, a, n; 32. nervous, L, b; 33. muscle, C, l; 34. epithelial, F, a, e, k; 35. connective, A, c, f, m

31.7. ORGAN SYSTEMS [pp.532-533]

1. circulatory (D); 2. respiratory (E); 3. urinary (excretory)(A); 4. skeletal (C); 5. endocrine (J); 6. reproductive (G); 7. digestive (I); 8. muscular (H); 9. nervous (B); 10. integumentary (F); 11. lymphatic (K)

31.8. HUMAN INTEGUMENTARY SYSTEM [pp.534-535]
31.9. NEGATIVE FEEDBACK IN HOMEOSTASIS

1. skin; 2. epidermis; 3. stratified squamous; 4. dermis; 5. connective tissue; 6. Keratinocytes; 7. keratin; 8. hair; 9. Melanocytes; 10. melanin; 11. ultraviolet; 12. melanocytes; 13. vitamin D; 14.

elastic; 15. cultured; 16. burns; 17. G; 18. C; 19. F;
20. B; 21. D; 22. A; 23. H; 24. E; 25. II; 26. III; 27. I;
28. Unbroken skin is an effective barrier against
pathogens; 29. melanocytes, sweat glands, oil
glands, any other differentiated structures; 30. A
process of homeostasis in which a change causes a
reverses the change; 31. A; 32. D; 33. D; 34. B; 35.
D; 36. C; 37. D; 38. C

SELF-TEST
1. d; 2. b; 3. c; 4. c; 5. d; 6. c; 7. c; 8. d; 9. a; 10. d; 11.
d; 12. b; 13. b; 14. c; 15. b; 16; c; 17. a; 18. b; 19. b;
20. c; 21. A; 22. G; 23. E; 24. J; 25. H; 26. I; 27. F; 28.
B; 29. D; 30. C

Chapter 32   Neural Control

32.1. IN PURSUIT OF ECSTASY [p.541]
**32.2. EVOLUTION OF NERVOUS SYSTEMS
[pp.542-543]**
1. A; 2. F; 3. G; 4. E; 5. D; 6. D; 7. H; 8. C; 9.
autonomic system; 10. somatic system; 11.
peripheral nerves; 12. spinal cord; 13. Brain; 14. B;
15. F; 16. D; 17. A; 18. J; 19. I; 20. H; 21. C; 22. E;
23. G

**32.3. NEURONS—THE GREAT
COMMUNICATORS [pp.544-545]**
1. dendrite (B); 2. input zone (A); 3. trigger zone
(C); 4. conducting zone (E); 5. axon (D); 6. output
zone (F); 7. A; 8. F; 9. C; 10. B; 11. D; 12. E; 13. C;
14. B; 15. C; 16. B; 17. E; 18. D; 19. Sensory
neurons; 20. Interneurons; 21. Motor neurons; 22.
Neurons have specialized extensions called axon
and dendrites; they are modified for electrical and
chemical processes for the transmission of signals;
23. Dendrites receive signals and axons transmit
signals.

32.4. THE ACTION POTENTIALS [pp.546-547]
1. F, sodium; 2. T; 3. F, sodium; 4. F, threshold
potential; 5. T; 6. -70mV; 7. T; 8. Na^+, K^+; 9. II; 10.
III; 11. II; 12. III; 13. II; 14. I; 15. IV; 16. I; 17. I; 18.
IV; 19. IV; 20. I;

**32.5. HOW NEURONS SEND MESSAGES TO
OTHER CELLS [pp.548-549]**
**32.6. A SMORGASBORD OF SIGNALS [pP.550-
551]**
1. F; 2. H; 3. G; 4. C. 5. A; 6. E; 7. B; 8. D; 9.
synaptic integration; 10. T; 11. synapse; 12.
receptors; 13. T; 14. B; 15. A.; 16. G; 17. E; 18. C;
19. F; 20. D; 21. a. stimulant, block Ach receptors;
b. stimulant, stop uptake of dopamine; c.
stimulant, increase serotonin, norepinephrine, and
dopamine; d. depressant, inhibit Ach output; e.
depressant, inhibit Ach output; f. analgesic, slow
clearing of synapse; g. analgesic, slow clearing of
synapse; h. hallucinogen, binds to serotonin
receptor—distort sensory data; i. hallucinogen,
alters levels of dopamine, serotonin,
norepinephrine, and GABA—relaxed and
inattentive; 22. Drug takes on a vital biochemical
function in the body's homeostasis; 23. Tolerance,

habituation, inability to stop, concealing use,
extreme or dangerous actions to get or use drug,
deterioration of relationships, anger about use,
prefer drug use over other activities.

**32.7. THE PERIPHERAL NERVOUS SYSTEM
[pp.552-553]**
1. A; 2. E; 3. D; 4. C; 5. B; 6. Increase speed of nerve
transmission; 7. They have opposite effects.
Sympathetic alerts body to face stress while
parasympathetic slows down activities for normal
resting conditions; 8. optic nerve; 9. vagus nerve;
10. pelvic nerve; 11. midbrain; 12. medulla
oblongata; 13. cervical nerves; 14. thoracic nerves;
15. lumbar nerves; 16. sacral nerves.

32.8. THE SPINAL CORD [pp.554-555]
1. spinal cord; 2. spinal nerve; 3. vertebra; 4.
intervertebral disk; 5. meninges; 6. B; 7. C; 8. F; 9.
A; 10. D; 11. E; 12. G; 13. c; 14. b; 15. d; 16. a; 17. c;
18. d; 19. d; 20. b; 21. d

32.9. THE VERTEBRATE BRAIN [pp.556-557]
1. D; 2. F; 3. G; 4. H; 5. C; 6. B; 7. A; 8. E; 9. a; 10. a;
11. a; 12. b; 13. b; 14. b; 15. c

**32.10. A CLOSER LOOK AT THE HUMAN
CEREBRUM [pp.558-559]**
32.11. EMOTION AND MEMORY[p.559]
1. h; 2. i; 3. b; 4. e; 5. d; 6. h; 7. f; 8. a; 9. h; 10. c; 11.
g; 12. Sperry demonstrated that signals that
cross the corpus callosum coordinate the functions of the
two cerebral hemispheres, each of which has
responded to visual signals from the opposite side
of the body; 13. Short; 14. long; 15. permanently;
16. skill; 17. prefrontal; 18. motor; 19. cerebellum;
20. basal ganglia; 21. Declarative; 22. sensory; 23.
gatekeeper; 24. hippocampus; 25. association; 26.
basal ganglia; 27. thalamus

**32.12. NEUROGLIA—THE NEURON'S
SUPPORT STAFF [pp.560-561]**
1. a. line the ventricles and produce cerebrospinal
fluid; b. phagocytic cells within the central nervous
system; c. form blood-brain barrier, take up
neurotransmitters, form lactate, and make nerve

growth factor; d. produce myelin sheath on axons in central nervous system.

SELF-TEST

1. a; 2. a; 3. b; 4. d; 5. a; 6. b; 7. c; 8. d; 9. b; 10. b; 11. d; 12. d; 13. c; 14. e; 15. B; 16. d; 17. b; 18. d; 19. c; 20. d; 21. c; 22. b; 23. b; 24. c; 25. a

---

Chapter 33 Sensory Perception

33.1. A WHALE OF A DILEMMA [p.565]

33.2. OVERVIEW OF SENSORY PATHWAYS [pp.566-567]

1. T; 2. T; 3. chemoreceptors; 4. chemoreceptors; 5. T; 6. fewer; 7. sensory adaptation; 8. somatic; 9. d; 10. c; 11. a; 12. b; 13. a; 14. e; 15. e; 16. b; 17. d; 18. b; 19. b; 20. a; 21. B; 22. A (e); 23. E (a); 24. C (c); 25. D (d); 26. B (b)

33.3. SOMATIC AND VISCERAL SENSATIONS [pp.568-569]

33.4. CHEMICAL SENSES[pp.570-571]

1. somatosensory cortex; 2. lips/mouth (fingers/hand); 3. fingers/hand (lips/mouth); 4. Free nerve endings; 5. pain; 6. Meissner's corpuscles; 7. pain; 8. endorphins (enkephalins); 9. enkephalins (endorphins); 10. sensory; 11. Olfactory (chemo-); 12. nose; 13. five; 14. olfactory; 15. Pheromones; 16. olfactory (chemo-); 17. vomeronasal; 18. chemoreceptors

33.5. DIVERSITY OF VISUAL SYSTEMS [pp.572-573]

33.6. A CLOSER LOOK AT THE HUMAN EYE [pp.574-575]

33.7. LIGHT RECEPTION AND VISUAL PROCESSING [pp.576-577]

33.8. VISUAL DISORDERS [p. 577]

1. lens; 2. cornea; 3. iris; 4. sclera; 5. lens; 6. visual accommodation; 7. rods; 8. cones; 9. fovea; 10. rhodopsin; 11. blue-green; 12. optic nerve; 13. blind spot (optic disk); 14. J (h)15. C (d); 16. M (g); 17. H (e); 18. A (a); 19. D (i); 20. I (k); 21. B (f); 22. G (j); 23. K (none); 24. E (e); 25. F (b); 26. L (none); 27. Farsightedness; 28. macula; 29. T; 30. lens; 31. cornea; 32. T; 33. red; 34. Roundworms; 35. E; 36. A; 37. D; 38. F; 39. C; 40. B

33.9. SENSE OF HEARING [pp.578-579]

33.10. ORGANS OF EQUILIBRIUM[p.580]

1. fluid in the canals; 2. T; 3. static equilibrium; 4. T; 5. T; 6. basilar membrane; 7. cochlea; 8. T; 9. ;10. A; 11. B; 12. I; 13. C; 14. G; 15. H; 16. E; 17. F; D; 19. B; 20. G; 21. C; 22. A; 23. E; 24. D; 25. F; 26. C; 27. G; 28. A; 29. F; 30. D; 31. B; 32. E; 33. Permanent damage to the nerve hairs can result in permanent hearing loss that cannot be repaired.

SELF-TEST

1. c ; 2. c ; 3. d ; 4. e ; 5. d ; 6. a ; 7. c ; 8. b; 9. b; 10. a ; 11. d; 12. c; 13. d; 14. c; 15. a; 16. e; 17. b; 18. a; 19. a; 20. c; 21. d; 22. b

---

Chapter 34 Endocrine Control

34.1. HORMONES IN THE BALANCE [p.585]

34.2. THE VERTEBRATE ENDOCRINE SYSTEM [pp.586-587]

1. Atrazine; 2. endocrine disruptor; 3. hermaphrodites; 4. feminized (abnormal); 5. kepone (DDT); 6. DDT (kepone); 7. E; 8. A; 9. B; 10. F; 11. C; 12. D; 13.a. 1, several releasing and inhibiting hormones, synthesizes ADH and oxytocin; b. 1, ACTH, TSH, LH, somatotropin (STH), prolactin (PRL); c. 1, stores and secrete two hypothalamic hormones, ADH and oxytocin; d. 2, cortisol, aldosterone, (sex hormones of opposite sex); e. 2, epinephrine, norepinephrine; f. 2, estrogens, progesterone; g. 2, testosterone; h. 1, melatonin; i. 1, thyroxine and triiodothyronine; j. 4, parathyroid hormone (PTH); k. 1, thymosins; l. 1, insulin, glucagon

34.3. THE NATURE OF HORMONE ACTION

[pp.588-589]

1. a; 2. b; 3. a; 4. b; 5. b; 6. a; 7. b; 8. b; 9. b; 10. b; 11. A second messenger carries the signal from the cell membrane to the inner parts of the cell when the hormone attaches to a receptor on the cell membrane surface. Only steroid (lipid) hormones can pass directly through the membrane and into the nucleus of the cell. All others must attach to receptors on cell surface and then need a means to transfer the signal to inner parts of the cell—second messenger.; 12. If a cell that is normally a target cell for a particular hormone lacks the receptor for that hormone, then the cell will be unable to respond to the signal. It will act as though no hormone was secreted; 13. A. 4, B. 5, C. 2, D. 1, E. 3; 14. A. 3, B. 1, C. 4, D. 2, E. 5,

34.4. THE HYPOTHALAMUS AND PITUITARY GLAND [pp.590-591]

1. A (H); 2. P (G); 3. A (A); 4. A (I); 5. A (D); 6. A (B); 7. P (E); 8. A (C); 9. A (F); 10. T; 11. hypothalamus; 12. hypothalamus; 13. anterior pituitary; 14. T; 15. anterior pituitary; 16. glandular, nerve; 17. prolactin and oxytocin

34.5. GROWTH HORMONE FUNCTION AND DISORDERS [p.592]
34.6. SOURCES AND EFFECTS OF OTHER VERTEBRATE HORMONES [p.593]

1. F; 2. A; 3. E; 4. C; 5. B; 6. D; 7. Injections of recombinant human growth hormone (rhGH) increase the growth rate of children with lower than normal GH. Treatments are expensive and some argue that short stature is not a defect to be treated; 8. Hormones can signal to organs and organs can release hormones. Most organ cells have multiple receptors for more than one hormone.

34.7. THYROID AND PARATHYROID GLANDS [pp.594-595]

1. negative feedback; 2. hypothalamus; 3. TSH (thyroid stimulating hormone); 4. thyroid; 5. iodine; 6. simple goiter; 7. hyperthyroidism; 8. parathyroid; 9. calcium; 10. bones; 11. kidneys; 12. rickets; 13. metamorphosis; 14. thyroid disruptors; 15. Iodized salt provides a steady source of iodine for the thyroid to prevent hypothyroidism and goiter; 16. Hyperthyroidism can produce symptoms of anxiety, insomnia, heat intolerance, weight loss, and tremors.

34.8. PANCREATIC HORMONES [pp.596-597]
34.9. THE ADRENAL GLANDS [pp.598-599]
34.10. GONADS [p.600]
34.11. THE PINEAL GLAND [p.601]

1. c, B; 2. e, G; 3. b, D; 4. d, A; 5. a, G; 6. g, C; 7. f, E; 8. a, B, 9. d, A; 10. c, B; 11. e, C; 12. b, D; 13. juvenile (child); 14. low; 15. T; 16. elevated; 17. T; 18. long-term; 19. A. 2, B. 4, C. 8, D. 6, E. 1, F. 3, G. 5; H. 7; 20. Responses may include blindness, increased blood pressure, increased risk of kidney disease, increased risk of amputation; 21. Western diets and sedentary lifestyles.; 22. Harm can be done to the cardiovascular, reproductive, and immune systems, as well as affect memory and growth; 23. SAD can occur when a person's melatonin clock gets out of rhythm; 24. Cancer and infectious diseases.

35.13. A COMPARATIVE LOOK AT A FEW INVERTEBRATES [p.613]

1. molt; 2. ecdysone; 3. steroid; 4. X-; 5. eye stalk; 6. insects; 7. molt-inhibiting hormone

SELF-TEST

1. a; 2. d; 3. b; 4. e; 5. d; 6. b; 7. a; 8. c; 9. L; 10. E; 11. N; 12. D; 13, F; 14. A; 15. I; 16. B; 17. O; 18. G; 19. J; 20. C; 21. M; 22. K; 23. H; 24. U; 25. Q; 26. P; 27. W; 28. S; 29. R; 30. T; 31. V

Chapter 35   Structural Support and Movement

35.1. MUSCLES AND MYOSTATIN [p.607]
35.2. INVERTEBRATE SKELETONS [pp.608-609]
35.3. THE VERTEBRATE ENDOSKELETON [pp.610-611]

1. c; 2. a; 3. b; 4. b; 5. a; 6. c; 7. B; 8. D; 9. E; 10. A; 11. C; 12. cranial bones; 13. vertebrae; 14. intervertebral disks; 15. clavicle; 16. scapula; 17. humerus; 18. radius; 19. ulna; 20. carpals; 21. metacarpals; 22. phalanges; 23. pelvic girdle; 24. femur; 25. patella; 26. tibia; 27. fibula; 28. tarsals; 29. metatarsals; 30. phalanges

35.4. BONE STRUCTURE AND FUNCTION [pp.612-613]

35.5. SKELETAL JOINTS—WHERE BONES MEET [pp.614-615]

1. F; 2. C; 3. E; 4. D; 5. H; 6. G; 7. A; 8. B; 9. c; 10. b; 11. c; 12. a; 13. c; 14. b; 15. c; 16. a; 17. C; 18. A; 19. F; 20. G; 21. E; 22. B; 23. D; 24. A; 25. D, E; 26. G; 27. B; 28. H; 29. C, F; 30. Ligaments holding bones together in a joint stretch too far or tears; 31. Crucuate ligaments are in the knee joint. They stabilize the knee. When damaged, the bones shift and may give out when trying to stand or walk.; 32. Arthritis is inflammation of a joint. Bursitis is when fluid filled sacs called bursae that cushion joints become inflamed.

35.6. SKELETAL-MUSCULAR SYSTEMS [pp.616-617]
35.7. HOW DOES SKELETAL MUSCLE CONTRACT? [pp.618-619]

1. muscle fiber; 2. tendon; 3. bone; 4. lever; 5. fixed; 6. joint; 7. force; 8. skeletal; 9. opposition; 10; resists (reverses); 11. Smooth; 12. stomach; 13. heart; 14. tendon; 15. triceps brachii (C); 16. pectoralis major (G); 17. serrattus anterior (I); 18. external oblique (M); 19. rectus abdominus (K); 20. adductor longus (D); 21. Sartorius (L); 22. quadriceps femoris (F); 23. tibialis anterior (A); 24. biceps brachii (O); 25. deltoid (N); 26. trapezius (P); 27. latissimus dorsi (J);

28. gluteus maximus (H); 29. biceps femoris (E); 30. gastrocnemius (B); 31. F; 32. D; 33. G; 34. E; 35. A; 36. C; 37. B; 38. actin; 39. T; 40. closer together; 41. stays the same; 42. smaller; 43. T; 44. actin; 45. T; 46. A. 3, B. 2, C. 5, D. 4, E. 1.

35.8. NERVOUS CONTROL OF MUSCLE CONTRACTION [pp.620-621]
35.9. MUSCLE METABOLISM [pp.622-623]

1. E; 2. D; 3. F; 4. A; 5. B; 6. C; 7. H; 8. G; 9. I; 10. L; 11. J; 12. K; 13. Troponin and tropomyosin cover up the binding sites on actin. When calcium is released from the sarcoplasmic reticulum, troponin and tropomyosin move out of the way to expose the myosin-binding sites.; 14. An isotonic contraction is one where you would observe motion. In an isometric contraction, the muscle is flexed and released, but no actual movement is observed; 15. Exercise tones muscle and builds stronger bones; 16. *C. tetani* prohibits the release of GABA, which is a muscle inhibitor. As a result, muscles over the whole body tense; 17. It impairs motor neuron function.

SELF-TEST

1. b; 2. d; 3. c; 4. d; 5. b; 6. c; 7. b; 8. a; 9. c; 10. c; 11. a; 12. b; 12. a; 14. a; 15. c; 16. a; 17. b; 18. d; 19. c; 20. b

---

Chapter 36   Circulation

36.1. A SHOCKING SAVE [p.627]
36.2. THE NATURE OF BLOOD CIRCULATION [pp.628-629]

1. Sudden cardiac arrest; 2. Cardiopulmonary resuscitation (CPR); 3. defibrillator; 4. closed; 5. Blood; 6. heart; 7. interstitial fluid; 8. heart; 9. rapidly; 10. capillary; 11. slows; 12. single; 13. pulmonary; 14. systemic; 15. pressure; 16. A. aorta; B. heart; C. gut cavity; D. dorsal blood vessel; E. ventral blood vessel; F. two of 5 hearts; 17. Open circulatory system: Blood is pumped into short tubes that open into spaces in the body's tissues, mixes with tissue fluids, then reenters the heart at openings in the heart wall.; 18. Closed circulatory system: Blood flow is confined within blood vessels that have continuously connected walls. Blood does not mix with interstitial fluid. The blood is moved by the action of five pairs of "hearts."

36.3. HUMAN CARDIOVASCULAR SYSTEM [pp.630-631]
36.4. COMPONENTS AND FUNCTIONS OF BLOOD [pp.632-633]
36.5. HEMOSTASIS [p.634]
36.6. BLOOD TYPING [pp.634-635]

1. body; 2. T; 3. T; 4. moving away from the heart; 5. T; 6. a. Solvent; b. Plasma proteins; c. Red blood

cells; d. Neutrophils; e. Lymphocytes; f. Defense against parasitic worms; g. Platelets; 7. Blood carries oxygen, nutrients, and other solutes to cells and transports the cells' metabolic wastes and secretions. It transports hormones to all parts of the body. It helps stabilize pH and regulate body temperature. It is the transport system for cells and proteins that help protect and repair tissues.; 8. Anti Rh antibodies can pass across the placenta and damage red blood cells in the fetal circulation resulting in anemia or death of the fetus; 9. T; 10. T; 11. plasma; 12. bone marrow; 13. T; 14. short lived; 15. neutrophils (macrophages, monocytes); 16. platelets; 17. C; 18. B; 19. F; 20. D; 21. A; 22. E; 23. stem (E); 24. red blood (D); 25. platelets (B); 26. neutrophils (A); 27. B (C); 28. T (C); 29. monocytes (A); 30. Acts as solvent for proteins, salts, dissolved gases, hormones, and nutrients. Also carries/supports white blood cells and red blood cells; 31. jugular; 32. superior vena cava; 33. pulmonary; 34. renal; 35. inferior vena cava; 36. iliac; 37. femoral; 38. femoral; 39. iliac; 40. abdominal; 41. renal; 42. brachial; 43. coronary; 44. pulmonary; 45. carotid; COLORING: all arteries are red except for the pulmonary trunk and arteries, which are blue. All veins are blue except for the pulmonary veins which are red. 46. A, C, E,

G; 47. A, B, C, D, E, F, G, H; 48. C, G; 49. G; 50. F, G.

36.7. THE HUMAN HEART [pp.636-637]
1. D; 2. E; 3. C; 4. A; 5. F; 6. B; 7. aorta; 8. left pulmonary veins; 9. left semilunar valve; 10. left ventricle; 11. inferior vena cava; 12. right atrioventricular valve; 13. superior vena cava; COLORING: Both vevna cavae, the right atrium, right ventricle, and the pulmonary trunk and arteries are blue. The left atrium, left ventricle, the aorta, and the pulmonary veins are red.

36.8. BLOOD VESSEL STRUCTURE AND FUNCTION [p.638]
36.9. BLOOD PRESSURE [p.639]
36.10. MECHANISMS OF CAPILLARY EXCHANGE [p.640]
36.11. VENOUS FUNCTION [p.641]
36.12. BLOOD AND CARDIOVASCULAR DISORDERS [pp.642-643]
1. aorta (arteries); 2. slows; 3. friction; 4. Arterioles; 5. vasodilation; 6. vasoconstriction; 7. Receptors; 8. kidneys; 9. vein; 10. artery; 11. arteriole; 12. capillary; 13. smooth muscle, elastic fibers; 14.

valve; 15. capillaries; 16. slowly, need; 17. T; 18. sixty; 19. T; 20. muscles in veins contract; 21. ultrafiltration; 22. T; 23. Blood pressure depends on total blood volume and how much is pumped out of the heart in a given time period; 24. At 125, ventricles are contracting. At 86, ventricles are relaxed and atria are contracting. 25. I; 26. F; 27. E; 28. G; 29. A; 30. H; 31. J; 32. D; 33. C; 34. B

36.13. INTERACTIONS WITH THE LYMPHATIC SYSTEM [pp.644-645]
1. C; 2. B; 3. D; 4. A; 5. tonsils; 6. thymus; 7. thoracic; 8. spleen; 9. lymph node(s); 10. lymphocytes; 11. bone marrow; 12. 1. Drainage channels for water and plasma proteins that leaked out of capillaries; 2. Delivers fats absorbed from intestine to blood; 3. Transports cell debris, pathogens, foreign cells to lymph nodes.

SELF-TEST
1. c; 2. e; 3. e; 4. a; 5. b; 6. e; 7. a; 8. e; 9. d; 10. a; 11. H; 12. M; 13. G; 14. B; 15. O; 16. F; 17. R; 18. N; 19. E; 20. L; 21. A; 22. Q; 23. J; 24. D; 25. K; 26. I; 27. C; 28. P

Chapter 37   Immunity

37.1. FRANKIE'S LAST WISH [p.649]
37.2. INTEGRATED RESPONSE TO THREATS [pp.650-651]
37.3. SURFACE BARRIERS [pp.652-653]
37.4. TRIGGERING INNATE DEFENSES [pp.654-655]

37. 5. INFLAMMATION AND FEVER [pp.656-657]

1. antigens; 2. 1000; 3. pathogen-associated molecular patterns (PAMPs); 4. complement; 5. innate; 6. adaptive; 7. skin; 8. mucus; 9. Lysozyme; 10. Gastric; 11. bacterial; 12. pathogens; 13. mucus membranes; 14. T; 15. T; 16. cytokines; 17. C; 18. D; 19. E; 20. G; 21. H; 22. A; 23. F; 24. B; 25.(A) d; 25.(B) b; 25.(C) a; 25.(D) e; 25.(E) c. E; 26. E; 27. A; 28. B; 29. D; 30. F; 31. C; 32. B; 33. A; 34. C; 35. F; 36. E; 37. D; 38. A fever increases the rate of enzyme activity resulting in more rapid metabolism, tissue repair, and activity of phagocytes. It may slow pathogen cell division.

37.6. ANTIGEN RECEPTORS [pp.658-659]
37.7. OVERVIEW OF ADAPTIVE IMMUNITY [pp.660-661]
37.8. THE ANTIBODY-MEDIATED IMMUNE RESPONSE [pp.662-663]
37.9. THE CELL-MEDIATED IMMUNE

RESPONSE [pp.664-665]
1. innate; 2. adaptive; 3. MHC marker (surface antigens); 4. non-self; 5. lymphocytes; 6. B cell (B lymphocyte); 7. T cell (T lymphocyte); 8. viruses (bacteria or fungus or protest); 9. cancer (tumor); 10. identity; 11. antigen; 12. antigen-presenting; 13. effector; 14. memory; 15. B; 16. T; 17. recognition of antigens; 18. T; 19. cytokines; 20. proteins; 21. T cells; 22. antigen presenting cells; 23. D; 24. C; 25. B; 26. A; 27. E; 28. C; 29. E; 30. F; 31. I; 32. B; 33. J; 34. D; 35. H; 36. G; 37. A; 38. G; 39. E; 40. A; 41. F; 42. B; 43. C; 44. D

37.10. WHEN IMMUNITY GOES WRONG [p.666-667]
37.11. HIV AND AIDS [pp.668-669]
37.12. VACCINES [pp.670-671]
1. I; 2. M; 3. B; 4. L; 5. E; 6. O; 7. N; 8. J; 9. C; 10. G; 11. H; 12. A; 13. K; 14. F; 15. D; 16. primary; 17. antigens; 18. T; 19. secondary; 20. reverse transcriptase; 21. too much; 22. T; 23. Edward Jenner

SELF-TEST
1. e; 2. b; 3. b; 4. a; 5. a; 6. e; 7. c; 8. c; 9. e; 10. a; 11. c; 12. e; 13. d; 14. a; 15. b; 16. d; 17. a; 18. c; 19. H; 20; D; 21. E; 22. J; 23. B; 24. G; 25. C; 26. I; 27. F; 28. A

Chapter 38 Respiration

38.1. CARBON MONOXIDE – A STEALTHY POISON [p.675]
38.2. THE NATURE OF RESPIRATION [pp.676-677]
38.3. INVERTEBRATE RESPIRATION [pp.678-679]

38.4. VERTEBRATE RESPIRATION [pp.680-681]
1. B; 2. E; 3. A; 4. C; 5. G; 6. D; 7. F; 8. C; 9. D; 10. B; 11. F; 12. G; 13. A; 14. E; 15. Cool, fast-flowing; 16. T; 17. first; 18. four; 19. T; 20. b; 21. b; 22. a; 23. a; 24. c; 25. d; 26. a; 27. c; 28. b; 29. c; 30. b

38.5. HUMAN RESPIRATORY SYSTEM [pp.682-683]
38.6. CYCLIC REVERSALS IN AIR PRESSURE GRADIENTS [pp.684-685]
1. alveoli (B); 2. bronchiole (J); 3. alveolar sac (L); 4. oral cavity (A); 5. pleural membrane (K); 6. intercostal muscles (H); 7. diaphragm (C); 8. nasal cavity (G); 9. pharynx (F); 10. epiglottis (M); 11. larynx (E); 12. trachea (I); 13. lung (D); 14. a; 15. a; 16. b; 17. b; 18. b; 19. b; 20. a; 21. b; 22. C; 23. G;

24. A; 25. F; 26. D; 27. B; 28. E; 29. A maneuver that raises the intra-abdominal pressure of a choking individual to dislodge an object in the trachea

38.7. GAS EXCHANGE AND TRANSPORT [pp.686-687]
38.8. RESPIRATION IN EXTREME ENVIRONMENTS [pp.688-689]
38.9. RESPIRATORY DISEASES AND DISORDERS [pp.690-691]
1. b; 2. a; 3. a (or c); 4. b; 5. a; 6. b; 7. c; 8. a; 9. K; 10. H; 11. C; 12. L; 13. E; 14. J; 15. A; 16. F; 17. G; 18. I; 19. B; 20. D; 21. F; 22. D; 23. A; 24. B; 25. E; 26. C

SELF-TEST
1. L; 2. H; 3. P; 4. Q; 5. F; 6. E; 7. U; 8. I; 9. R; 10. B; 11. K; 12. A; 13. C; 14. S; 15. M; 16. N; 17. G; 18. T; 19. J; 20. D; 21. b; 22. e; 23. a; 24. c; 25. c; 26. a; 27. a; 28. a; 29. a; 30. e; 31. e; 32. d; 33. d; 34. c; 35. c

---

Chapter 39 Digestion and Human Nutrition

39.1. YOUR MICROBIAL "ORGAN" [pp.695]
39.2. THE NATURE OF DIGESTIVE SYSTEMS [pp.696-697]
39.3. OVERVIEW OF THE HUMAN DIGESTIVE SYSTEM [pp.698-699]
39.4. DIGESTION IN THE MOUTH [p.699]
1. nutrition; 2. mechanically (chemically); 3. chemically (mechanically); 4. absorption; 5. circulatory; 6. incomplete; 7. circulatory; 8. complete; 9. opening; 10. Movements; 11. secretion; 12. absorption; 13. excretion; 14. Teeth; 15. incisors; 16. molars; 17. ruminants; 18. crown; 19. stomach; 20. small intestine; 21. accessory; 22. pancreas; 23. salivary amylase; 24. epiglottis; 25. esophagus; 26. salivary glands (H); 27. liver (F); 28. gallbladder (C); 29. pancreas (B); 30. anus (D); 31. large intestine (colon) (E); 32. small intestine (J); 33. stomach (A); 34. esophagus (G); 35. pharynx (I); 36. mouth (oral cavity) (K); 37. E; 38. G; 39. C; 40. J; 41. I; 42. D; 43. A; 44. F; 45. H; 46. B

39.5. FOOD STORAGE AND DIGESTION IN THE STOMACH [p.700]
39.6. STRUCTURE OF THE SMALL INTESTINE [p.701]
39.7. DIGESTION AND ABSORPTION IN THE SMALL INTESTINE [pp.702-703]
39.8. ABSORPTION AND STORAGE IN THE COLON [p.704]

39.9. METABOLISM OF ABSORBED ORGANIC COMPOUNDS [p.705]
1. C; 2. G; 3. F; 4. A; 5. D; 6. B; 7. H; 8. E; 9. a. mouth, small intestine (duodenum); small intestine (jejunum/ileum); monosaccharide (sugar); b. stomach, small intestine (duodenum); small intestine (jejunum/ileum); amino acids; c. small intestine (duodenum); small intestine (jejunum/ileum); fatty acids, monoglycerides; d. small intestine (duodenum); small intestine (jejunum/ileum); nucleotides (nucleotide base and monosaccharide); 10. acidic; 11. stomach; 12. T; 13. gastrin; 14. T; 15. osmosis; 16. T; 17. autonomic nerves; 18. water; 19. lipid; 20. T; 21. Because they are organs which to not directly function in digestion but produce enzymes, compounds, or hormones that assist in the process; 22. G; 23. J; 24. I; 25. C; 26. A; 27. H; 28. D; 29. E; 30. B; 31. F; 32. F; 33. A; 34. E; 35. B; 36. C; 37. D

39.10. HUMAN NUTRITIONAL REQUIREMENTS [pp.706-707]
39.11. VITAMINS, MINERALS, AND PHYTOCHEMICALS [pp.708-709]
39.12. WEIGHTY QUESTIONS, TANTALIZING ANSWERS [pp.710-711]
1. vegetables and fruits; 2. milk (dairy); 3. whole grains; 4. refined; 5. saturated; 6. trans; 7. sugar; 8. salt; 9. lactose intolerant; 10. carbohydrates; 11.

vitamins; 12. fiber; 13. gluten-free; 14. Essential fatty acids; 15. Essential amino acids; 16. vitamins; 17. minerals; 18. Leptin; 19. ghrelin; 20. obesity; 21. gastric bypass surgery; 22. E; 23. I; 24. H; 25. J; 26. F; 27. G; 28. C; 29. D; 30. A; 31. B; 32. a. 2070; b. 2900; c. 1230; 33. Consulting text Figure 40.14 (men's column, 6 feet tall) yields 178 pounds as his ideal weight. 195-178=17 pounds overweight. Maintaining 178 pounds at a low activity level would require an intake of 1,780 kilocalories/day; 34. Because BMI does not take note of where the weight is carried on the body (e.g. "beer belly"); 35. Genetics can determine which individuals are more at risk for weight gain due to specific genetic predispositions but cannot account for the national trend in obesity which is more due to lifestyle such as overeating and lack of exercise; 36. Individuals suffering from anorexia nervous are not able to eat enough calories to maintain a weight within fifteen percent of normal. This can lead to malnourishment and even death. Individuals suffering from bulimia often "binge and purge" by eating too much and then vomiting to avoid absorbing the calories. This can lead to brittle teeth and increases the risk of esophageal cancer.

SELF-TEST
1. b; 2. b; 3. a; 4. e; 5. d; 6. c; 7. c; 8. b; 9. c; 10. D; 11. M; 12. B; 13. A; 14. L; 15. G; 16. E; 17. J; 18. H; 19. N; 20. C; 21. F; 22. I; 23. K

Chapter 40   Maintaining the Internal Environment

40.1. TRUTH IN A TEST TUBE [p.715]
40.2. REUGLATING FLUID VOLUME AND COMPOSITION [p.716-717]
1. diabetes mellitus; 2. bacterial infection; 3. protein; 4. LH (luteinizing hormone); 5. pregnant; 6. menopause; 7. ten; 8. interstitial; 9. blood; 10. extracellular; 11. urinary; 12. volume; 13. homeostasis; 14. D; 15. J; 16. F; 17. B; 18. I; 19. A; 20. E; 21. C; 22. H; 23. G; 24. less; 25. losing; 26. T; 27. uric acid, urea; 28. T; 29. larger; 30. gain; 31. T; 32. Mammals gain water through solids, liquids, and metabolism; 33. Mammals lose water through urine, feces, and evaporation.

40.3. THE HUMAN URINARY SYSTEM [pp.718-719]
40.4. HOW URINE FORMS [pp.720-721]
40.5 REGULATING THIRST AND URINE FORMATION [pp.722-723]
40.6. ACID-BASE BALANCE [p.723]
1. kidney; 2. ureter; 3. urinary bladder; 4. urethra; 5. kidney cortex; 6. kidney medulla; 7. ureter; 8. kidneys; 9. nephrons; 10. Bowman's capsule; 11. glomerular; 12. renal corpuscle; 13. proximal; 14. loop of Henle; 15. collecting duct; 16. ureter; 17. urine; 18. ureter; 19. urinary bladder; 20. urethra; 21. reflex; 22. skeletal (sphincter); 23. voluntary; 24. E; 25. D; 26. K; 27. G; 28. M; 29. A; 30. H; 31. B; 32. F; 33. J; 34. C; 35. L; 36. I; 37. c; 38. a; 39. b; 40. c; 41. a; 42. B; 43. E; 44. D; 45. C; 46. A; 47. pH; 48. bicarbonate (HCO_3^-); 49. T; 50. little; 51. too little; 52. High; 53. The hypothalamus in the brain monitors osmoreceptors to detect dehydration, at which point, antidiuretic hormone will be produced by the posterior pituitary to reabsorb more water through the kidneys.

40.7. WHEN KIDNEYS FAIL [p.724]
40.8. HEAT GAINS AND LOSSES [p.725]
40.9. RESPONSES TO HEAT AND COLD[pp.726-729]
1. C; 2. E; 3. A; 4. B; 5. D; 6. F; 7. E; 8. D; 9. B; 10. A; 11. C; 12. G; 13. F; 14. E; 15. a; 16. c; 17. c; 18. b; 19. a; 20. c; 21. c; 22. b; 23. b; 24. D; 25. Drink plenty of water and avoid excessive exercise. Take breaks to monitor how you feel. Wear light-colored clothing that allows evaporation. Avoid direct sunlight, wear a hat, and use sunscreen; 26. In hemodialysis, a dialysis machine is connected to a patient's blood vessel and the blood is filtered through an artificial kidney. Peritoneal dialysis is done overnight where dialysis solution is pumped into a patient's abdominal cavity and drained the following morning; 27. Brown adipose tissue is located under the skin and feathers/fur can be used to trap air, which helps hold in body heat. Panting and sweating in mammals can also be utilized to disperse extra heat from the body. Behaviors such as hibernation and dormancy are also used during stressful thermal periods.

SELF-TEST
1. c; 2. e; 3. d; 4. b; 5. d; 6. b; 7. d; 8. e; 9. b; 10. a; 11. d; 12. d; 13. b; 14. b; 15. c; 16. b; 17. d; 18. c; 19. d; 20. d; 21. d.

Chapter 41   Animal Reproductive Systems

41.1. INTERSEX CONDITIONS [p.731]
41.2. MODES OF REPRODUCTION [pp.732-733]
41.3. REPRODUCTIVE SYSTEM OF HUMAN MALES [pp.734-735]
41.4. SPERM FORMATION [pp.736-737]

1. C; 2. F; 3. B; 4. I; 5. G; 6. H; 7. J; 8. A; 9. E; 10. D; 11. prostate gland (G); 12. urinary bladder (F); 13. Urethra (I); 14. Penis (A); 15. Testis (H); 16. ejaculatory duct (C); 17. seminal vesicle (C); 18. bulbourethral gland (E); 19. vas deferens (K); 20. epididymis (D); 21. scrotum (B); 22. I; 23. E; 24. G; 25. B; 26. H; 27. J; 28. C; 29. D; 30. A; 31. F; 32. Risk factors for prostate cancer include advancing age, a diet rich in animal fats, smoking, genetics, and a sedentary lifestyle; 33. Once a month, after a warm shower or bath, men should examine their testes to detect lumps, enlargement, or hardening.

41.5. REPRODUCTIVE SYSTEM OF HUMAN FEMALE [pp.738-739]
41.6. PREPARATIONS FOR PREGNANCY [pp.740-741]
41.7. PMS TO HOT FLASHES [p.742]

1. ovary (I); 2. oviduct (fallopian tube) (D); 3. clitoris (H); 4. labium minora (E); 5. labium majora (C); 6. uterus (b); 7. myometrium (A); 8. endometrium (G); 9. vagina (F); 10. E; 11. K; 12. G; 13. C; 14. A; 15. B; 16. I; 17. F; 18. D; 19. H; 20. J;

21. L; 22. B; 23. J; 24. H; 25. A; 26. G; 27. C; 28. I; 29. E; 30. D; 31. F; 32.A. ovarian follicle; 32.B. secreted protein; 32.C. primary oocyte; 32.D. first polar body; 32.E. secondary oocyte; 32.F. corpus luteum; 33. Ovarian cancer is so deadly because it often goes undetected until it has enlarged or spread; 34. Cervical cancer can be detected early by a Pap smear.

41.8. WHEN GAMETES MEET [pp.742-743]
41.9. PREVENTING OR SEEKING PREGNANCY [pp.744-745]
41.10. SEXUALLY TRANSMITTED DISEASES [pp.746-747]

1. T; 2. oxytocin; 3. T; 4. one; 5. *zona pellucidum*; 6. diploid zygote; 7. after fertilization; 8. million; 9. F; 10. E; 11. I; 12. J; 13. C; 14. H; 15. A; 16. D; 17. B; 18. G; 19. E; 20. H; 21. D; 22. I; 23. B; 24. G; 25. A; 26. F; 27. C; 28. The least effective forms of birth control include the rhythm method, withdrawal, and spermicides (if used alone); 29. The most effective forms are abstinence, tubal ligation, vasectomy, hormone contraceptives and intrauterine devices; 30. Condoms also help prevent STDs.

SELF-TEST

1. b; 2. c; 3. e; 4. a; 5. a; 6. b; 7. d; 8. c; 9. d; 10. e; 11. a; 12. a; 13. d; 14. b; 15. b; 16. a; 17. d; 18. c; 19. c; 20. a

---

Chapter 42   Animal Development

42.1. MIND BOGGLING BIRTHS[p.751]
42.2. STAGES OF REPRODUCTION AND DEVELOPMENT[pp.752-753]

1. Gamete formation; 2. fertilization; 3. zygote; 4. Cleavage; 5. blastomeres; 6. Gastrulation; 7. Ectoderm; 8. endoderm; 9. mesoderm; 10. F; 11. B; 12. C; 13. A; 14. E; 15. D; 16. F; 17. D; 18. A; 19. G; 20. B; 21. H; 22. C; 23. Cytoplasmic localization means that cytoplasmic contents of the egg are not evenly distributed. As a result, as cleavage occurs, blastomeres have different contents. This leads to differences in cell differentiation.

42.3. FROM ZYGOTE TO GASTRULA [pp.754-755]
42.4. SPECIALIZED TISSUES AND ORGANS FORM [pp.756-757]
42.5. AN EVOLUTIONARY VIEW OF DEVELOPMENT [p.758]

1. induction; 2. morphogens; 3. gradient; 4. cytoplasmic localization; 5. homeotic; 6. position; 7. physical; 8. surface-to-volume; 9. architectural; 10. phyletic; 11. changes the developmental pattern; 12. T; 13. induction; 14. cell differentiation; 15. morphogens; 16. concentration gradient; 17. bones and skeletal muscles; 18. T; 19. Physical; 20. phyletic; 21. One group of cells has an effect on adjacent cells and regulates their developmental pattern; 22. Apoptosis is when cells self-destruct at an appropriate time to create a particular physical shape. During development, there are times when cells are needed initially but then need to disappear at a later time. They do so by self-destruction. One example is webbing between digits on the hand that disappears later so that fingers are separated.; 23. (1) Molecules in different areas of unfertilized egg induce local expression of master genes creating chemical gradients. (2) Depending on location in

gradient, other master genes are activated or inhibited, forming additional gradients. (3) Depending on position within gradient, homeotic genes are expressed regulating development of specific body parts.

42.6. OVERVIEW OF HUMAN DEVELOPMENT [p.759]
42.7. EARLY HUMAN DEVELOPMENT [pp.760-761]
42.8. EMERGENCE OF THE VERTEBRATE BODY PLAN [pp.762-763]
42.9. STRUCTURE AND FUNCTION OF THE PLACENTA [p.763]
42.10. EMERGENCE OF DISTINCTLY HUMAN FEATURES [pp.764-765]

1. implantation; 2. chorionic villi; 3. placenta; 4. eighth; 5. fetus; 6. b; 7. a; 8. b; 9. b; 10. a; 11. b; 12 a; 13. b; 14. b; 15. b; 16. a; 17. A; 18. I; 19. F; 20. G; 21. B; 22. E; 23. C; 24. D; 25. H; 26. J; 27. L; 28. K; 29. D; 30. B; 31. A; 32. C; 33. c; 34. a; 35. b; 36. c; 37. a; 38. c; 39. b; 40. c; 41. c; 42. The placenta exchanges all materials between the blood of the mother and the blood of the fetus. Materials include nutrients and oxygen to the fetus and wastes and CO_2 to the mother's blood.; 43. Since the blood of the mother and the fetus never mix, the mother's system is not exposed to the different proteins of the fetus that could be treated as antigens. Therefore, no antibodies are made.; 44. I; 45. N; 46. B; 47. H; 48. D; 49. J; 50. M; 51. A; 52. K; 53. E; 54. C; 55. O; 56. F; 57. L; 58. G.

42.11. MISCARRIAGES, STILLBIRTHS, AND BIRTH DEFECTS [pp.766-767]
42.12. BIRTH AND LACTATION [pp.768-769]

1. folate; 2. iodine; 3. morning sickness; 4. organs; 5. teratogens; 6. teratogens; 7. toxins; 8. pathogens; 9. toxoplasmosis; 10. developmental problems; 11. stillborn; 12. Rubella; 13. alcohol; 14. caffeine; 15. hot tub; 16. paxil; 17. heart deformities; 18. acne; 19. cranial or facial deformities; 20. G; 21. E; 22. J; 23. A; 24. H; 25. C; 26. B; 27. D; 28. I; 29. F; 30. C; 31. D; 32. A; 33. E; 34. B

SELF-TEST

1. a; 2. b; 3. c; 4. a; 5. c; 6. e; 7. d; 8. b; 9. e; 10. a; 11. c; 12. b; 13. d; 14. e

---

Chapter 43   Animal Behavior

43.1. ALARMING BEE BEHAVIOR [p.773]
43.2. BEHAVIORAL GENETICS [pp.774-775]
43.3. INSTINCT AND LEARNING [pp.776-777]

1. c; 2. b; 3. e; 4. f; 5. a; 6. d; 7. g; 8. h

43.4. ENVIRONMENTAL EFFECTS ON BEHAVIORAL TRAITS [p.778]
43.5. MOVEMENTS AND NAVIGATION [p.779]
43.6. COMMUNICATION SIGNALS [pp.780-781]

1. b; 2. i; 3. j; 4. f; 5. g; 6. e; 7. h; 8. a; 9. c; 10. d; 11. k; 12. l

43.7. MATES, OFFSPRING, AND REPRODUCTIVE SUCCESS [pp.782-783] 43.8. LIVING IN GROUPS [pp.784-785]

43.9. WHY SACRIFICE YOURSELF? [pp.786-787]

1a. midwife toad; 1b. hanging flies; 1c. bison, lion, elk, elephant seals; 1d. sage grouse; 2. a; 3. d; 4. c; 5. b; 6. e; 7. i; 8. g; 9. j; 10. f; 11. h; 12. c; 13. d; 14. a; 15. b; 16. eusocial; 17. honeybees, termites and ants; 18. queen; 19. workers; 20. drones; 21. sterile; 22. altruistic; 23. relatives; 24. inclusive fitness; 25. family; 26. queens; 27. genes

SELF-TEST

1. c; 2. d; 3. b; 4. d; 5. a; 6. c; 7. c; 8. a; 9. d; 10. d

Chapter 44   Population Ecology

44.1. A HONKING MESS [p.791]
44.2. POPULATION DEMOGRAPHICS [pp.792-793]
44.3. POPULATION SIZE AND EXPONENTIAL GROWTH [pp.794-795]

1. k; 2. h; 3. d; 4. i; 5. b; 6. f; 7. a; 8. g; 9. j; 10. l; 11. c; 12. e; 13. Population ecology is the study of the characteristics of populations of a species, how populations interact with each other, with populations of other species, and with their physical environment. 14. Three characteristics that define a population: 1. it includes only members of one species; 2. the population occupies a particular limited area, 3. members breed with each other more than with members of another population.; 15. f; 16. g; 17. b; 18. c; 19. a; 20. h; 21. d; 22. e; 23a. it increases; 23b. it decreases; 23c. it must increase; 24. 100,000; 25a. 100,000; 25b. 300,000

44.4. LIMITS ON POPULATION GROWTH [pp.796-797]
44.5. LIFE HISTORY PATTERNS [pp.798-799]
44.6. EVIDENCE OF EVOLVING LIFE HISTORY PATTERNS [pp.800-801]

1. a; 2. c; 3. b; 4. e; 5. d; 6. f; 7. g; 8. b; 9. c; 10. a; 11. a; 12. c; 13. b

44.7. HUMAN POPULATION GROWTH [pp.802-803]
44.8. ANTICIPATED GROWTH AND CONSUMPTION [pp.804-805]

1. 200,000; 2. brain; 3. languaage; 4. agriculture; 5. hunter-gatherer; 6. grasses; 7. wheat; 8. rice; 9. fossil; 10. fertilizers; 11. pesticides; 12. disease; 13. cholera; 14. typhoid fever; 15. sanitation; 16. medicine; 17. vaccines; 18. antibiotics; 19. death; 20. 7 billion; 21. 9 billion; 22. total fertility rate; 23. replacement fertility rate; 24. grows; 25. broader; 26. b; 27. d; 28. a; 29. c

SELF-TEST
1. d; 2. a; 3. b; 4. d; 5. a; 6. a; 7. a; 8. c; 9. b; 10. a; 11. c; 12. a; 13. a

---

Chapter 45   Community Ecology

45.1. FIGHTING FOREIGN ANTS [p.809]
45.2. WHICH FACTORS SHAPE COMMUNITY STRUCTURE? [p.810]
45.3. MUTUALISM [p.811]

1. community; 2. habitat; 3. richness; 4. evenness; 5. abiotic; 6. dynamic; 7. disturbances; 8. symbiosis; 9. mutualism; 10. commensalism; 11. parasitism;12. h; 13. d; 14. g; 15. b; 16. c; 17. a; 18. f; 19. e

45.4.COMPETITIVE INTERACTIONS [pp.812-813]
45.5. PREDATOR-PREY INTERACTIONS [pp.814-815]
45.6. AN EVOLUTIONARY ARMS RACE [pp.816-817]
45.7 PARASITES AND PARASITOIDS [pp.818-819]

1. b; 2. e; 3. f; 4. c; 5. a; 6. d; 7. b; 8. c; 9. a; 10. e; 11. d; 12. parasite; 13. host; 14. endoparasite; 15. ectoparasite; 16. kill; 17. offspring; 18. predation; 19. mates; 20. sterility; 21. sex ratio; 22. adaptations; 23. negative or harmful; 24. brood parasite; 25. parenting; 26. lowers; 27. parasitoids; 28. biological or ecological; 29. chemical

45.8. ECOLOGICAL SUCCESSION [pp. 820-821]
45.9. SPECIES INTRODUCTION, LOSS AND OTHER DISTURBANCES [pp. 822-823]

1. b; 2. a; 3. b; 4. b; 5. a; 6. a; 7. d; 8. b; 9. e; 10. c; 11. f; 12. a; 13. g; 14. Three factors that affect succession are: 1. physical factors such as climate and soil, 2. random events determining species arrival, and 3. frequency and extent of disturbances.; 15. Species diversity is affected by environmental disturbances. Richness is greatest when communities are not disturbed frequently or severely.

45.10. BIOGEOGRAPHIC PATTERNS IN COMMUNITY STRUCTURE [pp. –824-825]

1a. tropical latitudes intercept more intense sunlight, receive more rainfall and have a longer growing season, 1b. tropical communities have been evolving longer than temperate ones; 1c. species richness may be self-reinforcing; 2. island C; 3. Closer islands have greater immigration, therefore greater diversity is expected. Larger islands are likely to have more varied habitats to accommodate more diverse organisms. Larger islands also are larger targets for immigrating organisms,

SELF-TEST
1. b; 2. b; 3. c; 4. a; 5. b; 6. b; 7. d; 8. d; 9. a; 10. d; 11. a; 12. c

Chapter 46   Ecosystems

46.1. TOO MUCH OF A GOOD THING [p.829]

46.2. THE NATURE OF ECOSYSTEMS [pp.830-831]

46.3. THE NATURE OF FOOD WEBS [pp.832-833]

46.4. ENERGY FLOW THROUGH ECOSYSTEMS [pp. 834-835]

1. H; 2. C; 3. K; 4. F; 5. M; 6. O; 7. B; 8. I; 9. N; 10. A; 11. P; 12. G; 13. L; 14. D; 15. J; 16. E; 17. Producer; 18. Herbivore; 19. Carnivore; 20. Decomposer; 21. Energy transfers in aquatic ecosystems are due to a variety of factors; the cell walls of most alga is composed of lignin instead of cellulose. This makes the alga much more digestible than its plant counterparts. Aquatic ecosystems usually have a higher percentage of cold blooded organisms than land ecosystems. Cold blooded organisms don't need as much energy in order to maintain a constant internal temperature.; 22. Eutrophication is the enrichment of an aquatic ecosystem through the accumulation of nutrients. Human activity causes nitrogen and phosphorus from fertilizers, detergents and animal wastes to enter ecosystems.

46.5. BIOGEOCHEMICAL CYCLES [p. 836]

46.6. THE WATER CYCLE [pp.836-837]

1. biogeochemical; 2. essential; 3. chemical; 4. geological; 5. reservoirs; 6. water; 7. water; 8. carbon; 9. nitrogen; 10. atmosphere; 11. sedimentary; 12. seafloor; 13. uplifting; 14. crust; 15. slower; 16. e; 17. f; 18. a; 19. d; 20. c; 21. b; 22. h; 23. g; 24. 80%; 25. 50%; 26. 50%

46.7. CARBON CYCLE [pp. 838-839]

46.8. GREENHOUSE GASES AND CLIMATE CHANGE [pp.840-841]

1. h; 2. e; 3. c; 4. a; 5. f; 6. d; 7. b; 8. g; 9. Solar radiation warms the earth. Heat is re-radiated and captured by greenhouse gases that re-emit the heat energy, some of which is trapped in the atmosphere. 15. 10. The gas layer helps keep the Earth warm enough to support life, but rising temperatures may cause glaciers to melt, a rise in seal levels and the disruption of global weather patterns.;11. the burning of fossil fuels by humans as a source of energy

46.9. NITROGEN CYCLE [p.842]

46.10. DISRUPTION OF THE NITROGEN CYCLE [p.843]

46.11 PHOSPHORUS CYCLE [p.844]

1. F; 2. E; 3. C; 4. G; 5. A; 6. B; 7. D; 8. All organisms need nitrogen to synthesize protein, DNA and RNA. Most nitrogen exists in a gaseous form unusable by higher plants. Certain bacteria and cyanobacteria are able to convert nitrogen gas into water soluble forms usable by plants and, in turn, the animals that feed on them.;9. Nitrous oxide is a greenhouse gas that contributes to global warming. It destroys the ozone layer allowing increased ultraviolet radiation to reach earth. Nitrates lead to eutrophication of waters and may be contributory in some cases of human cancer..

SELF- TEST

1. c; 2. c; 3. d; 4. c; 5. a; 6. a; 7. b; 8. a; 9. a; 10. a; 11. b; 12. d; 13. a; 14. a

---

Chapter 47   The Biosphere

47.1. WARMING WATER AND WILD WEATHER [p.849]

47.2. GLOBAL AIR CIRCULATION PATTERNS [pp.850-851]

47.3. THE OCEAN, LANDFORMS, AND CLIMATES [pp.852-853]

1. Climate; 2. Wind currents; 3. ocean currents; 4. sunlight; 5. warm; 6. equator; 7. global air; 8. moisture; 9. warms; 10. ascends (rises); 11. moisture; 12. descends (sinks); 13. moisture; 14. ascends (rises); 15. moisture; 16. descends (sinks); 17. easterlies; 18. westerlies; 19. C; 20. A; 21. D; 22. B; 23. E; 24. A; 25. D; 26. B; 27. C; 28. E

47.4. BIOMES [pp.854-855]

47.5. DESERTS [pp.856-857]

1. H; 2. F; 3. A; 4. G; 5. I; 6. B; 7. J; 8. E; 9. D; 10. C; 11. A; 12. E; 13. C; 14. B; 15. D

47.6. GRASSLANDS [p.858]

47.7. DRY SHRUBLANDS, AND WOODLANDS [p.859]

47.8. BROADLEAF FORESTS [pp.860-861]

47.9. CONIFEROUS FORESTS [p.862]

47.10. TUNDRA [p.863]

1. D; 2. B; 3. C; 4. A; 5. E; 6. F; 7. G; 8. J; 9. H; 10. I; 11. K

47.11. FRESHWATER ECOSYSTEMS [pp.864-865]

47.12. COASTAL ECOSYSTEMS [pp.866-867]

1. D; 2. B; 3. H; 4. C; 5. A; 6. G; 7. E; 8. F; 9. Oligotrophic lakes: deep, steeply banked; large deep-water volume relative to surface-water

volume; highly transparent; water blue or green; low nutrient content; oxygen abundant through all levels throughout the year; not much phytoplankton; green algae and diatoms dominant; aerobic decomposers favored in profundal zone; low biomass in profundal zone. Eutrophic lakes: shallow with broad littoral; small deep-water volume relative to surface-water volume; limited transparency; water green to yellow or brownish-green; high nutrient content; oxygen depleted in deep water during summer; abundant, thick masses of phytoplankton; cyanobacteria dominant; anaerobic decomposers in profundal zone; high biomass in profundal zone.; 10. Spring overturn:

winds cause vertical currents that forces oxygen-rich water in the surface layers to move down and nutrient-rich water from the lake's depth to move upwards.Fall overturn: the upper layer cools and sinks; this causes the oxygen-rich water to move downward while the nutrient-rich water moves upward.

47.13. CORAL REEFS [pp.868-869]

47.14. THE OPEN OCEAN [pp.870-871]

47.15. OCEAN-SEA INTERACTIONS [pp.872-873]

1. G; 2. F; 3. E; 4. J; 5. C; 6. I; 7. D; 8. B; 9. A; 10. H; 11. K; 12. Benthic; 13. Pelagic; 14. Neritic; 15. Oceanic

SELF-TEST

1. a; 2. d; 3. b; 4. d; 5. e; 6. b; 7. c; 8. b; 9. d; 10. c; 11. c; 12. a; 13. a; 14. a

---

Chapter 48   Human Impacts on the Biosphere

48.1. A LONG REACH [p.877]

48.2. THE EXTINCTION CRISIS [pp.878-879]

48.3. CURRENTLY THREATENED SPECIES [pp.880-881]

48.4. HARMFUL LAND USE PRACTICES [pp.882-883]

1. extinction; 2. species; 3. 99; 4. Five; 5. geologic; 6. Permian; 7. volcanic activity; 8. asteroid; 9. 6th; ; 10. humans;

11. B, E, H; 12. F; 13. C, I, J;   14. A, D, G; 15.Poor Agricultural practices can encourage soil erosion; exposed soil (plowed land) with no plant cover can be blown away by wind; drought increases problem; overgrazing has same effect. .;16. Acid rain, pesticide residues, fertilizer runoff, and emissions of greenhouse gasses degrade habitats.; 17. Deforestation encourages flooding due to runoff; no tree roots to hold soil back can cause landslides in hilly areas; runoff removes soil nutrients.; 18a. Mesozoic, Cenozoic; 18b. Cenozoic; 18c. Paleozoic, Mesozoic, Cenozoic; 19a. Paleozoic; 19b. Mesozoic; 19c. Mesozoic; 20. Cenozoic

48.5. TOXIC POLLUTANTS [pp.884-885]

48.6. OZONE DEPLETION AND POLLUTION [p.886]

48.7. GLOBAL CLIMATE CHANGE [p.887]

1. C; 2. F; 3. B; 4. E; 5. A; 6. D; 7. pollutants; 8. point source; 9. nonpoint sources; 10. sulfur dioxide; 11. nitrogen oxide; 12. acid rain; 13. accumulate; 14. biomagnification; 15. waterways; 16. wildlife; 17. ultraviolet; 18. mutations; 19. CFCs; 20. ozone hole; 21. global climate change; 22. Glaciers; 23. sealevels; 24. salt; 25. patterns; 26. rainfall; 27. flowering plants; 28. Migration; 29. heat stroke; 30. infectious diseases.

48.8. CONSERVATION BIOLOGY [pp.888-889]

48.9. REDUCING NEGATIVE IMPACTS [pp.890-891]

1. Biodiversity is measured at three levels: genetic diversity within species, species diversity, and ecosystem diversity.; 2. Survey the range of biodiversity, investigate the evolutionary and ecological origins of biodiversity, and identify ways to maintain and use biodiversity in ways that benefit human populations.; 3. biodiversity; 4. conservation biologist; 5. ecoregions; 6. hot spot; 7. conservation; 8. preserve; 9. biodiversity; 10. Ecological restoration; 11. Sustainability; 12. resources; 13. dead zone; 14. recycling; 15. energy consumption

SELF-TEST

1. a; 2. c; 3. d; 4. d; 5. b; 6. d; 7. b; 8. d; 9. d; 10. c
